Principles of
Adaptive Filters
and Self-learning
Systems

A. Zaknich

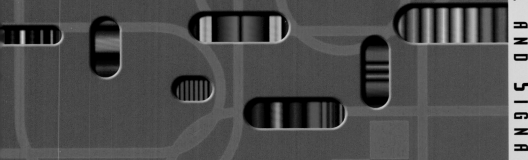

 Springer

Advanced Textbooks in Control and Signal Processing

Advanced Textbooks in Control and Signal Processing

Series Editors

Professor Michael J. Grimble, Professor of Industrial Systems and Director
Professor Emeritus Michael A. Johnson, Professor of Control Systems and Deputy Director

Industrial Control Centre, Department of Electronic and Electrical Engineering, University of Strathclyde, Graham Hills Building, 50 George Street, Glasgow G1 1QE, U.K.

Other titles published in this series:

Genetic Algorithms
K.F. Man, K.S. Tang and S. Kwong

Neural Networks for Modelling and Control of Dynamic Systems
M. Nørgaard, O. Ravn, L.K. Hansen and N.K. Poulsen

Modelling and Control of Robot Manipulators (2nd Edition)
L. Sciavicco and B. Siciliano

Fault Detection and Diagnosis in Industrial Systems
L.H. Chiang, E.L. Russell and R.D. Braatz

Soft Computing
L. Fortuna, G. Rizzotto, M. Lavorgna, G. Nunnari, M.G. Xibilia and R. Caponetto

Statistical Signal Processing
T. Chonavel

Discrete-time Stochastic Processes (2nd Edition)
T. Söderström

Parallel Computing for Real-time Signal Processing and Control
M.O. Tokhi, M.A. Hossain and M.H. Shaheed

Multivariable Control Systems
P. Albertos and A. Sala

Control Systems with Input and Output Constraints
A.H. Glattfelder and W. Schaufelberger

Analysis and Control of Non-linear Process Systems
K. Hangos, J. Bokor and G. Szederkényi

Model Predictive Control (2nd Edition)
E.F. Camacho and C. Bordons

Digital Self-tuning Controllers
V. Bobál, J. Böhm, J. Fessl and J. Macháček

Control of Robot Manipulators in Joint Space
R. Kelly, V. Santibáñez and A. Loría
Publication due July 2005

Robust Control Design with MATLAB®
D.-W. Gu, P.Hr. Petkov and M.M. Konstantinov
Publication due July 2005

Active Noise and Vibration Control
M.O. Tokhi
Publication due November 2005

A. Zaknich

Principles of Adaptive Filters and Self-learning Systems

With 95 Figures

 Springer

Anthony Zaknich, PhD

School of Engineering Science, Rockingham Campus,
Murdoch University, South Street, Murdoch, WA 6150, Australia

and

Centre for Intelligent Information Processing Systems,
School of Electrical, Electronic and Computer Engineering,
The University of Western Australia,
35 Stirling Highway, Crawley, WA 6009, Australia

Instructors Solutions Manual in PDF can be downloaded from the book's page at springeronline.com

British Library Cataloguing in Publication Data
Zaknich, Anthony
    Principles of adaptive filters and self-learning systems.
    (Advanced textbooks in control and signal processing)
    1. Adaptive filters 2. Adaptive signal processing 3. System
    analysis
    I. Title
    621.3'815324
ISBN-10: 1852339845

Library of Congress Control Number: 2005923608

Advanced Textbooks in Control and Signal Processing series ISSN 1439-2232
ISBN-10 1-85233-984-5
ISBN-13 978-1-85233-984-5
Springer Science+Business Media
springeronline.com

Typesetting: Camera ready by author
Production: LE-TEX Jelonek, Schmidt & Vöckler GbR, Leipzig, Germany
Printed in Germany
69/3141-543210 Printed on acid-free paper SPIN 10978566

Franica, Nikola, Iris, Nelli and Pi Pi

Series Editors' Foreword

The topics of control engineering and signal processing continue to flourish and develop. In common with general scientific investigation, new ideas, concepts and interpretations emerge quite spontaneously and these are then discussed, used, discarded or subsumed into the prevailing subject paradigm. Sometimes these innovative concepts coalesce into a new sub-discipline within the broad subject tapestry of control and signal processing. This preliminary battle between old and new usually takes place at conferences, through the Internet and in the journals of the discipline. After a little more maturity has been acquired by the new concepts then archival publication as a scientific or engineering monograph may occur.

A new concept in control and signal processing is known to have arrived when sufficient material has evolved for the topic to be taught as a specialised tutorial workshop or as a course to undergraduate, graduate or industrial engineers. *Advanced Textbooks in Control and Signal Processing* are designed as a vehicle for the systematic presentation of course material for both popular and innovative topics in the discipline. It is hoped that prospective authors will welcome the opportunity to publish a structured and systematic presentation of some of the newer emerging control and signal processing technologies in the textbook series.

This new advanced course textbook for the control and signal processing series, *Principles of Adaptive Filtering and Self-learning Systems* by Anthony Zaknich, presents a bridge from classical filters like the Wiener and Kalman filters to the new methods that use neural networks, fuzzy logic and genetic algorithms. This links the classification-based adaptive filtering methods to the innovative non-classical techniques, and both are presented in a unified manner. This eliminates the dichotomy of many textbooks which focus on either classical methods or non-classical methods.

The textbook is divided into six parts: Introduction, Modelling, Classical Filters and Spectral Analysis, (Classical) Adaptive Filters, Non-Classical Adaptive Systems and finally Adaptive Filter Applications. As befits an advanced course textbook there are many illustrative examples and problem sections. An outline Solutions Manual complete with a typical course framework and with specimen examination papers is also available to tutors to download from springeronline.com.

Solid foundations for a possible adaptive filtering course are laid in the Introduction (Part I) with an overview chapter and a linear systems and stochastic processes chapter of nearly 60 pages. All the main basic terms and definitions are found in this introductory part.

Signal models and optimization principles are covered in Part II. In the two chapters of this part are found concepts like the pseudo-inverse, matrix singular value decompositions, least squares estimation and Prony's method.

Filters proper emerge in Part III which covers the classical Wiener filter, the Kalman filter and power spectral density analysis methods. The chapter on the Kalman filter is nicely presented since it includes examples and an assessment of the advantages and disadvantages of the Kalman filter method.

In Part IV, adaptation and filtering are united to yield a set of chapters on adaptive filter theory. Since many of the techniques are used by control engineers it is pleasing to have a chapter devoted to adaptive control systems (Chapter 11). In fact the way that the author keeps linking the specifics of filtering theory to the broader fields of filter implementation, practical applications and control systems is a real strength of this book.

Neural networks, fuzzy logic and genetic algorithms are the constituent techniques of the non-classical methods presented in Part V. Each technique is given a chapter-length presentation and each chapter is full of reviews, perspectives and applications advice. In all three chapters links are made to similar applications in the field of control engineering. This gives credence to the idea that twin adaptive filtering and digital control systems courses would be powerful reinforcing strategy in any advanced systems postgraduate qualification.

The final part of the book comprises two chapters of adaptive filter applications (Part VI). Whilst the range of applications presented is not exhaustive, fields like speech encoding, event detection, data transmission and discussing both classical and non-classical filter solution methods are covered.

In summary Anthony Zaknich's is a particularly welcome entry to the *Advanced Textbooks in Control and Signal Processing* series. Graduate students, academics and industrial engineers will find the book is a constructive introduction to adaptive filtering with many of the chapters appealing to a wider control, electronic and electrical engineering readership.

M.J. Grimble and M.A. Johnson
Industrial Control Centre
Glasgow, Scotland, U.K.
January 2005

Preface

This book can be used as a textbook for a one semester undergraduate or postgraduate introductory course on adaptive and self-learning systems for signal processing applications. The topics are introduced and discussed sufficiently to give the reader adequate background to be able to confidently pursue them at depth in more advanced literature. Subject material in each chapter is covered concisely but with adequate discussion and examples to demonstrate the important ideas. Key chapters include exercises at the end of the chapter to help provide a deeper topic understanding. It is strongly recommended that the exercises be attempted first before making reference to the answers, which are available in a separate Solutions Manual. The Solutions Manual also includes a possible course outline using this book as the textbook, plus sample assignments and representative past examination papers with solutions that may aid in the design and conduct of such a course.

Topics are presented in a progressive sequence starting with a short introduction to adaptive filters and systems, linear systems and stochastic process theory. Unavoidably the first Chapter refers to some advanced concepts that are more fully described in later Chapters. In the first reading of this first Chapter it may be best to gloss over these and just accept a general understanding of what is described to gain initial familiarity with the terminology and ideas.

The introductory part of the book is followed by the detailed developments of system and signal modelling theory, classical Wiener filter theory, Kalman filter theory, spectral analysis theory, classical adaptive linear and nonlinear filter theory, adaptive control systems, nonclassical adaptive systems theory, through to adaptive filter application issues. Although the book concentrates on the more established adaptive filter theory, introductions to artificial neural networks, fuzzy logic and genetic algorithms are also included to provide a more generic perspective of the topic of adaptive learning. A significant further offering of the book is a method to seamlessly combine a set of both classical and/or nonclassical adaptive systems to form a powerful self-learning engineering solution method that is capable of solving very complex nonlinear problems typical of the underwater acoustic signal processing environment, as well as other equally difficult application domains.

The concepts of system adaptation and self-learning are quite general and can conjure up all sorts of ideas. In this book, these concepts have a very specific meaning. They signify that a system can be configured in such a way that allows it to in some sense progressively organise itself towards a learned state in response to input signals. All learning has to be with respect to some appropriate context and suitable constraints. A human designer hoping to achieve some meaningful functionality will of necessity initially supply the required context and constraints. The systems of interest here start out with predetermined structures of some sort. However, these structures have sufficient inherent flexibility to be able to adapt their parameters and components to achieve specific solutions formed from classes of relationships predetermined by those structures. Sometimes all the signals involved will come exclusively out of the system's environment and sometimes some of the signals will be supplied by a human supervisor, but in all cases the system must be able to eventually achieve coherent solutions within that context by its own adaptive or learning processes.

The difference between an adaptive system and a learning system is in principle slight. It is to do with history and convention, but more importantly it is to do with the degree of flexibility allowed by the system model. Classical linear and nonlinear adaptive filters typically have less flexibility in the way they can change themselves and are generally referred to as adaptive. On the other hand nonclassical adaptive systems such as Artificial Neural Networks (ANN), Adaptive Fuzzy Logic (FL), Genetic Algorithms (GA), and other machine learning systems have a much greater flexibility inherent within their structures and therefore can be seen more as learning systems.

The field of nonclassical learning systems is often referred to as Computational Intelligence (CI). However, the word intelligence can also conjure up unintended meanings. Intelligent methods are often referred to as model-free and are mostly based on the example signals (or data) rather than on the constraints imposed by the model itself (Haykin and Kosko 2001). In this special context, "more intelligent" implies more able to extract system information from the example data alone and be less dependent on *a priori* environmental and system information. It is fair to say that no limited physical system can be absolutely model free. Although some models like ANNs can be made to be very flexible, having a huge number of possible configurations or states, it is really a matter of degree. Before the advent of recent finite data based statistical learning theories (Cherkassky and Mulier 1998, Vapnik 1998, 2001) it was commonplace to limit the flexibility of learning machines down to a sufficient degree in order to force some regularization or smoothness in a local sense. This is somewhat like classical adaptive systems do by keeping the number of their model parameters (model order) to as low as necessary in order to achieve good generalization results for the chosen problem. The learning has to have a degree of local smoothness such that close input states are close to their corresponding output states; else generalization of learning would be impossible. The higher the order of the model with respect to the order of the problem the more difficult it is for the adaptive system to maintain adequate performance.

A unique situation is applicable to GAs with respect to CI and machine learning in that they have been able to consistently create numerous and varied programmed solutions automatically, starting from a high level statement of what needs to be done (Koza *et al* 2003). Using a common generic approach they have produced parameterized topologies for a vast number of complex problems. In that sense GAs are exhibiting what Turing called Machine Intelligence (MI). To him MI was some mechanical process that exhibited behaviour, if done by humans, would be assumed to be intelligent behaviour. Since Turing's time in the 1940s and 50s the term machine learning has tended to move away from this goal-orientated meaning more toward specific approaches for attempting to achieve automatic machine solutions to problems. These approaches tend to use somewhat arbitrary methodologies, many of which are statistical in nature. It is because of this de-emphasis on broad intelligence that narrower methodologies like adaptive filters, ANNs and FL can also been included within the definition of CI or machine learning. Although Turing did not have a way of doing it, he did suggest that true MI might be achieved by an evolutionary process whereby a computer program undergoes progressive modifications under the guidance of a natural selection process. Since many successful natural systems seem to have developed by natural evolutionary process it should be no surprise that GAs are also beginning to produce very impressive results, especially as computer processing speed and capacity increases. Genetic Programming (GP), a generalisation of GAs, can quite effectively use all the other CI methodologies as substructures in its evolutionary constructs. Therefore GP has a very special generic position in the scheme of adaptive and self-learning systems and probably deserves much more attention than it has received to date.

The final Chapter of this book introduces a model for seamlessly combining any set of adaptive filters or systems to form an engineering solution method for complex nonlinear adaptive filtering problems. The model is referred to as the Sub-Space Adaptive Filter (SSAF) model (Zaknich 2003b). This model, when constructed with a set of piecewise linear adaptive filters spread throughout the data space of the problem, can be adjusted by a single smoothing parameter continuously from the best piecewise linear adaptive filter in each sub-space to the best approximately piecewise linear adaptive filter over the whole data space. A suitable smoothing value in between ensures that all the adaptive filters remain decoupled at the centre of their individual operating spaces and at the same time neighbouring linear adaptive filters merge together smoothly at their common boundaries. The SSAF allows each piecewise linear adaptive filter section to be adapted separately as new data flows through it. By doing this the SSAF model represents a learning/filtering method for nonlinear processes that provides one solution to the stability-plasticity dilemma associated with standard adaptive filters and feedforward neural networks. This is achieved by simply keeping the piecewise linear adaptive filter models decoupled. As a complex nonlinear adaptive filter model, the SSAF adapts only the one piecewise linear adaptive filter that is in the current operating data space, leaving all the others unaffected. As a learning system, it is possible to achieve local learning with the SSAF without affecting previous global learning. This is done by the same process of only adjusting the one piecewise linear adaptive filter at a time. In principle the SSAF structure can smoothly integrate decoupled sets of any type of adaptive structures, linear or nonlinear or combinations of both. However, by keeping all the adaptive filters linear this makes it simpler to design and to deal with.

The SSAF does require human design input to determine the number of piecewise linear adaptive filters to use and where to place them in the data space.

However, the smoothing factor can then be systematically optimised using typical training data. It would be possible to automate the entire SSAF model construction, of adding and pruning piecewise linear adaptive filter centres, based on the data and keep it optimised as the data statistics change. However this is the subject of ongoing research and is not reported on in this book. The SSAF model presented here is a very useful approach, which can be applied to many practical problems with judicious human design. It is a generic structure that can meaningfully use and integrate all or any of the adaptive and self-learning systems covered in this book, as well as many others, to tailor make special adaptive solutions for individual problems.

Anthony Zaknich
Perth
Western Australia
January 2005

Acknowledgements

Since knowledge truly lives in brains and is subsequently broadcast from mind to minds I do recognise that all the people with whom I have engaged throughout my professional and academic careers have contributed significantly to the production of this book. "That which has been is that which will be; and that which has been done is that which will be done; and there is nothing new under the sun" (Ecclesiastes 1:9). Still, occasionally, if only by quantum tunnelling, a very rare soul must be able to happen upon a unique idea that the rest of us can then exploit in every conceivable way, to either add to or to reconfigure previous knowledge.

Nevertheless it is fitting and proper to identify and heartily thank those that have, in various ways, helped and contributed most significantly. These are, Professor Yianni Attikiouzel, Dr Chris deSilva, Dr Mike Alder, Dr Thomas Hanselmann, Stanley McGibney, Dr James Young, Brad Finch, Jordan Kosek and the many undergraduate students that have been subjected to and have test driven earlier versions of the document. Then there are the anonymous reviewers who have provided valid and expert suggestions for the significant improvement of the book. Professor Derek Humpage who, before his untimely death on the 31st of October 2003, always unreservedly supported and encouraged my work, also deserves a very special mention. Other people that deserve a mention for miscellaneous input and assistance are Associate Professor Thomas Bräunl, Peyman Kouchakpour, Nicholas Pelly, Sandra Snook and Linda Barbour. Finally, without the help of Professor Lakhmi C. Jain I'm sure I would not have gained timely access to the publisher for the submission of my initial book proposal.

This book certainly offers some unique contributions of my own but it also draws from many other fine textbooks by taking relevant ideas and representing them in a context suitable to the specific aims of the book. Of all the textbooks that have been referenced Monson Hayes' "Statistical digital signal processing and modelling" was one of the best that I do highly recommend to anyone. Also, Simon Haykin's book "Adaptive filter theory" has been especially helpful. Of all the authors on this and allied subjects I have found Simon Haykin to be the one that I can most identify with, both in respect to relevance of subject matter and emphasis. Haykin's work has always drawn from a more generic framework of concepts that rightly places adaptive filters in the same category as artificial neural networks and other nonlinear optimising systems. Åström and Wittenmark's book "Adaptive control" has been an invaluable source for the identification and understanding of the key adaptive control ideas and their very interesting history.

Alcoa World Alumina through the Technology Delivery Group has kindly approved of the use of two images in the Artificial Neural Networks Chapter to demonstrate some practical applications. Dr Gerald Roach and John Cornell of Alcoa were very helpful to me during the work that I performed for Alcoa in relation to those applications.

In the final analysis, this book represents my perspective on the subject and as such I accept full responsibility for any errors or oversights.

Contents

PART I. INTRODUCTION

This introduction to the subject of adaptive filters and self-learning systems consists of two chapters including a general introduction to adaptive systems and an overview of linear systems and stochastic processes. Some of the more significant types of adaptive and self-learning systems of interest to engineering design are overviewed. These include; linear adaptive filters, nonlinear adaptive filters, adaptive controllers, Artificial Neural Networks (ANN), Fuzzy Logic (FL), and Genetic Algorithms (GA). Linear adaptive filters, nonlinear adaptive filters, and adaptive controllers are categorised as classical adaptive systems as they represent a culmination of initial research in these areas. On the other hand ANNs, FL and GAs can be regarded as nonclassical Computational Intelligence (CI) approaches to adaptive systems because they tend to go beyond classical methods. However, it is fair to say that the boundary between classical and nonclassical can be a little blurred in some cases, where development proceeded in concert using simpler underlying concepts. Sufficient discussion is provided to give a reasonable introductory understanding of them and to show some of the fundamental relationships between them. All of these adaptive systems may be used for many different types of functions and applications related to signal processing and control as is indicated in these introductory and in later Chapters.

A short history is given of all of these approaches starting with linear estimation theory upon which they are either directly founded or at least reliant on for signal processing applications. Linear estimation theory technically began with Galileo Galilei in 1632 but it was Gauss who was given credit for it, based on his very significant invention of the method of least squares in 1795. Through a series of research efforts starting in the late 1930s through to the 1940s by Kolmogorov, Klein and Wiener mean squares estimation was studied in relation to stochastic prediction and filtering processes. Work during the 1950s by Widrow and Hoff resulted in the very successful and now famous Least Mean Squares (LMS) algorithm that can be used to adapt a linear filter. During the same time period work begun by Placket eventually resulted in the family of Recursive Least Squares (RLS) algorithms (a special case of the Kalman filter) for linear filter adaptation. From then on significant progress has been made on linear and subsequently nonlinear adaptive filter theory, eventually opening the way for the more current nonclassical methods of CI in the form of ANNs, FL and GAs.

The fundamentals of adaptive filters are discussed to show the basic commonalities and issues involved. All adaptive filter systems require some kind of desired signal to guide the adaptation process. Clearly, if the desired signal for a given input signal were always known it would obviate the need for any kind of a processing system at all. However, in practice, such a signal is not always known either in part or in whole. Therefore there are many ways that adaptive filters can be configured to work with physically available information that is in some way correlated with the theoretically desired signal. A study of specific ways to deal with this issue for prediction, modelling, inverse-modelling and interference cancelling problems provides considerable insight into how to approach other new problems. Not all adaptive algorithms are suitable for use under all practical conditions so it is necessary to understand their individual strengths and limitations for proper choice in specific applications.

The second Chapter reviews the basic aspects of linear systems and stochastic processes as a necessary background for later topics. A good grounding is provided in discrete-time signal processing concepts, including, the discrete Fourier Transform and its practical uses, along with a concise review of digital sampling theory. The relationship between continuous-time and discrete-time signals and transformations is investigated through a study of the Laplace Transform and the z-Transform and their connections. This culminates in an overview of discrete-time linear shift invariant system theory and properties of some special filter forms.

A basic summary of linear algebra appropriate to discrete signal processing is provided to define operational elements and functions for the various algorithmic structures and operations required for adaptive filtering. To this same end a very basic overview of random signals and processes is also provided with formal definitions of fundamental stochastic processes, functions and operators.

1. Adaptive Filtering

Adaptive filters represent a significant part of the subject of statistical signal processing upon which they are founded. Historically, the parametric approach has been the main engineering approach to signal processing and is based on *a priori* models derived from scientific knowledge about the problem. At the other extreme, the alternative nonparametric approach is based on the use of more general models trained to replicate desired behaviour using statistical information from representative data sets. Adaptive filters are actually based on an approach which is somewhere in between these two extremes. When *a priori* knowledge of a dynamic process and its statistics is limited then the use of adaptive filters can offer performance improvements over the more conventional parametrically based filter designs. Furthermore, they can offer other signal processing benefits that would not be possible otherwise. Consequently, adaptive filters have found application in diverse fields including communications, controls, robotics, sonar, radar, seismology and biomedical engineering to name but a few.

Filtering in the most general terms is a process of noise removal from a measured process in order to reveal or enhance information about some quantity of interest. Any real data or signal measuring process includes some degree of noise from various possible sources. The desired signal may have added noise due to thermal or other physical effects related to the signal generation system, or it may be introduced noise due the measuring system or a digital data sampling process. Often the noise is a wide-sense stationary random process (has a constant finite mean and variance, and an autocorrelation function dependent only on the difference between the times of occurrence of the samples), which is known and therefore may be modelled by a common statistical model such as the Gaussian statistical model. It may also be random noise with unknown statistics. Otherwise, it may be noise that is correlated in some way with the desired signal itself. The so-called filtering problem can be identified and characterised more specifically by the terms filtering, smoothing, prediction (Haykin 1996) and deconvolution (Hayes 1996).

1.   Filtering, strictly means the extraction of information about some quantity of interest at the current time t by using data measured up to and including the time t.

2. Smoothing, involves a delay of the output because it uses information extracted both after and before the current time t to extract the information. The benefit expected from introducing the delay is more to do with accuracy than filtering.

3. Prediction, involves forecasting information some time into the future given the current and past data at time t and before.

4. Deconvolution, involves the recovery of the filter characteristics given the filter's input and output signals.

Filters can be classified as either linear or nonlinear types. A linear filter is one whose output is some linear function of the input. In the design of linear filters it is necessary to assume stationarity (statistical-time-invariance) and know the relevant signal and noise statistics *a priori*. The linear filter design attempts to minimise the effects of noise on the signal by meeting a suitable statistical criterion. The classical linear Wiener filter, for example, minimises the Mean Square Error (MSE) between the desired signal response and the actual filter response. The Wiener solution is said to be optimum in the mean square sense, and it can be said to be truly optimum for second-order stationary noise statistics (fully described by constant finite mean and variance). For nonstationary signal and/or noise statistics, the linear Kalman filter can be used. Very well developed linear theory exists for both the Wiener and Kalman filters and the relationships between them.

When knowledge of the signal and noise statistics is unavailable *a priori* it is still possible to develop a useful filter by using a recursive algorithm to adjust the filter parameters based on the input data stream. This is what an adaptive filter does. If the signal and noise statistics are stationary then the adaptive filter would be expected to eventually converge to the optimum Wiener solution. If they are nonstationary then the adaptive filter tracks them if they vary at a sufficiently slow rate. The adaptation rate must be faster than the rate of change in statistics to maintain tracking. The parameters of an adaptive filter are updated continuously as the data flows through it; therefore the adaptive filter is strictly a nonlinear system. However, it is common to distinguish linear and nonlinear adaptive filters. A linear adaptive filter is one whose output is some linear combination of the actual input at any moment in time between adaptation operations. A nonlinear adaptive filter does not necessarily have a linear relationship between the input and output at any moment in time. Many different linear adaptive filter algorithms have been published in the literature. Some of the important features of these algorithms can be identified by the following terms (Haykin 1996),

1. Rate of convergence - how many iterations to reach a near optimum Wiener solution.

2. Misadjustment - measure of the amount by which the final value of the MSE, averaged over an ensemble of adaptive filters, deviates from the MSE produced by the Wiener solution.

3. Tracking - ability to follow statistical variations in a nonstationary environment.

4. Robustness - implies that small disturbances from any source (internal or external) produce only small estimation errors.

5. Computational requirements - the computational operations per iteration, data storage and programming requirements.

6. Structure - of information flow in the algorithm, e.g., serial, parallel etc., which determines the possible hardware implementations.

7. Numerical properties - type and nature of quantization errors, numerical stability and numerical accuracy.

The filters described so far may be referred to as classical adaptive filters in so far as they draw upon theory and methods extending from classical Wiener filter theory. A nonclassical approach to adaptive filtering is one that does not rely so much on linear modelling techniques. Artificial neural networks, fuzzy logic, and genetic algorithms have come to prominence in more recent years and are described more as learning systems and belong to the family of CI methods. These employ a range of nonlinear learning techniques that are not dependent on such strict assumptions about either the process model or process statistics. Nevertheless, they can still often be adapted in whole or in part by some form of a gradient descent algorithm (Principe *et al* 2000) that attempts to minimise a mean square error function, not unlike the classical adaptive filters.

1.1 Linear Adaptive Filters

A linear adaptive filter system filters a sequence of input data by controlling its adjustable parameters via an adaptive process. The choice of filter structure is a very important part of the system. There are three main types of structures commonly used (Haykin 1996),

1. Transversal structure (tapped delay line) - similar to the linear FIR filter structure.

2. Lattice predictor - a modular structure with a lattice appearance.

3. Systolic array - a parallel computing network ideally suited for mapping important linear algebra computations such as matrix multiplication, triangulation, and back substitution.

Of these the transversal structure, although not necessarily the most efficient, is very successfully employed for many practical systems. It forms the basis of the

Finite Impulse Response (FIR) discrete-time filter (Loy 1988). The terms associated with this filter structure are defined more thoroughly in later Chapters but for now it is sufficient to say that given an input sequence set of discrete real numbers $\{x[n]\}$, where n is an integer index value, the output sequence $y[n]$ of a Mth order FIR filter is defined by Equation 1.1 and depicted in Figure 1.1. The index value n represents the current discrete-time instant, and $n - k$ represents the previous kth instant, i.e., delayed by k instants.

$$y[n] = \sum_{k=0}^{M} b[k]\, x[n-k] \tag{1.1}$$

where:
   $b[k]$ are the fixed filter coefficients that define the filter's characteristics.

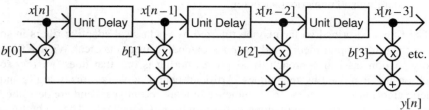

Figure 1.1. FIR Transversal Structure

   To design a real FIR filter, Equation 1.1 must be converted into realisable blocks, including a means for obtaining a delayed version of the input sequence $x[n]$, a means for multiplying input signal values in the delay line by the filter coefficients, $b[k]$, and a means for adding the scaled sequence values. The FIR filter is completely defined once the coefficients of the filter are known. For example, if the filter coefficients are the set $\{b[k]\} = \{3,-1,2,1\}$ then this represents a third order ($M = 3$) FIR filter, having a tap length of four. Here, a tap is simply a tap-off point in a serial delay line. The equation for this example filter can be expanded into a four-point (4-tap) difference equation defined by Equation 1.2.

$$y[n] = \sum_{k=0}^{3} b[k]\, x[n-k]$$

$$y[n] = b[0]\, x[n] + b[1]\, x[n-1] + b[2]\, x[n-2] + b[3]\, x[n-3] \tag{1.2}$$

e.g.,

$$y[n] = 3x[n] + -x[n-1] + 2x[n-2] + x[n-3]$$

The general direct-form realisation of a FIR filter using basic computational elements is depicted in Figure 1.1 with the tapping points shown with black dots. Notice that the input sequence $x[n]$ flows through the delay line continuously and uniformly step by step producing another output value $y[n]$ at each integer step, indefinitely. This filter can be made into an adaptive filter by the addition of a suitable adaptation mechanism that is capable of sensibly adapting the coefficients $b[n]$ progressively at each time step based on some real-time data information.

1.1.1 Linear Adaptive Filter Algorithms

No unique algorithmic solution exists for linear adaptive filtering problems. There are various algorithms and approaches that may be used depending on the requirements of the problem. However, there are two main approaches to the development of recursive adaptive filter algorithms (Haykin 1996),

1. **The Stochastic Gradient Approach** - uses a tapped delay line or transversal structure. The relation between the tap weights and the mean square error between the desired and actual filter output is a multi-dimensional paraboloid (quadratic) error function with a uniquely defined minimum, representing the optimum Wiener solution. This solution can be found by the well-established optimisation method called steepest descent, which uses the gradient vector to gradually descend step by step to the minimum of the error function. The so-called Wiener-Hopf equations, in matrix form, define this optimum Wiener solution. A simpler way to do this is with the Least Mean Squares (LMS) algorithm, invented by B. Widrow and M. E. Hoff Jr in 1959. It is a modified system of Wiener-Hopf equations and is used to adapt the filter weights toward the minimum. This algorithm estimates the gradient of the error function from instantaneous values of the correlation matrix of the tap inputs and the cross-correlation vector between the desired response and the tap weights. The LMS algorithm is very simply and elegantly defined by Equation 1.3.

$$\mathbf{w}[k+1] = \mathbf{w}[k] + 2\eta \, e[k] \, \mathbf{x}[k] \qquad (1.3)$$

where:

η = learning rate parameter.
$e[k]$ = scalar error (desired output minus the actual output).
$\mathbf{x}[k]$ = $[x_1, x_2,...., x_p]^T$, the tap vector at time instance k.
$\mathbf{w}[k]$ = $[w_1, w_2,..., w_p]^T$, the tap weight matrix at time instance k.

A problem with the LMS algorithm is that it is slow to converge and is dependent on the ratio of the largest to smallest eigenvalue of the correlation matrix of the tap inputs. The higher the ratio, the slower the convergence. Nevertheless, it is very popular and the most widely used learning algorithm, which under the right conditions can perform very adequately. Its tracking behaviour is said to be model-independent and consequently it exhibits good tracking behaviour.

A lattice structure can also be used with the gradient approach in which case the resulting adaptive filtering algorithm is called the Gradient Adaptive Lattice (GAL).

2. **Least Squares Estimation (LSE)** - minimises an objective, or error, function that is defined as the sum of weighted error squares, where the error or residual is defined as the difference between the desired and actual filter output as before. LSE can be formulated in two important ways, with

block estimation or recursive estimation approaches. In block estimation the input data sequence is arranged in blocks of equal time length and processing proceeds block by block. In recursive estimation the processing proceeds sample by individual time sample. Recursive estimation is more popular because it typically requires less data storage overhead than block estimation. Recursive Least Squares (RLS) can be seen as a special case of the well known Kalman filter, which itself is a form of LSE. The Kalman filter uses the idea of state, where state represents a measure of all the relevant inputs applied to the filter up to and including a specific instance in time. In the most general terms the Kalman filtering algorithm can be defined by Equation 1.4.

$$s[k+1] = s[k] + \mathbf{K}(k)\, \mathbf{i}[k] \tag{1.4}$$

where:

$\mathbf{K}(k)$ = the Kalman gain matrix at instance k.
$\mathbf{i}[k]$ = the innovation vector at instance k.
$s[k]$ = the state at instance k.

The innovation vector $\mathbf{i}[k]$ in Equation 1.4 contains the new information (being the observed new data at time k less its linear prediction based on observations up to and including time k-1) that is presented to the filter at the time of the processing for the instance k. As there is a one-to-one correspondence between the Kalman and RLS variables it is possible to learn useful ideas for RLS from the vast Kalman filter literature. There are three main categories of RLS depending on the specific approach taken (Haykin 1996),

1.   **The Standard RLS Algorithm** - uses a tapped delay line or transversal structure. Both the RLS and Kalman algorithms rely on the matrix inversion lemma, which results in lack of numerical robustness and excessive numerical complexity. The next two categories address these problems.

2.   **Square-root RLS Algorithms** - linear adaptive filter approaches based on QR-decomposition of the incoming data matrix and they represent the square-root forms of the standard RLS algorithm. The QR-decomposition can be performed by the Householder transformation and the Givens rotation, which are both numerically stable and robust data-adaptive transformations.

3.   **Fast RLS Algorithms** - by exploiting the redundancy in the Toeplitz structure of the input data matrix (a matrix where all the elements along each of its diagonals have the same value) and through use of linear least squares prediction in both the forward and backward direction the standard and square-root RLS algorithms can be reduced in computational complexity from $O(M^2)$ to $O(M)$, where M is the number of adjustable weights and $O(.)$ denotes "the order of." This reduction in computational

complexity is welcomed for hardware realisations. There are two types of fast RLS algorithms depending on the structure used (Haykin 1996),

1. **Order-recursive Adaptive Filters** - make linear forward and backward predictions using a lattice like structure. These can be realised in numerically stable forms.

2. **Fast Transversal Filters** - where the linear forward and backward predictions are done with separate transversal filters. These suffer from numerical stability problems and they require some form of stabilisation for practical implementations.

The tracking behaviour of the family of RLS algorithms, unlike the LMS algorithm, are model-dependent and therefore their tracking behaviour can be inferior to the stochastic gradient approach unless care is taken to choose an accurate model for the underlying physical process producing the input data.

1.2 Nonlinear Adaptive Filters

The linear adaptive filters discussed above are all based on the minimum mean square error criterion, which results in the Wiener solution for wide sense stationary statistics. This means that these filters can only relate to the second-order statistics of the input data and are strictly only optimum for Gaussian, or at least symmetrical, statistics. It is a fortunate happenstance that these types of filters have been found to be useful for statistics that deviate from this Gaussian ideal. If the input data has non-Gaussian statistics, where the Wiener solution is not guaranteed to be optimum, it is necessary to incorporate some form of nonlinearity in the structure of the adaptive filter to deal adequately with the higher-order statistical information. Although this will improve the learning efficiency it will be at the expense of more complex mathematical analysis of the problem. One important type of nonlinear adaptive filter is the adaptive Volterra filter.

1.2.1 Adaptive Volterra Filters

The adaptive Volterra filter can be seen as a kind of polynomial extension to the linear adaptive filter. It includes a zero order Direct Current (DC) offset term, a first-order linear term, and then a number of higher order terms starting with the second-order quadratic terms, third-order cubic terms and so on to some chosen order. In practice the Volterra filter is often implemented only up to quadratic or cubic order and rarely higher because of the huge increase in computational complexity beyond that, especially for high input dimensions.

1.3 Nonclassical Adaptive Systems

Three types of nonclassical adaptive systems that do not rely on linear modelling techniques are Artificial Neural Networks (ANNs), Fuzzy Logic (FL) and Genetic Algorithms (GAs). All these types of systems can be classified as nonlinear learning structures. Some forms of ANNs are similar to the classical adaptive systems in that they do have a set of parameters that are optimised based on the minimisation of a scalar quadratic error function. To some extent, fuzzy logic systems also have the same kind of similarity as they can be integrated with ANNs to produce hybrid adaptive systems. On the other hand genetic algorithms are different in their form and function although they do have various types of adaptation, or learning mechanisms designed to search, if not for optimal then at least better states.

1.3.1 Artificial Neural Networks

ANNs are a type of massively parallel computing architecture based on brain-like information encoding and processing models. The particular class of supervised training or learning ANNs have a similar external form as the linear adaptive filter. That is, there is a desired output behaviour that the ANN tries to learn as it is exposed to input training data and then it tries to generalise that behaviour after training. In this form ANNs offer the following advantages for adaptive filter applications,

1.   Ability to learn the model of almost any continuous (and preferable differentiable) nonlinear physical process given sufficient input-output data pairs generated from that process.

2.   Ability to accept weak statistical assumptions about the process.

3.   Ability to generalise its learning to new data after initial training.

4.   VLSI implementation in a massively parallel structure that is fault tolerant to hardware failure of some of the circuits because of the inherent redundancy.

ANNs, although theoretically able to model linear processes, are less useful for this purpose and should not be used for linear modelling. The well-established linear design methods are easier to use and analyse. The major disadvantage of ANNs is that it is much harder to specify and analyse their application to specific problems. Since the process model is developed from a limited set of training input-output data pairs a degree of uncertainty may exist about the bounds of applicability of the ANN solution. The training data may not be fully representative of the process and may not contain rare but very significant samples that are critical to the system's success.

1.3.2 Fuzzy Logic

Initially fuzzy logic was conceived of as a better method for sorting and handling data but has since proven to be good for control applications because it effectively mimics human control logic. It uses an imprecise but very descriptive language to deal with input data more like a human operator does and it is very robust and forgiving of operator and data input error. FL can be implemented in hardware, software, or a combination of both.

FL is based on the idea that although people do not require precise numerical information input, they are still capable of highly adaptive control functionality. Therefore, it is reasonable to assume that if feedback controllers could be programmed to accept noisy, imprecise inputs, they may be much more effective and perhaps easier to implement. FL uses a simple rule-based approach, such as "IF A AND B THEN C," to control problems as opposed to a strict system model based approach. In that sense a FL model is empirically-based, built by a designer's experience rather than on his/her technical understanding of the system. The design is based on imprecise terms such as "too cool," "add heat," etc. that are descriptive rather than numerically specific. For instance, if a person was trying to regulate the temperature of a shower they might just increase the hot water tap a little if they felt it was too cool and then adjust again if it still was not satisfactory. FL is capable of mimicking this type of behaviour but at a much higher rate than a human can do it.

1.3.3 Genetic Algorithms

Genetic algorithms represent a learning or adaptation method based on search that is analogous to biological evolution and can be described as a kind of simulated evolution. The interest in GAs lies in the fact that evolution is known to be a successful and robust method for biological adaptation. GAs can be seen as general optimisation mechanisms that are not guaranteed to find strictly "optimum" solutions but they often succeed in finding very suitable solutions. They are very useful not only for machine learning problems including function approximation and learning network topologies but also for very many other types of complex problems. In their most common form GAs work with hypotheses that may be described by symbolic expressions, computer programs, specific model parameters, collections of rules, and so on. They are useful in applications where hypotheses contain complex interacting parts, where the impact of each part on overall hypothesis fitness may be difficult to understand or model. GAs can also take advantage of parallel computer hardware since they lend themselves to computational subdivision into parallel subparts.

When the hypotheses are specifically computer programs the evolutionary computing process is called Genetic Programming (GP), where GP is a method for automatically creating computer programs. It starts from a high-level statement of what needs to be done and uses the Darwinian principle of natural selection to breed a population of improving programs over many generations (Koza *et al* 2003). Given a collection or population of initial hypotheses the search for an acceptable hypothesis proceeds from one generation to the next by means of operations inspired by processes in biological evolution such as random mutation and crossover. A measure of "fitness" is required to evaluate the relative worth of the hypotheses in each generation. For each generation the most "fit" hypotheses are selected probabilistically as seeds for producing the next generation by mutating and then recombining their components.

What makes GAs very special is that very little design effort is required to make transitions to new problem solutions within a given domain or even new problems from a completely different domain. In this sense GAs can truly be classified as intelligent in a broader sense because they can provide solutions through a generic approach that can rival solutions produced by human intelligence.

1.4 A Brief History and Overview of Classical Theories

It is both interesting and instructive to consider a brief history of the related areas of linear estimation theory, linear adaptive filters, adaptive signal processing applications and adaptive control. The following historical summary is according to Haykin (Haykin 1996) and the view of Åström and Wittenmark (Åström and Wittenmark 1995). It is not in any sense complete but it is sufficient to provide a suitable structure to relate the fundamentally important discoveries and techniques in these areas. Some, but not all, of the most significant techniques mentioned below are more fully developed and analysed in later Chapters.

1.4.1 Linear Estimation Theory

Galileo Galilei, in 1632, originated a theory of estimation, which he developed to minimise various functions of errors. However, it was Gauss who was given credit for the development of linear estimation theory. This was based on his invention of the method of least squares that he developed in 1795 to study the motion of heavenly bodies. Legendre invented the method of least squares independently of Gauss and actually published before Gauss in 1805 and was therefore subsequently given equal credit for the invention.

In the late 1930s and 1940s Kolmogorov and, Krein and Wiener originated the first studies of minimum mean square estimation in connection with stochastic processes. In 1939 Kolmogorov (Kolmogorov 1939) developed a comprehensive treatment of the linear prediction problem for discrete-time stochastic processes. In 1945 Krein (Krein 1945) subsequently extended Kolmogorov's results to continuous-time by using a bilinear transformation. By 1949 Wiener (Wiener 1949), working independently of either Kolmogorov or Krein, had formulated the continuous-time linear prediction problem but in a different context to the other two. He derived an explicit formula for the optimum predictor as well as solving the filtering problem of estimating a process corrupted by added noise. This required the solution of the integral equation known as the Wiener-Hopf equation, which was developed in 1931.

In 1947 Levinson (Levinson 1947) formulated the Wiener filtering problem in discrete-time in the form of a transversal filter structure as expressed by matrix Equation 1.5.

$$\mathbf{R}\mathbf{w}_0 = \mathbf{p} \tag{1.5}$$

where:

$\mathbf{R}$ = autocorrelation matrix of the tap inputs.

$\mathbf{w}_0$ = tap-weight vector of the optimum Wiener filter solution.

$\mathbf{p}$ = the cross-correlation vector between the tap inputs and the desired output.

For the special case of stationary inputs $\mathbf{R}$ takes the form of a Toeplitz structure, which allowed Levinson to derive a recursive procedure for solving the matrix Equation 1.5. Later in 1960 Durbin (Durbin 1960) rediscovered Levinson's procedure when he used it for recursive fitting of autoregressive time-series models.

Both Wiener and Kolmogorov assumed that the stochastic process was stationary and that there would be an infinite amount of data available. Other researchers in the 1950s generalised the Wiener and Kolmogorov filter theory for finite observation intervals and for nonstationary processes. However, these solutions were found to be complex and difficult to apply to the prevailing application of satellite orbit estimation. In 1960 Kalman (Kalman 1960) achieved considerable fame with his Kalman filter algorithm, which seemed to be very suitable for the dynamical estimation problems of the new space age. Kalman's original filter was developed for discrete-time processes. A year later in 1961, in conjunction with Bucy, he also developed it for the continuous-time case (Kalman and Bucy 1961).

Over the period from 1968 to 1973 Kailath reformulated the solution to the linear filtering problem by using the so called "innovations" approach, which was first introduced by Kolmogorov in 1941. The term "innovation" conveyed the idea of new information that is statistically independent of past samples of the process, i.e., orthogonal to the linear estimate given all the past data samples.

1.4.2 Linear Adaptive Filters

From earlier work in the 1950s the LMS algorithm for adaptive transversal filters emerged in 1959. It was developed by Widrow and Hoff for their ADALINE pattern recognition system (Widrow and Hoff 1960). The LMS algorithm is a stochastic gradient algorithm and is closely related to the concept of stochastic approximation developed by Robins and Monro (Robins and Monro 1951). The GAL algorithm was developed by Griffiths around 1977 and is only structurally different from the LMS algorithm. In 1981 Zames (Zames 1981) introduced the so called H^∞ norm (or minimax criterion) as a robust index of performance for solving problems in estimation and control. Subsequently, it was shown by Hassibi et al (Hassibi et al 1996) that the LMS algorithm is optimum under this new H^∞ criterion and thereby proving that its performance is robust. In 1965 Lucky (Lucky 1965) introduced a zero-forcing algorithm alternative to the LMS algorithm for the adaptive equalisation of communication channels, which also used a minimax type of performance criterion.

The family of RLS algorithms saw its beginnings with the work of Placket (Placket 1950). After much work by many researchers, in 1974 Godard (Godard 1974) presented the most successful application of the Kalman filter theory used to derive a variant of the RLS algorithm. It wasn't until 1994 that Sayed and Kailath (Sayed and Kailath 1994) exposed the exact relationship between the RLS algorithm and Kalman filter theory opening the way for the full exploitation of the vast literature on Kalman filtering for solving linear adaptive filtering problems. They showed that QR-decomposition-based RLS and fast RLS algorithms were simply special cases of the Kalman filter.

1.4.3 Adaptive Signal Processing Applications

Five significant applications of linear adaptive signal processing are,

1. Adaptive equalisation.

2. Speech coding.

3. Adaptive spectrum analysis.

4. Adaptive noise cancellation.

5. Adaptive beamforming.

Adaptive equalisation of telephone channels to minimise data transmission intersymbol interference was first developed by Lucky in 1965 (Lucky 1965). He used his minimax criterion based zero-forcing algorithm to automatically adjust the tap weights of a transversal equaliser by minimising what he called the peak distortion. This pioneering work by Lucky spearheaded many other significant contributions to the adaptive equalisation problem. In 1969, Gerosho and Proakis, and Miller independently reformulated the adaptive equaliser problem using a mean square-error criterion. In 1978 Falconer and Ljung (Falconer and Ljung 1978) developed a simplifying modification to a Kalman based algorithm, for adaptive tap adjustment, derived by Godard in 1974. This simplification reduced the computational complexity of Godard's algorithm to that comparable with the LMS algorithm. Satorius, Alexander and Pack in the late 1970s and early 80s showed the usefulness of lattice-based algorithms for adaptive equalisation.

Linear Predictive Coding (LPC) was introduced and developed for the problem of speech coding in the early 1970s by Atal and Hanauer. In LPC the speech waveform is represented directly in terms of time-varying parameters related to the transfer function of the vocal tract and excitation characteristics. The predictor coefficients are determined by minimising the mean square error between actual and predicted speech samples. Although a lattice structure for the linear prediction problem was developed by a number of investigators it was Saito and Itakura who were credited with the invention in 1972. They were able to show that the filtering process of a lattice predictor model and an acoustic tube model of speech were identical.

From the time when Schuster invented the periodogram for analysing the power spectrum of a time-series in 1898 until 1927 it was the only numerical method available for spectrum analysis. In 1927 Yule (Yule 1927) introduced a new approach based on the concept of a finite parameter model for a stationary stochastic process. This new approach was developed to combat the problem of the periodogram's erratic behaviour when applied to empirical time-series observed in nature such as sunspot activity. Yule's model was a stochastic feedback model in which the present sample of the time-series is assumed to consist of a linear combination of past samples plus an error term. This approach was called autoregressive spectrum analysis. Burg rekindled interest in the autoregressive method in the 1960s and 70s with his maximum-entropy method of power spectrum estimation directly from the available time-series. In 1971 Van den Bos (Van den Bos 1971) was able to show that the maximum-entropy method is equivalent to least squares fitting of an autoregressive model to the known autocorrelation sequence. The maximum-entropy method involved the extrapolation of the autocorrelation function of the time series in such a way that the entropy of the corresponding probability is maximised at each step of the extrapolation.

In 1967 Kelly of Bell Telephone Laboratories was given credit for inventing an adaptive filter for speech echo cancellation, which used the speech signal itself in the adaptation processes. Work on echo cancellers only started around 1965. Another type of adaptive noise canceller was the line canceller used for removing the mains power frequency interference from instrument and sensor preamplifier circuits. This was invented by Widrow and his co-workers at Stanford University. An early version of the device was built in 1965 and described in Widrow's paper in 1975 (Widrow *et al* 1975).

Initial contributions to adaptive array antennas were made by Howells in the late 1950s and by Applebaum in 1966. Howells developed a sidelobe canceller that became a special case of Applebaum's adaptive antenna array system. Applebaum's algorithm was based on maximising the Signal-to-Noise Ratio (SNR) at the array output for any type of noise. This classic work was reprinted in the 1976 special issue of IEEE Transactions on Antennas and Propagation (Applebaum and Chapman 1976). Another major work related to adaptive array antennas was put forward independently by Widrow and his co-workers in 1967. Their theory was based on the LMS algorithm and their paper, (Widrow *et al* 1967), the first publication in the open literature on adaptive array antenna systems, was considered to be another classic of that era. In 1969 Capon (Capon 1969) proposed a different method for solving the adaptive beamforming problem based on variance (average power) minimisation. Finally, in 1983, McWhirter (McWhirter 1983) developed a simplified version of the Gentleman-Kung systolic array for recursive least squares estimation, which is very well suited for adaptive beam forming applications.

1.4.4 Adaptive Control

Much of the history that is related to adaptive filters is also relevant to adaptive control systems as they incorporate much of the same theory. In fact, the main difference between the two is mostly a matter of application rather than underlying principles of operation. It is helpful to view signal processing and control theory, generally, as divergent branches of application of the same underlying theory. In some ways there should be more reintegration of the two fields for the sake of economy of understanding.

Historically, adaptive control has been very difficult to define explicitly, because it is seen to be superficially similar to feedback control. Both feedback control and adaptive control involve changing behaviour to conform to new circumstances. Attempts to draw distinctions between the two have not always been successful but it is now commonly agreed that a constant-gain feedback is not an adaptive system. From a pragmatic view point adaptive control can be seen as a special type of nonlinear feedback control in which the states of the process are separated into two categories related to the rate of change involved. In this view the slowly changing states are seen as the parameters and the fast ones are the ordinary feedback states. This definition precludes linear constant parameter regulators and gain scheduling from being called adaptive. Constant parameter regulators do not change their parameters and gain scheduled systems don't have any feedback once the parameters are changed to a new state.

There was extensive research on adaptive control applied to autopilots for high performance aircraft in the early 1950s. The dynamics of high-performance aircraft undergo major changes when they fly from one operating point to another (Levine 1996). This autopilot control problem was investigated by Whitaker *et al* (Whitaker *et al* 1958) using Model Reference Adaptive Control (MRAC). Early enthusiasm for more sophisticated regulators, which work well over a wider range of conditions, diminished through bad hardware, nonexistent theory, a general lack of insight, and finally an in flight test disaster. However, in the 1960s important underlying theory for adaptive control was introduced through the development of state space and stability theory based on Lyapunov and other important results in stochastic control theory. Correct proofs for stability of adaptive systems under very restrictive assumptions were developed in the late 1970s and early 1980s. Nevertheless, controversies over the practicality of adaptive control were still raging, mostly based on the sensitivity and potential instability of earlier designs. From this early work new and interesting research began into the robustness of adaptive control and into controllers that are universally stabilising. By the mid 1980s the field of robust adaptive control was opened based on new designs and analysis. In the late 1980s and early 1990s the focus of adaptive control research turned to extending the results of the 1980s to certain classes of nonlinear plants with unknown parameters. This led to new classes of MRAC with improved transient and steady-state performance.

Adaptive control has traditionally been classified into the MRAC and Adaptive Pole Placement Control schemes (APPC). In MRAC both the poles and zeros of the plant model are changed and in APPC only the poles are changed so that the closed-loop plant has the same input-output properties as those of the reference model.

1.5 A Brief History and Overview of Nonclassical Theories

The three main types of nonclassical adaptive or learning systems are ANN, FL and GAs. These form the foundation of what is now called the computational intelligent systems that have slowly developed into viable and accepted engineering solution methods over the past six decades. Although their origins are not much more recent than the classical adaptive filtering theories they have found broader commercial application only in more recent times.

1.5.1 Artificial Neural Networks

The history of ANNs has two significant periods. The period before 1970 represents the period of initial investigation and the period after 1970 opened the modern era of ANNs.

William James, in 1890, was the first to publish about brain structure and function in connection with psychological theories and neuropsychological research (James 1890). The first theorists to conceive the fundamentals of neural computing were W. S. McCulloch and W. A. Pitts in 1943 (McCulloch and Pitts 1943). They derived theorems related to the then current neural models. Their work proved that networks consisting of neurons could represent any finite logical expression but they did not demonstrate any learning mechanisms. It was Donald Hebb, in 1949, who was the first to define a method of neural network learning (Hebb 1949). Rosenblatt, in 1958, defined the ANN structure called the Perceptron that engineers recognised as a "learning machine" (Rosenblatt 1958). This work laid the foundations for both supervised and unsupervised training algorithms that are seen today in both the Multi-Layer Perceptron (MLP) and Kohonen networks respectively.

The advent of silicon based integrated circuit technology and consequent growth in computer technology in the 1960s was instrumental in the general surge in artificial neural computer systems. The ADALINE introduced by Widrow and Hoff was similar to the Perceptron but it used a much better learning algorithm, called the LMS algorithm, which can also be used for adaptive filters. The extension of the LMS algorithm is used in today's MLP. As the 1960s drew to a close there was growing optimism for the advance of ANN technology. However, funding and research activity in ANNs took a major dive after the publication of Minsky and Papert's book "Perceptrons" in 1969, which was mistakenly thought to have criticised the whole field of ANNs rather than just the simple Perceptron.

The decade of the 1970s saw a much reduced but stable activity in ANN research by a smaller number of researchers including Kohonen, Anderson, Grossberg and Fukushima. After the low period of the 1970s, several very significant publications appeared between 1982 and 1986 that advanced the state of ANN research. John J. Hopfield published a most significant single paper in 1982 (Hopfield 1982) and a follow-on paper in 1984 (Hopfield 1984) identifying ANN structures that could be generalised and that had a high degree of robustness. The Parallel Distributed Processing (PDP) Research Group published the first two volumes of their "Parallel Distributed Processing" in 1986 followed by a third volume in 1988. The most significant contribution of the PDP volumes was the derivation and subsequent popularisation of the Backpropagation-of-error learning algorithm for MLPs. Closely following that, important ANNs based on Radial Basis Functions (RBFs) (Powell 1985) (Broomhead and Lowe 1988) including Donald Specht's Probabilistic Neural Network (PNN) (Specht 1988) and General Regression Neural Network (GRNN) (Specht 1988, 1991) were introduced.

A significant resurgence in interest in ANNs occurred in the 1980s as computers got bigger, faster and cheaper. This ubiquitous computing power allowed the development of many mathematical tools to express analytically, the complex equilibrium state energy landscapes necessary to study ANN architectures. Because of this increased and enthusiastic research activity, especially in conjunction with statistics, many new and useful learning theories have now been proposed and implemented. One of the most important of these is Vapnik's "Statistical Learning Theory" (Cherkassky and Mulier 1998).

1.5.2 Fuzzy Logic

The basic foundations of fuzzy logic were conceived by Lotfi Zadeh in 1965 as an extension of classic set theory (Zadeh 1965). He presented it not as a control methodology, but as a way of processing data by allowing partial set membership rather than specific or crisp set membership/non-membership. Due to inadequacy of computing systems at the time this approach to set theory was not applied to control systems until the 1970s. U.S. manufacturers were not quick to embrace this technology, whereas the Europeans and Japanese began to aggressively build commercial products with it almost from the outset.

Ebraham Mamdani applied FL to control a simple steam engine for the first time in 1974 at the University of London (Mamdani 1974). It was not for another six years that the first industrial application appeared for the control of a cement kiln by F. H. Smidth of Denmark. Fuji Electric of Japan applied FL to the control of a water purification plant in the 1980s and Hitachi later developed an automatic train control system. This led to the FL boom in Japan in the early 1990s with the production of household electronics products using FL. Since then, FL has been applied to a wide range of growing applications including decision support systems, investment consultation, fault diagnosis, medical diagnosis, transport scheduling, management strategy, social and environmental systems (Tanaka 1997).

1.5.3 Genetic Algorithms

In 1948 Alan Turing identified an approach to machine intelligence based on genetical or evolutionary search by which a combination of genes is sought based on survival value. He didn't specify how to conduct the search or mention the concept of population recombination but he did introduce the idea that a number of child-machines should be experimented with to see how well they learn and then choose from the best of these (Turing 1950). Here, the structure of the machine represented hereditary material. Changes of the machine represented mutations, and natural selection (fitness) was based on the experimenter's judgement. It was left to John Holland between 1962 and 1975 to introduce the crucial concepts of maintaining large populations and sexual recombination within them (Holland 1962, 1995).

Since the 1950s there has been a great variety of experimentation with evolution-based computational approaches, which has included optimisation of numerical parameters in engineering design. In 1966 Fogel, Owens and Walsh (Fogel, Owens and Walsh 1966) first developed evolutionary programming, which was a method of evolving finite-state machines. This method was followed up and further developed by numerous researchers including John Koza (Koza 1992). Koza applied the search strategy of GAs to hypotheses consisting of computer programs, which has now come to be known as Genetic Programming (GP).

1.6 Fundamentals of Adaptive Networks

An adaptive network can be used to model either a linear system whose parameters are unknown (or changing with time) or a nonlinear system whose model is unknown (or also changing with time). A linear adaptive system will eventually converge to a linear solution over sufficient time and range of input signals. It will then continue to adapt only if the system or noise statistics change. For a nonlinear process, a linear adaptive system can only adapt to a linear approximation at the current operating point. It is possible however, to keep a historical record of the set of linear models for each small region around a set of operating points and then apply an appropriate model as the set point changes. This is called schedule or switching control with multiple models. A nonlinear adaptive network will adapt to a more accurate model at the current operating point, but like the linear adaptive network it cannot generalise this to new operating points, unless a historical record is kept. To ensure a more robust control of nonlinear systems it is desirable to have some historical information about the system over the expected range of operating points in parallel with a fast adaptive network to make up for any differences.

The basic system structure that is applicable to both adaptive and some learning networks is depicted in Figure 1.2. In the most general terms the vector of the noisy input signal at discrete instance k is $\mathbf{x}_k$ and the vector error signal is $\mathbf{e}_k = (\mathbf{d}_k - \mathbf{y}_k)$, where $\mathbf{d}_k$ is the vector of the noiseless desired output response and $\mathbf{y}_k$ is the actual vector network response given an input of $\mathbf{x}_k$. The network is adapted or trained by presenting it with successive input and desired vector pairs and then using the error signal to adapt the network in such a way as to consistently reduce the error according to some specific learning rule. A least squares error rule is commonly used for both linear networks and nonlinear learning and adaptation. When the network is trained or adapted the error arrives at some statistical minimum. As the statistics of the input vectors change with time the network can be continually adapted to maintain a minimum error, otherwise the network parameters are fixed after training. Either way the network then represents an estimate of the noiseless model of the underlying process or function at that point. The process function is represented by the set of input and desired vector pairs used to train the network.

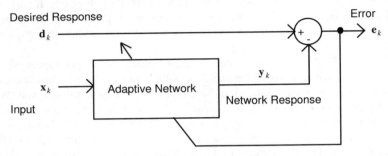

Figure 1.2. Basic Adaptive Structure

There are two main considerations related to the basic adaptive structure shown in Figure 1.2. Firstly, if the desired responses are known why is the network needed at all? Secondly, the adaptation mechanism may be simple or complex depending on the network and consequently, the convergence may take considerable time or computation. In the first case, although the desired responses are not usually known explicitly it is often possible to derive them or find responses that are correlated to them. Since there is no general solution to this problem it is necessary to look at specific examples to gain insight into application issues. The most common generic configurations according to (Lim and Oppenheim 1988) are for,

1.   Adaptive prediction.

2.   Adaptive forward modelling.

3.   Adaptive inverse modelling.

4.   Adaptive interference cancelling.

These can best be represented by Figures 1.3 to 1.6 respectively.

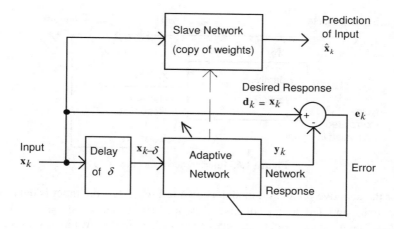

Figure 1.3. Adaptive Prediction

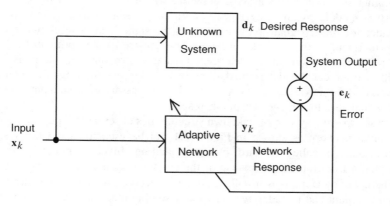

Figure 1.4. System Forward Modelling

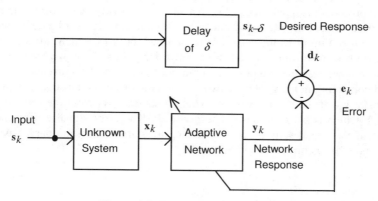

Figure 1.5. Inverse System Modelling

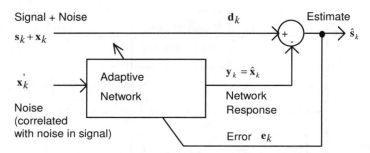

Figure 1.6. Interference Cancelling

For the adaptive prediction model, shown in Figure 1.3, the input is delayed by δ time units and fed to an adaptive network. The input serves as the desired response. The network weights are adapted and when they converge this produces a best estimate of the present input given an input delayed by δ. After convergence the weights are copied into a slave network, which is then taken to be the best predictive model. Wiener developed optimum linear least squares filtering techniques for linear signal prediction. When the signal's autocorrelation function is known, Wiener's theory yields the impulse response of the optimum filter. The autocorrelation function can be determined using a correlator, or alternatively the predictive filter can be determined directly by adaptive filtering. For nonlinear problems and for non-Gaussian noise it is strictly necessary to use adaptation with a nonlinear network to achieve acceptable results.

In cases where a system of unknown structure has observable inputs and outputs an adaptive network as shown in Figure 1.4 can be used to model the system's response. This is called forward system modelling. Inverse modelling involves developing a filter that is the inverse of the unknown system, as shown in Figure 1.5. The delay by δ units is usually included to account for the propagation delay through the plant and the adaptive network, assuming that both are causal systems, i.e., their output depends only on inputs up to and including the current time.

Separating a signal from additive noise, also called interference cancelling, is a common problem in signal processing. An adaptive network as shown in Figure 1.6 can be used to subtract the noise out of the signal. This gives a better result than applying an optimum Kalman or a Wiener filter, both of which introduce some inevitable phase distortion. The adaptive network solution is only viable when there is an additional reference input $\mathbf{x'}_k$ containing noise that is correlated with the original corrupting noise $\mathbf{x}_k$. The network filters the reference noise $\mathbf{x'}_k$, to produce an estimate $\mathbf{y}_k$, of the actual noise $\mathbf{x}_k$. Then, it subtracts $\mathbf{y}_k$ from the primary input $(\mathbf{s}_k + \mathbf{x}_k)$, which acts as the desired response $\mathbf{d}_k$. The error signal $\mathbf{e}_k$ becomes the estimate of the signal $\mathbf{s}_k$ if, $\mathbf{s}_k$, $\mathbf{x'}_k$, $\mathbf{x}_k$ and $\mathbf{y}_k$ are statistically stationary, have zero means, and $\mathbf{s}_k$ is uncorrelated with $\mathbf{x'}_k$ and $\mathbf{x}_k$.

1.7 Choice of Adaptive Filter Algorithm

In the most general terms an adaptive algorithm tries to minimise an appropriate objective or error function that involves the input, reference and filter output signals. An objective function must be non-negative and ideally have an optimum value of zero. The adaptive algorithm can be seen to consist of three main parts, the definition of the minimisation algorithm, the definition of the objective function and the definition of the error signal (Dinz 1997).

The most commonly used minimisation methods used for adaptive filters are Newton's method, quasi-Newton methods and the steepest-descent gradient method (Principe 2000). Gradient methods are easy to implement but the Newton method usually requires less iterations to achieve convergence. A good compromise between these two are the Quasi-Newton methods which have reasonable computational efficiency and good convergence. However, the Quasi-Newton methods are susceptible to instability problems. In all these methods the gain or convergence factor must be chosen carefully based on good knowledge of the specific adaptation problem.

The error function can be formed in many ways but the most common ways include the Mean Square Error (MSE), Least Squares (LS), Weighted Least Squares (WLS), and Instantaneous Squared Value (ISV). Strictly speaking the MSE is approximated by the other more practical methods since the MSE is a theoretical value requiring an infinite amount of data. ISV is the easiest to implement but it has noisy convergence properties. The LS method is suitable for stationary data, whereas WLS is valuable for slowly varying data statistics. The choice of error signal is crucial to algorithm complexity, convergence properties, robustness and control of biased or multiple solutions.

There is a great diversity of adaptive applications with their own peculiarities. Every application must be carefully evaluated and understood before a suitable adaptive algorithm can be chosen, because a solution to one application may not be suitable for another. The choice of algorithm must take into account not only the specifics of the application environment but also issues of computational cost, performance, and robustness. Often it can be instructive to apply the simple but robust LMS or Backpropagation-of-error algorithm to the problem first, to study, evaluate and compare the benefits of an adaptive solution to the problem. Further and more detailed design decisions can then be made based on those findings. All adaptive system forms can be implemented to accept and process either real or complex input signals depending on the requirements.

2. Linear Systems and Stochastic Processes

A review of linear systems theory and stochastic processes is presented here to provide a reference and a summary of the fundamental ideas required for following chapters that draw upon linear methods. It is assumed that the reader already has a basic familiarity with these concepts and thus reading through them will cement them more firmly in mind. More advanced readers may skip this Chapter but it is recommended that they at least skim through it because there are some useful summary panels that help provide a good overview of important concepts.

Every physical system is broadly characterised by its ability to accept an input such as voltage, current, pressure etc. and to produce an output response to this input. The analysis of most systems can be reduced to the study of the relationship between certain input excitations and the resulting outputs. The two main types of systems are lumped parameter and distributed parameter systems. Lumped parameter systems are those made-up of a finite number of physical discrete elements, each of which are able to store or dissipate energy (for example capacitors, inductors and resistors) or if it is a source, to deliver energy. These systems can be described by ordinary differential equations. Distributed parameter systems consist of elements that cannot be described by simple lumped elements because of propagation time delays in signals traversing the elements. However, each infinitesimal part of distributed elements can be modelled in lumped parameter form. These systems can be described in terms of partial differential elements, for example transmission lines. Figure 2.1 shows the taxonomy of the type of systems of interest to signal processing. The main system types relevant to this book are the stochastic and the deterministic systems, which include both continuous-time and discrete-time linear and nonlinear systems, and in particular discrete-time linear and nonlinear time varying systems.

A system is defined as an entity that manipulates one or more signals to accomplish a function, thereby yielding new signals. In its simplest form a system takes an input signal, performs some signal processing and then presents a required output signal. Some examples of important engineering systems include communications systems, control systems, electrical and electronic circuits, electrical power systems, remote sensing systems, and biomedical signal processing systems. System analysis is usually performed to determine system response to input or excitation signals over desired ranges. Some reasons why this may be done are to establish a performance specification, aid in component selection, uncover and study system deficiencies, explain unusual or unexpected system operation or produce quantitative data related to system operation for various needs. A mathematical system model is essential for the analysis of the system under various ambient conditions and to determine the effects of changing parameters of system components. In most practical circumstances a final system design requires an intelligent mixture of analytic and experimental approaches. An initial analytic study provides a basis for interpretation of experimental results and a design direction, which then establishes suitable variable ranges for experimental design.

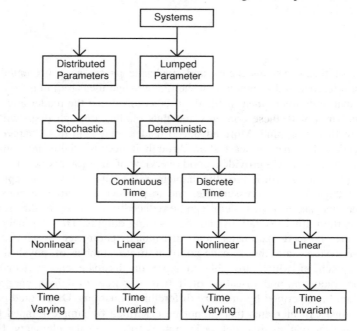

Figure 2.1. A System Representation

A signal is defined to be a function of one or more variables that conveys information on the nature of a physical phenomenon. A signal can also be seen as a varying quantity or variable that represents a form of energy propagation within or through a system. Signals can be one-dimensional, two-dimensional or multi-dimensional, depending on the number of variables. A signal can be naturally occurring or it can be synthesised or simulated. The variation of signal amplitude as a function of some independent variable or variables (usually time) is defined as the signal waveform. Signals can be generally classified as deterministic (predictable), random (unpredictable) or chaotic (predictable only in the short term). Random or stochastic signal theory is an essential part of linear systems analysis because all real systems and signals are invariably contaminated with random noise of some type or other. A chaotic signal emanates from a chaotic system, defined by a coupled system of nonlinear differential equations, whose parameters are fixed. Because chaotic systems are defined to be very sensitive to initial conditions they are only predictable in the short term. Any small uncertainty in the initial conditions of a chaotic system grows exponentially with time. Therefore a chaotic signal is a random-like signal with only slight predictability. Examples of chaotic signals and systems include fluid turbulence and radar sea clutter.

2.1 Basic Concepts of Linear Systems

A mathematical system model of a physical process relates the input signals to the output signals of the system. A linear system relates the inputs to the outputs by a linear transformation, $LT(.)$, as shown in Figure 2.2. Systems may be continuous-time systems or discrete-time systems. Continuous-time system input signals, impulse responses, and output signals are usually denoted by symbols $x(t)$, $h(t)$, and $y(t)$ respectively, where t is the continuous-time variable. The corresponding discrete-time symbols are either $x(n)$, $h(n)$, and $y(n)$, or alternatively and preferably, using square brackets, $x[n]$, $h[n]$, and $y[n]$, where n is the discrete-time variable. The square brackets are very often used for discrete-time variables but round brackets are also perfectly acceptable and can be used interchangeably as appropriate.

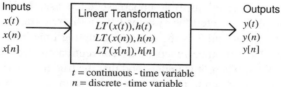

$$t = \text{continuous - time variable}$$
$$n = \text{discrete - time variable}$$

Figure 2.2. Linear System

The output of a noiseless linear continuous-time system $y(t)$ is equal to the convolution of the input signal $x(t)$ with the system's impulse response $h(t)$, as defined by Equation 2.1. The Fourier transform of the impulse response $h(t)$ is defined as the frequency response of the system $H(f)$ or $H(\omega)$, where f is the frequency variable in Hertz and ω is the frequency variable in radians per second.

$$y(t) = x(t) * h(t) = \int_{-\infty}^{\infty} x(\tau)h(t-\tau)d\tau$$

$$x(t) * h(t) \Leftrightarrow X(f)H(f)$$

(2.1)

where:

   *   denotes convolution.

   $X(f)$ is the frequency response of $x(t)$.

   $H(f)$ is the frequency response of $h(t)$.

The class of real physical systems are causal systems. A causal system produces an output that at any time t_1 is a function of only those input values that have occurred for times up to and including t_1 (Gabel and Roberts 1987). Noncausal systems produce outputs as a function of inputs ahead of t_1. These are often useful for computational purposes in the process of eventually producing real outputs for real systems.

The state of a system is defined by a minimal set of variables known at t_1 such that for all inputs for $t > t_1$ it is possible to calculate the system outputs for $t > t_1$. Continuous-time systems are systems for which the input, output and state are all functions of a continuous real-time variable t, e.g. $f(t)$. Sets of variables for continuous-time signals and systems can be represented as vector quantities as follows,

$$\mathbf{x}(t) = \begin{bmatrix} x_1(t) \\ x_2(t) \\ \vdots \\ x_p(t) \end{bmatrix} = [x_1(t) \quad x_2(t) \quad .. \quad x_p(t)]^T$$

Vectors are denoted by bold lowercase letters and are single column matrices as exemplified above. Scalar values and time domain signals are denoted by plane lowercase letters. Uppercase plain letters can denote scalars or signal or system frequency domain representations. Bold uppercase letters usually denote matrices.

Uniformly sampled discrete-time systems have a discrete time variable $t_n = t = nT$, where n is an integer and T is the uniform sampling time interval, i.e., $f(t_n) \equiv f(nT) \equiv f(n) \equiv f[n]$. Discrete-time functions by convention are represented using an integer time variable n, where it is understood that the true time variable is actually nT. Sets of variables for discrete-time signals and systems can also be represented as vector quantities as follows,

$$\mathbf{x}[n] = \begin{bmatrix} x_1[n] \\ x_2[n] \\ \vdots \\ x_p[n] \end{bmatrix} = [x_1[n] \quad x_2[n] \quad .. \quad x_p[n]]^T$$

Linear systems possess the property of supposition, i.e., for a scalar rational constant α and signals x_1 and y_1,

if $x_1 \rightarrow y_1$ and $x_2 \rightarrow y_2$ then $x_1 + x_2 \rightarrow y_1 + y_2$
and if $x \rightarrow y$ then $\alpha x \rightarrow \alpha y$

A linear system is defined by a linear transformation LT of inputs x into outputs y if the LT satisfies,

$LT(\alpha x_1 + \beta x_2) \rightarrow \alpha LT(y_1) + \beta LT(y_2)$,
where: α and β are arbitrary constants.

This type of system is referred to as Linear Shift-Invariant (LSI) if it is a discrete-time system and Linear Time-Invariant (LTI) if is a continuous-time system. LTI systems are characterised by linear differential systems equations with constant parameters, i.e.,

$$\text{if } x(t) \rightarrow y(t) \text{ then } x(t + \tau) \rightarrow y(t + \tau)$$

LSI systems are characterised by difference equations with constant parameters, i.e.,

$$\text{if } x[n] \rightarrow y[n] \text{ then } x[n + k] \rightarrow y[n + k]$$

Constant parameter systems have parameters that do not change with time and variable parameter systems have parameters that do change or adapt with time.

2.2 Discrete-time Signals and Systems

Uniformly sampled discrete-time signals and systems defined by x have a discrete time integer index variable n, and are typically denoted by $x(n)$ or $x[n]$. Discrete-time systems are most often characterised by difference equations. The most general form of a difference equation for a LSI system is as follows,

$$a_0 y[n] = b_0 x[n] + b_1 x[n-1] + \dots + b_q x[n-q] - a_1 y[n-1] - \dots - a_p y[n-p]$$

$$a_0 y[n] = \sum_{m=0}^{q} b_m x[n-m] - \sum_{m=1}^{p} a_m y[n-m]$$

The constant indices p and q determine the order of the system (p, q)

i.e., $p = 0, 1, 2, \dots$, and $q = 0, 1, 2, \dots$, and $a_0 = 1$ (by convention)

The associated transfer function is,

$$H(z) = \frac{Y(z)}{X(z)} = \frac{B(z)}{A(z)} = \frac{\displaystyle\sum_{m=0}^{q} b_m z^{-m}}{\displaystyle\sum_{m=1}^{p} a_m z^{-m}}$$

The variables a_m and b_m are the filter coefficients that need to be solved for particular filter realisations. When upper bound constant indices p and q are both nonzero the system described by the difference equation (and associated transfer function) is said to be recursive or an Autoregressive Moving-Average (ARMA) model (pole-zero filter). If $q = 0$ and $p > 0$ the model is called Autoregressive (AR), (all-pole filter). If $p = 0$ and $q > 0$ the system is nonrecursive, which is also referred to as a Finite Impulse Response (FIR) system or Moving Average (MA) model (all-zero filter). Although recursive systems can in theory implement FIR systems, they more typically have Infinite Impulse Responses (IIR). Often recursive systems are spoken of as though they were all IIR, but that is not necessarily the case.

The output of a noiseless LSI discrete-time system $y[n]$, is equal to the convolution sum of the input signal $x[n]$, with the system's impulse response $h[n]$, as defined by Equation 2.2.

$$y[n] = x[n] * h[n] = \sum_{m=-\infty}^{+\infty} x[m]h[n-m]$$

$$x[n] * h[n] \Leftrightarrow X(k)H(k)$$

(2.2)

The impulse response is defined as the output of the system given a unit sample $\delta[n]$ input as defined by Equation 2.3.

$$\delta[n] = \begin{cases} 1, & n = 0 \\ 0, \text{otherwise} \end{cases}$$

(2.3)

The finite impulse response of a nonrecursive system is defined by Equation 2.4. The variables $b[m]$ (or b_m) in Equation 2.4 are the FIR filter's coefficients, which are equal to the FIR filter's impulse response, i.e., $b[n] \equiv b_m \equiv h[n]$.

$$y[n] = \sum_{m=0}^{q} b[m]\delta(n-m)$$

(2.4)

An example of a simple recursive system is defined by Equation 2.5.

$$y[n] = a_1 y[n-1] + x[n]$$

(2.5)

The impulse response of a recursive system is likely to be infinite and for the example Equation 2.5 it is $h[n] = a_1^n u[n]$, where $u[n]$ is the unit step defined by Equation 2.6.

$$u[n] = \begin{cases} 1, & n \geq 0 \\ 0, \text{otherwise} \end{cases}$$

(2.6)

Another most important signal often used for the Fourier decomposition of signals is the periodic complex exponential defined by Equation 2.7 which is also known as Euler's identity.

$$e^{jn\omega_0} = \cos(n\omega_0) + j\sin(n\omega_0)$$

(2.7)

where:
ω_0 is a constant.

The properties of stability and invertibility are important properties for LSI systems. Stability implies that an output $y[n]$ is bounded in amplitude for whenever the input is bounded. This type of system is said to be stable in the Bounded-Input-Bounded-Output (BIBO) sense, i.e., for any bounded input $|x[n]| \leq A < \infty$, the output is bounded, $|y[n]| \le B < \infty$. For LSI systems, stability is guaranteed whenever the unit sample response is absolutely summable, i.e., as defined by Equation 2.8.

$$\sum_{n=-\infty}^{+\infty} \left| h[n] \right| < \infty \tag{2.8}$$

A system is invertible if the input to the system may be uniquely determined from observation of the output. The invertibility of a LSI system is intimately related to its phase characteristics. A LSI system is causal, invertible and stable if it is of minimum phase or equivalently minimum delay (Proakis and Dimitris 1996).

2.3 The Discrete Fourier Transform (DFT)

The numerical calculation of the frequency spectrum of a continuous-time signal is an important engineering tool for both system design and analysis. The numerical calculation of the Fourier transform (Kammler 2000) of a continuous-time signal $h(t)$ involves two sampling processes. Firstly, the Fourier transform integral defined by Equation 2.9 must be approximated by a summation.

$$H(j\omega) = H(j2\pi f) = \int_{-\infty}^{+\infty} h(t)e^{-j\omega t}\,dt = \int_{-\infty}^{+\infty} h(t)e^{-j2\pi t}\,dt \tag{2.9}$$

This implies that the continuous-time function $h(t)$, must be represented by a discrete sequence $h[n]$. Secondly, $H(j\omega)$ must be represented by a discrete set of frequency samples $H(k)$. The sampling process in both time and frequency results in periodicity in both domains, which creates certain problems such as aliasing that must be avoided. Aliasing is avoided by band-limiting the signal to a bandwidth less than or equal to half the sampling frequency F_s, which is referred to as the Nyquist sampling rate. Given a N-point real, discrete-time signal $h(nT)$ of finite-duration NT, a corresponding periodic signal $h_p(nT)$ with period NT can be formed as defined by Equation 2.10.

$$h_p(nT) = \sum_{m=-\infty}^{+\infty} h(nT + mNT) \tag{2.10}$$

Taking the periodic signal defined by Equation 2.10, its corresponding periodic Discrete Fourier Transform (DFT), $H_p(k)$ is defined as follows,

$$H_p(k) = \sum_{n=0}^{N-1} h_p[n]e^{\frac{-j2\pi kn}{N}} \quad , \quad \text{The DFT is circular or periodic with period } N.$$

The index k represents the discrete frequencies $f_k = f(k) = \dfrac{F_s k}{N}$.

The discrete frequency domain variable is represented by the frequency integer index k such that the discrete frequency represented by k is $f_k = f(k) = F_s k / N$. The continuous-time Fourier transform $H(f(k))$ at these discrete frequencies is only approximately related to the DFT as follows,

$$H(f_k) = \int_{-\infty}^{\infty} h(t) e^{-j2\pi k_k t} dt$$

$$H(f_k) \approx \sum_{n=0}^{N-1} h_p[n] e^{-j2\pi f_k nT} T = T \sum_{n=0}^{N-1} h_p[n] e^{-\frac{j2\pi kn}{N}}$$

$$H(f_k) = H(f(k)) \approx T H(k)$$

The Fourier transform and DFT are only approximately related to each other due to the fact that the DFT is circular. If F_s, the sampling frequency, is made greater than twice the highest frequency existing in the signal the approximation gets better and better as F_s is increased further. However, the DFT $H_p(k)$ and what is called the Discrete-time Fourier Transform (DTFT) $H_p(e^{j\theta})$ are directly related as defined by Equation 2.11.

$$H_p(k) = H_p(e^{j\theta}) \Big|_{\theta = \frac{2\pi k}{N}} \tag{2.11}$$

The DTFT is sampled at N frequencies that are equally spaced between 0 and 2π, to produce the DFT. The Inverse DFT (IDFT) $h_p[n]$ is defined as follows,

$$h_p[n] = \frac{1}{N} \sum_{k=0}^{N-1} H_p(k) e^{\frac{j2\pi kn}{N}} \text{, The IDFT is circular or periodic with period } N.$$

$$h_p[n] = IDFT\left[H_p(k)\right] = \frac{1}{N} DFT\left[H_p^*(k)\right]$$

The power or energy of a time domain signal representation must be equal to the power or energy of its corresponding frequency domain representation. This most useful property is known as Parseval's theorem, which states that the sum of the squares of a signal, $x[n]$, is equal to the integral of the square of its DTFT, as defined by Equation 2.12.

$$\sum_{n=-\infty}^{\infty} |x[n]|^2 = \frac{1}{2\pi} \int_{-\pi}^{+\pi} |X(e^{j\theta})|^2 d\theta \tag{2.12}$$

2.3.1 Discrete Linear Convolution Using the DFT

The DFT and IDFT can be used to perform linear convolution of two discrete-time signals, $x[n]$ of length N_x and $h[n]$ of length N_h. If $N \geq N_x + N_h$, then the N-point circular convolution of the two sequences is certain to be the same as their linear convolution and it can be computed as follows,

$$x[n] * h[n] \rightarrow x_p[n] * h_p[n]$$

$$x_p[n] * h_p[n] = \sum_{m=0}^{N} x_p[m] h_p[n-k]$$

$$= IDFT_N \left[DFT_N \left[x[n] \right] DFT_N \left[h[n] \right] \right]$$

A N-point DFT is fairly inefficient to compute because it requires N^2 complex multiplications and $(N^2 - N)$ complex additions. For large N, the computation efficiency can be improved dramatically by using a Fast Fourier Transform (FFT) algorithm. For a radix-2 decimation-in-time N-point FFT, if N is power of 2, it only requires $(N/2)\log_2 N$ complex multiplications and $(N)\log_2 N$ complex additions.

2.3.2 Digital Sampling Theory

The digital signal sampling theorem is known as the Uniform Sampling Theorem. A band-limited analogue continuous-time signal can be uniquely and completely described by a set of uniformly time spaced discrete analogue amplitude samples taken at a sampling frequency rate of F_s, the Nyquist sampling rate, provided no signal energy exists at a frequency equal to or greater than $F_s/2$. This is strictly only true if the samples are of infinite amplitude precision and sampled with a Dirac impulse function having a zero time width. Of course this is unrealisable, so the theory must be treated as an approximation to real sampling systems.

The initial signal measurement process can be modelled by Equation 2.13.

$$m(t) = s(t) + n(t) \tag{2.13}$$

where:

$m(t)$ is the measured signal including noise.
$s(t)$ is the true signal (if it is stochastic there is a random nature to $s(t)$).
$n(t)$ is a random noise component (the most common type is Gaussian noise).

If the signal $s(t)$ is to be digitally sampled at uniform intervals of T seconds (a sampling rate of $F_s = 1/T$ Hz) for a total of N samples then,

$m(kT) = s(kT) + n(kT)$, where $k = 1,....., N$ the sampling integer index number.
or for short,
$m[k] = s[k] + n[k]$

In theory, if $m[k]$ is the exact amplitude value of the signal plus noise at precisely time $t = kT$ then some useful things can be said about the sampling process. Given that the highest frequency component (highest Fourier series component) in the signal $m(t)$ is say f_h, then according to the sampling theorem, a choice of $F_s \geq 2 f_h$ is sufficient to fully capture all the $m(t)$ signal information in the N uniformly sampled amplitude points of infinite amplitude precision. Fortunately a band-limited analogue continuous-time signal has a lot of redundancy and the signal amplitudes between the sample points can, at least in theory, be fully recovered through the use of a special interpolation equation implicit in the sampling and recovery processes. This discussion strictly only applies to $m(t)$ not $s(t)$, but for theoretical purposes it can be assumed that $m(t) = s(t)$. However, in practice this is never exactly the case. In any practical case, the amplitude $m[k]$ is never exact because the Analogue to Digital Converter (ADC) that must be used to do the sampling has a finite accuracy and resolution. The sample is never taken at exactly time $t = kT$ because all ADCs have finite aperture times (or sample capture times). Therefore, for various reasons, there is always some small error in the digitised signal $m[k]$ compared to the actual signal $m(t)$. This error can be minimised if the resolution of the ADC is high (has a high number of conversion bits) and the aperture time is kept small (aperture time $\leq T/8$). Cheaper ADCs have much high errors of these two types.

When a signal is digitally sampled the sampled signal has a periodic frequency spectrum $M_p(f)$, which is the signal's baseband spectrum $M(f)$ repeated in higher bands at integer multiples of F_s. If F_s is less than twice f_h, the signal's baseband frequency spectrum is inevitably aliased or corrupted by the overlap of the higher bands. Refer to and compare example Figures 2.3, 2.4 and 2.5.

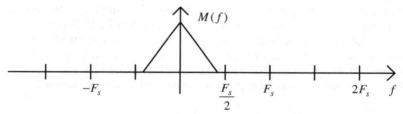

Figure 2.3. Signal Baseband Spectrum

The ideal sampling process introduces exact signal spectrum replicas centred at integer multiples of F_s as depicted in Figure 2.4.

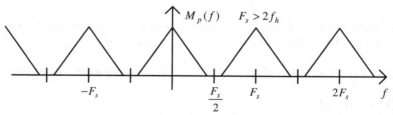

Figure 2.4. Ideal Sampled Spectrum

If the sampling frequency is lower than the Nyquist frequency aliasing occurs, which produces signal errors not removable by signal processing, except under special circumstances. As can be seen in Figure 2.5 aliasing means that the signal spectrum replicas overlap, producing a corruption to the baseband frequency spectral region.

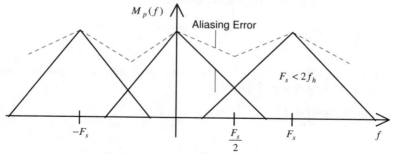

Figure 2.5. Aliased Sampled Spectrum

If the same digital input signal is to be converted back to an analogue signal it is output via a Digital to Analogue Converter (DAC) and passed through another lowpass filter called the anti-imaging filter, which cuts off all frequencies above f_h or $F_s/2$ to remove the higher spectral bands associated with the periodic digital signal spectrum. In this way the original baseband spectrum can be recovered and consequently the original continuous-time analogue signal. Refer to example Figures 2.6 and 2.7.

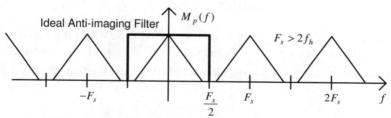

Figure 2.6. Sampled Spectrum and Ideal Anti-imaging Filter

Figure 2.6 shows a perfect brick wall lowpass anti-imaging filter, which is impossible to design in practice. Consequently, a real lowpass filter must be used, which inevitably introduces some errors in the perfectly recovered baseband signal spectrum shown in Figure 2.7.

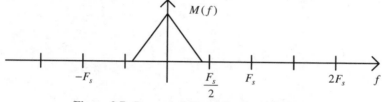

Figure 2.7. Recovered Signal Baseband Spectrum

A proof of the sampling theory discussed above may be derived as follows. Assume a band-limited continuous-time analogue signal $m(t)$ for which the Fourier transform $M(j\omega) = 0$ for $|\omega| \geq 2\pi F_s/2$. A sampled signal $\hat{m}(t)$ is generated by sampling the signal $m(t)$ using an ideal signal sampler $p(t)$ defined by Equation 2.14, which is also known as the comb function.

$$p(t) = \sum_{n=-\infty}^{\infty} \delta(t - nT) \qquad (2.14)$$

The sampled signal is defined by Equations 2.15 and 2.16 and the sampling processes are represented by Figure 2.8.

$$\hat{m}(t) = p(t)m(t) \qquad (2.15)$$

$$\hat{m}(t) = \sum_{n=-\infty}^{\infty} m(nT)\delta(t - nT) \qquad (2.16)$$

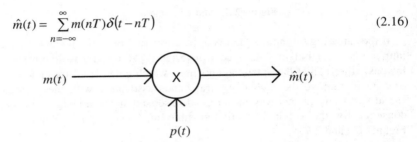

Figure 2.8. Sampling Process

Equation 2.16 represents the convolution of signal $m(nT)$ with the single impulse function $\delta(nT)$. Taking the Fourier transform $FT[\ ]$ of the sampled signal $\hat{m}(t)$ results in the mathematical development defined by Equations 2.17 to 2.20.

$$M_p(j\omega) = FT[\hat{m}(t)] = FT\left[\sum_{n=-\infty}^{\infty} m(nT)\delta(t - nT)\right] \qquad (2.17)$$

$$M_p(j\omega) = \sum_{n=-\infty}^{\infty} m(nT)\ FT[\delta(t - nT)] \qquad (2.18)$$

$$M_p(j\omega) = \sum_{n=-\infty}^{\infty} m(nT) \sum_{n=-\infty}^{\infty} e^{-jn2\pi F_s} \qquad (2.19)$$

$$M_p(j\omega) = \frac{1}{T} \sum_{n=-\infty}^{\infty} M(j\omega + jn2\pi F_s) \qquad (2.20)$$

From Equation 2.20 it can be seen that the sampled spectrum $M_p(j\omega)$ is simply the baseband spectrum $M(j\omega)$ repeated at integer multiples of F_s, as stated previously.

2.3.2.1 Analogue Interpolation Formula

If $H(f)=0$ for all $|f| < \dfrac{F_s}{2}$ (band-limited) then the function $h(t)$ is completely and fully determined by its samples $h[n]$ through the analogue interpolation Equation 2.21.

$$h(t) = T_s \sum_{n=-\infty}^{\infty} h[n] \frac{\sin(2\pi f_c(t - nT_s))}{\pi(t - nT_s)}$$ (2.21)

where:

$$2f_c = \frac{1}{T_s} = F_s$$

2.4 The Fast Fourier Transform (FFT)

The FFT has been and still is one of the most useful and successful linear signal processing algorithms ever developed (Cooley and Tukey 1965). Its value and success is derived from the fact that it can be used to compute the DFT very efficiently, thereby contributing to many practical real-time digital signal processing implementations. It can be easily developed starting from the DFT of a sequence $\{x[n]\}$ as defined by Equation 2.22.

$$X(k) = \sum_{n=0}^{N-1} x[n] W_N^{-kn}, \qquad 0 \le k \le N-1$$ (2.22)

where:

$$W_N = e^{j2\pi/N}$$

An example, for $N=4$, of the matrix form of Equation 2.22 is as follows,

$$\begin{bmatrix} X(0) \\ X(1) \\ X(2) \\ X(3) \end{bmatrix} = \begin{bmatrix} W_4^0 & W_4^0 & W_4^0 & W_4^0 \\ W_4^0 & W_4^1 & W_4^2 & W_4^3 \\ W_4^0 & W_4^2 & W_4^4 & W_4^6 \\ W_4^0 & W_4^3 & W_4^6 & W_4^9 \end{bmatrix} \begin{bmatrix} x[0] \\ x[1] \\ x[2] \\ x[3] \end{bmatrix}$$

$$\begin{bmatrix} X(0) \\ X(1) \\ X(2) \\ X(3) \end{bmatrix} = \begin{bmatrix} W_4^0 & W_4^0 & W_4^0 & W_4^0 \\ W_4^0 & W_4^1 & W_4^2 & W_4^3 \\ W_4^0 & W_4^2 & W_4^0 & W_4^2 \\ W_4^0 & W_4^3 & W_4^2 & W_4^1 \end{bmatrix} \begin{bmatrix} x[0] \\ x[1] \\ x[2] \\ x[3] \end{bmatrix}$$

$$\begin{bmatrix} X(0) \\ X(1) \\ X(2) \\ X(3) \end{bmatrix} = \begin{bmatrix} W_4^0 & W_4^0 & W_4^0 & W_4^0 \\ W_4^0 & W_4^1 & -W_4^0 & -W_4^1 \\ W_4^0 & -W_4^0 & W_4^0 & -W_4^0 \\ W_4^0 & -W_4^1 & -W_4^0 & W_4^1 \end{bmatrix} \begin{bmatrix} x[0] \\ x[1] \\ x[2] \\ x[3] \end{bmatrix}$$

Due to periodicity $W_N^{k+N} = W_N^k = W_N^{k \bmod N}$.

Due to symmetry $= W_N^k = -W_N^{k+N/2}$, for even N.

The Fast Fourier Transform (FFT) is actually a name given to the whole family of algorithms for computing the DFT with fewer multiplications than N^2, required by the DFT defining Equation 2.22. In one of its simplest forms the FFT proceeds as follows. Assume the sequence length N is a power of 2 and it is split into two equal lengths $N/2$. Now, suppose that the two subsequences can be transformed separately and then combined to form the required DFT. If this can be done, then only $0.5N^2$ multiplications would be required to compute the DFT plus the computational cost of the combining operations. It is then possible to keep splitting the subsequences in two until there are only subsequences of length one left. Since the DFT of a sequence of length one is itself, the need to do any transformations can thereby be avoided entirely, leaving only a set of combining operations to do. It takes $\log_2 N$ iterations to get down to subsequences of one, which means that $\log_2 N$ combining operations are required. It turns out that it is possible to do this in such a way as to require only N multiplications for each combining operation, reducing the computational burden from N^2 to $(N)\log_2 N$.

The key to this method is the combining operation, which can be developed as along the following lines. Let the N point sequence $\{x[n]\}$ be split into its even and odd parts as follows,

$$\{x_e[n]\} = \{x[0], x[2], ..., x[N-2]\}$$
$$\{x_o[n]\} = \{x[1], x[3], ..., x[N-1]\}$$

The DFT of the even sequence $\{x_e[n]\}$ is defined by $X_e(k)$ and the DFT of the odd sequence $\{x_o[n]\}$ is defined by $X_o(k)$. The DFT $X(k)$ of the sequence $\{x[n]\}$, being linear, can now be rewritten in terms of the defining Equation 2.23.

$$X(k) = \sum_{n=0}^{N/2-1} x_e[n] W_N^{-2nk} + \sum_{n=0}^{N/2-1} x_o[n] W_N^{-(2n+1)k} \qquad (2.23)$$

where:

$$W_N = e^{j2\pi/N}$$

For an even N, $W_N^2 = W\big|_{N=\frac{N}{2}} = W_{N/2}$ therefore,

$$X(k) = \sum_{n=0}^{N/2-1} x_e[n] W_{N/2}^{-nk} + \sum_{n=0}^{N/2-1} x_o[n] W_{N/2}^{-nk}$$

$$= X_e(k) + W_N^{-k} X_o(k)$$

For $k > N/2$, the periodic property of the DFT can be used and therefore,

$$X_e(k + N/2) = X_e(k)$$
$$X_o(k + N/2) = X_o(k)$$

For any even N, $W_N^{k+N/2} = -W_N^k$, hence the required combining method is,

$$X(k) = \begin{cases} X_e(k) + W_N^{-k} X_o(k), & \text{for } 0 \le k \le N/2 \\ X_e(k - N/2) - W_N^{-(k-N/2)} X_o(k - N/2), & \text{otherwise} \end{cases}$$

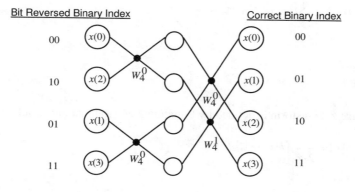

Bit Reversed Binary Index   Correct Binary Index

Where, the FFT Butterfly operator is defined as:

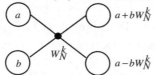

Figure 2.9. FFT Parallel Computations for $N = 4$

The price for this approach to the FFT is that the input sequence must be presented in a permuted order to obtain the result in the correct order, but this has a very low computational overhead. This permutation is called decimation in time and can be done simply if the sequence index numbers are firstly expressed in a binary form. The bits of binary index number are then reversed to produce the new index for that sequence position. For $N = 4$ there are two stages of parallel operations. The input sequence is bit reversed ordered and the FFT is performed efficiently using the so-called FFT butterfly operator as demonstrated in Figure 2.9.

2.5 The z-Transform

The z-Transform is a generalisation of the discrete-time Fourier transform in the same way that the continuous-time Laplace transform is a generalisation of the continuous-time Fourier transform (Haykin and Van Veen 1999). In fact, the two-sided or bilateral z-Transform can be derived directly from the bilateral Laplace transform. The bilateral Laplace transform, a generalisation of the Fourier transform Equation 2.9, and the most general form of the Laplace transform, is defined by Equation 2.24.

$$H(s) = H(\sigma + j\omega) = \int_{-\infty}^{+\infty} h(t)e^{-st}dt = \int_{-\infty}^{+\infty} h(t)e^{-(\sigma + j\omega)t}dt \qquad (2.24)$$

For a signal $h[n]$ the bilateral z-Transform is defined by Equation 2.25.

$$H(z) = \sum_{n=-\infty}^{+\infty} h[n]z^{-n} \qquad (2.25)$$

where:

$z = re^{j\theta}$
$\theta = \omega T$

The inverse z-Transform is defined by Equation 2.26 (Rabiner and Gold 1975).

$$h[n] = \frac{1}{2\pi j} \oint_{C_1} H(z)z^{n-1}dz \qquad (2.26)$$

where:
C_1 is a closed path in region of convergence encompassing origin in z-plane.

The Discrete-time Fourier Transform (DTFT) is the z-Transform as defined by Equation 2.25 for the special case when $z = e^{j\theta}$, i.e. $r = 1$.

2.5.1 Relationship between Laplace Transform and z-Transform

In the time-domain an infinite-length signal $h[n]$ sampled uniformly at an interval of T_s seconds can be represented by,

$$h(nT_s) = \sum_{n=-\infty}^{\infty} h(t)\delta(t - nT_s) = \sum_{n=-\infty}^{\infty} h(nT_s)\delta(t - nT_s)$$

The Laplace transform $H(s)$ of this sampled signal $h[n]$ is,

$$H(s) = \int_{-\infty}^{\infty} \left(\sum_{n=-\infty}^{\infty} h(nT_s)\delta(t - nT_s) \right) e^{-st} dt$$

$$H(s) = \sum_{n=-\infty}^{\infty} h(nT_s) \int_{-\infty}^{\infty} \left(\delta(t - nT_s) \right) e^{-st} dt$$

$$H(s) = \sum_{n=-\infty}^{\infty} h(nT_s) e^{-snT_s}$$

There are two forms of the z-Transform commonly used, the bilateral and the unilateral transforms. The bilateral transform is defined over infinite discrete-time from $n = -\infty$ to $n = +\infty$, and is suitable for general signals, causal and noncausal. Whereas, the unilateral is defined from $n = 0$ to $n = +\infty$ and is only suitable for use with causal signals.

2.5.1.1 Bilateral z-Transform

If $z = e^{-sT_s}$ is taken as the complex z-domain variable then an infinite length of $h[n]$ can be represented by what is called the two-sided or bilateral z-Transform,

$$H(z) = \sum_{n=-\infty}^{\infty} h(nT_s) z^{-n} = \sum_{n=-\infty}^{\infty} h[n] z^{-n} \text{, dropping the } T_s \text{ for convenience.}$$

The relationship between the discrete-time domain and the discrete-frequency domain is,

$$\text{Discrete} - \text{Time Domain} \Leftrightarrow \text{Discrete} - \text{Frequency Domain}$$

$$h[n] = \sum_{k=-\infty}^{\infty} h[k]\delta[n-k] \Leftrightarrow H(z) = \sum_{n=-\infty}^{\infty} H[n] z^{-n}$$

$$H(z) = H(e^{sT_s}) = H(e^{(\sigma + j\omega)T_s}) = H(e^{\sigma T_s} e^{j\omega T_s}) = H(re^{j\omega T_s}) = H(re^{j\theta})$$

where:

$r = e^{\sigma T_s}$

$s = \sigma + j\omega$, the complex Laplace-domain variable.

$\theta = \omega T_s$ radians, ω is in radians per second and T_s is in seconds.

For example, the z-Transform of the sequence,

$$x[n] = 0.5\delta[n] + 0.8\delta[n-2] - 8\delta[n-5]$$

is,

$$X(z) = \sum_{n=-\infty}^{\infty} x[n]z^{-n} = \sum_{n=-\infty}^{\infty} \left(0.5\delta[n] + 0.8\delta[n-2] - 8\delta[n-5]\right)z^{-n}$$

$$X(z) = 0.5z^{-0} + 0.8z^{-2} - 8z^{-5} = 0.5 + 0.8z^{-2} - 8z^{-5}$$

2.5.1.2 Unilateral z-Transform

For causal signals and systems the one-sided or unilateral z-Transform is often used by changing the lower limit in the summation of the bilateral z-Transform as follows,

$$X(z) = \sum_{n=0}^{\infty} x[n]z^{-n} = x[0] + x[1]z^{-1} + x[2]z^{-2} + ... + x[n]z^{-n} + ...$$

In most practical cases signals begin at (or after) $n = 0$, taken as the reference point. Furthermore, the impulse response $h[n]$ of any causal system is zero for $n < 0$. Therefore, whether describing signals or LTI systems with the z-Transform, the unilateral version is usually adequate. Care must be taken with respect to its region of convergence in the z-domain. An instructive example is to determine the unilateral z-Transform of a unit step, $u[n] \xrightarrow{z} U(z)$,

$$X(z) = \sum_{n=0}^{\infty} u[n]z^{-n} = \sum_{n=0}^{\infty} z^{-n}$$

This transform only converges for $\left|z^{-1}\right| < 1$, therefore, $X(z) = \dfrac{1}{1-(z^{-1})} = \dfrac{z}{z-1}$.

2.5.1.3 Region of Convergence (ROC) for the z-Transform

The function $X(z)$ is the z-Transform of $x[n]$ but since it is a power series it may not converge for all z. Just like the Laplace transform the z-Transform also has a Region of Convergence (ROC). The ROC defines the values of z for which the z-Transform sum will converge. Take the example, $x[n] = a^n u[n]$. The bilateral z-Transform is,

$$X(z) = \sum_{n=-\infty}^{\infty} a^n u[n]z^{-n} = \sum_{n=0}^{\infty} a^n z^{-n} = \sum_{n=0}^{\infty} (a^n z^{-1})^n$$

This power series converges to,

$$X(z) = \frac{1}{1-az^{-1}} = \frac{z}{z-a}, \text{ iff } \left|az^{-1}\right| < 1, \text{ i.e., iff } |z| > |a|, \text{ which is its ROC}$$

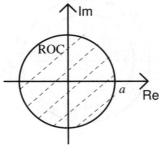

z - plane

Another example is, $x[n] = -a^n u[-(n+1)]$. The z-Transform of this signal is,

$$X(z) = \sum_{n=-\infty}^{\infty} -a^n u[-n-1]z^{-n} = \sum_{n=-\infty}^{-1} -a^n z^{-n} = -\sum_{n=1}^{\infty} a^{-n} z^n = -\sum_{n=1}^{\infty} \left(\frac{z}{a}\right)^n$$

This power series converges to,

$$X(z) = -\left(\frac{\frac{z}{a}}{1-\frac{z}{a}}\right) = \frac{-1}{\frac{a}{z}-1} = \frac{1}{1-\frac{a}{z}} = \frac{1}{1-az^{-1}}, \text{ iff } \left|a^{-1}z\right| < 1, \text{ i.e.,}$$

iff $|z| < |a|$, which is its ROC.

2.5.1.4 Region of Convergence (ROC) for General Signals

1. If the signal $x[n]$ is right handed (causal), the ROC is the area outside a circle, $|z| > |a|_{max}$, the magnitude of the largest pole. This is a necessary and sufficient condition for convergence.

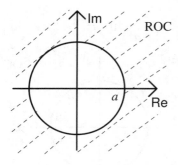

2. If the signal $x[n]$ is left handed (anti-causal), the ROC is the area inside a circle, $|z| < |b|_{min}$, the magnitude of the smallest pole.

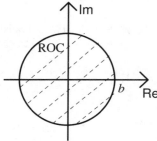

3. If the signal $x[n]$ is two-side or the sum of a left and right sided signal, the ROC is either a donut, $|a|_{max} < |z| < |b|_{min}$, or else the individual ROCs don't overlap, producing the null set.

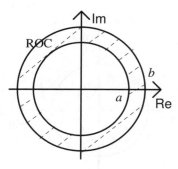

4. If the signal $x[n]$ is of finite duration, then the ROC is the entire z-plane, except possibly for $z = 0$ and $z = \infty$.

2.5.2 General Properties of the DFT and z-Transform

When $r = 1$, then $z = e^{j\theta}$ and the z-Transform becomes the DTFT as defined by Equation 2.27.

$$H(e^{j\theta}) = H(z)\Big|_{z=e^{j\theta}} = \sum_{n=-\infty}^{+\infty} h[n]e^{-j\theta n} = \sum_{n=-\infty}^{+\infty} h[n]z^{-n} \tag{2.27}$$

where:

$\theta = \omega T$ radians

The z-Transform is a continuous function of θ with a period of 2π and can be used for digital filter design, amongst other things. A short summary of the properties of the DFT and z-Transform are listed below (Hayes 1996, Stanley *et al* 1984),

Property	Sequence	DFT	z-Transform
Transform	$x[n]$	$X(k)$	$X(z)$
Delay	$x[n-m]$	$e^{-j2\pi nk/N}X(k)$	$z^{-m}X(z)$
Modulation	$e^{j2\pi mn}x[n]$	$X(k - m)$	$X(e^{j(\theta - 2\pi n)})$
Conjugation	$x^*[n]$	$X^*(N-k)$	$X^*(z^*)$
Time Reversal	$x[N-n]$	$X(N-k)$	$X(z^{-1})$
Convolution	$x[n]*h[n]$	$X(k)H(k)$	$X(z)H(z)$
Multiplication	$x[n]h[n]$	$X(k)*H(k) / N$	-
Multiplication by α^n	$\alpha^n x[n]$	-	$X(z/\alpha)$
Multiplication by n	$nx[n]$	-	$-z\, d/dz\, X(z)$

A Summary of some common series closed-form formulations is as follows,

$$\sum_{n=0}^{N-1} a^n = \frac{1-a^N}{1-a}$$

$$\sum_{n=0}^{N-1} na^n = \frac{(N-1)a^{N+1} - Na^N + a}{(1-a)^2}$$

$$\sum_{n=0}^{N-1} n = \frac{1}{2}N(N-1)$$

$$\sum_{n=0}^{N-1} n^2 = \frac{1}{6}N(N-1)(2N-1)$$

A Summary of some common z-Transform pairs is as follows,

Sequence	Transform	Convergence
$\delta[n]$	1	All z
$\alpha^n[u[n]-u[n-N]]$	$\dfrac{1-\alpha^N z^{-N}}{1-\alpha z^{-1}}$	$\|z\|>0$
$\alpha^n u[n]$	$\dfrac{1}{1-\alpha z^{-1}}$	$\|z\|>\alpha$
$-\alpha^n u[-n-1]$	$\dfrac{1}{1-\alpha z^{-1}}$	$\|z\|<\alpha$
$\alpha^{\|n\|}$	$\dfrac{1-\alpha^2}{(1-\alpha z^{-1})(1-\alpha z)}$	$\alpha<\|z\|<\dfrac{1}{\alpha}$

2.6 Summary of Discrete-time LSI Systems

Discrete-time LSI systems can be generally described by the block diagram of Figure 2.10 and the following panel of expressions and equations. Figure 2.10 shows the relationship between the system z-Transform $H(z)$ and the input and output signal z-Transforms, $X(z)$ and $Y(z)$ respectively.

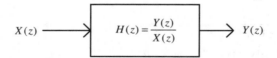

$$X(z) \longrightarrow \boxed{H(z) = \frac{Y(z)}{X(z)}} \longrightarrow Y(z)$$

Figure 2.10. z-Transformed Linear System

The following panel provides a concise summary of important equations and relationships for discrete-time LSI systems including the relationship between z-Transforms and difference equations. The relationship between the DFT and the DTFT (z-Transform when $r = 1$) is exemplified in Figure 2.11, which shows a diagram of a simple DFT $H(k)=H(z)\big|_{z=e^{j\frac{2\pi k}{N}}}$, for $N = 12$. Notice carefully the relationships between all the variables for the DTFT and corresponding DFT representations.

$$z = re^{j\theta}, \theta = \omega T, \theta_k = \frac{2\pi k}{N}, \text{ for } k = 0,1....N-1$$

$z^{-1} = e^{-j\omega T} = e^{-j\theta}$, is a unit delay when $r = 1$

where : T is the sampling interval, i.e., $T = 1/F_s$

$X(z)$ is the z – Transform (or DTFT when $r = 1$) of sequence $x[n]$.

$Y(z)$ is the z – Transform (or DTFT when $r = 1$) of sequence $y[n]$.

$H(z)$ is the z – Transform (or DTFT when $r = 1$) of impulse response $h[n]$.

$$H(z) = \frac{Y(z)}{X(z)} = \frac{B(z)}{A(z)} = \frac{b_0 z^{-q}}{a_0 z^{-p}} \left(\frac{z^q + \frac{b_1}{b_0} z^{q-1} + ... + \frac{b_q}{b_0}}{z^p + \frac{a_1}{a_0} z^{p-1} + ... + \frac{a_p}{a_0}} \right), \text{ usually } a_0 = 1,$$

Therefore,

$$H(z) = \frac{b_0 + b_1 z^{-1} + ... + b_q z^{-q}}{a_0 + a_1 z^{-1} + ... + a_p z^{-p}} = \frac{\sum_{m=0}^{q} b_m z^{-m}}{a_0 + \sum_{m=1}^{p} a_m z^{-m}},$$

which can be represented as the difference equation,

$$a_0 y[n] + a_1 y[n-1] + ... + a_p y[n-p] = b_0 x[n] + b_1 x[n-1] + ... + b_q x[n-q]$$

$$a_0 y[n] = \sum_{m=0}^{q} b_m x[n-m] - \sum_{m=1}^{p} a_m y[n-m],$$

where :

the constant indices p and q represent the order (p,q) of the system p and $q \leq N-1$.

In terms of poles and zeros in the z - plane,

$$H(z) = \frac{b_0}{a_0} z^{-q+p} \frac{(z-z_1)(z-z_2)....(z-z_q)}{(z-p_1)(z-p_2)....(z-p_p)},$$

$$z_m = \text{zeros}, p_m = \text{poles}$$

$$H(z) = \frac{b_0}{a_0} \frac{(1-z_1 z^{-1})(1-z_2 z^{-1})....(1-z_q z^{-1})}{(1-p_1 z^{-1})(1-p_2 z^{-1})....(1-p_p z^{-1})}$$

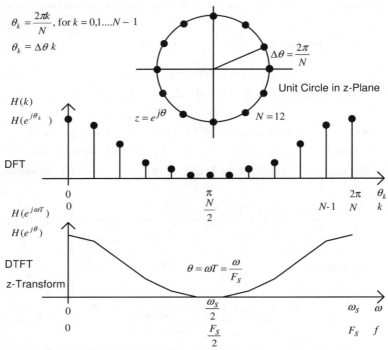

Figure 2.11. DFT and DTFT (z-Transform for $r = 1$) Relationship

2.7 Special Classes of Filters

Three filter classes that are of special significance are linear phase filters, allpass filters and minimum phase filters. Linear phase filters are important in speech and image processing because they introduce no signal distortion due to the filter's phase characteristics. If a causal filter's phase characteristic is linear then all frequencies are delayed through the filter by the same amount resulting in no distortion, only a bulk signal delay. The delay τ is equal to the derivative with respect to frequency (radians per second), or slope, of the phase characteristic (in radians), i.e., $\tau = \dfrac{d\theta(\omega)}{d\omega}$. A generalised linear phase filter has the form defined by Equation 2.28 (Hayes 1996),

$$H(e^{j\theta}) = A(e^{j\theta})e^{j(\beta - \alpha\theta)} \tag{2.28}$$

where:

$A(e^{j\theta})$ is a real valued function of θ, and α and β are constants.

To realise a linear phase causal filter using a finite-order linear constant coefficient difference equation it must be a FIR filter. However, not all FIR filters are linear phase filters. For a FIR filter to be linear phase its impulse response must be either conjugate symmetric (Hermitian) or conjugate antisymmetric (anti-Hermitian) as defined by Equations 2.29 and 2.30 respectively.

$$h^*[n] = h[N-1-n], \quad \text{Hermitian} \tag{2.29}$$

or

$$h^*[n] = -h[N-1-n], \quad \text{anti-Hermitian} \tag{2.30}$$

These constraints require that the zeros of the filter function $H(z)$ occur in conjugate pairs as defined by Equation 2.31, i.e., if $H(z)$ has a zero at $z = z_1$ then there must also be a zero $z = \dfrac{1}{z^*_1}$.

$$H^*(z^*) = \pm z^{N-1} H\left(\frac{1}{z}\right) \tag{2.31}$$

Allpass filters have a constant magnitude frequency response and are useful for phase equalisation. The $H(z)$ must be of the form defined by Equation 2.32, i.e., if $H(z)$ has a zero (or pole) at $z = \alpha_k$ there must also be a zero (or pole) at $z = \dfrac{1}{\alpha^*_k}$.

$$H^*(z^*) = z^{-n_0} A \prod_{k=1}^{N} \frac{z^{-1} - \alpha_k^*}{1 - \alpha_k z^{-1}}, \tag{2.32}$$

where:

A is a constant.

A minimum phase filter is a stable causal filter that has a rational filter transfer function with all its poles and zeros inside the unit circle of the z-plane, i.e., $|p_k| < 1$ and $|z_k| < 1$. A minimum phase filter has a stable and causal inverse, $1/H(z)$, which also has a minimum phase. For a rational LSI system characterised by the transfer function $H(z) = \dfrac{B(z)}{A(z)}$ it is causal, invertible and stable if all its poles (roots of $A(z)$) and zeros (roots of $B(z)$) are inside the unit circle of the z-plane. In this case the filter is a minimum phase system as its phase characteristic is always inside the bound of $\pm\pi$ radians (Oppenheim and Schafer 1975). An alternative condition for minimum phase is $H^{-1}(z)H(z) = 1$. For such a system or signal, $\ln(|H(e^{j\omega T_s})|)$ and $\arg(H(e^{j\omega T_s}))$ are Hilbert transform pairs.

Maximum phase systems have all their zeros outside the unit circle, and mixed phase systems have a mixture of zeros in and out of the unit circle. Allpass systems are defined by $|H(e^{j\omega T_s})| = 1$ for all ω. Any rational system $H(z)$ corresponding to a causal system can be expressed as $H(z) = H_{\min}(z)H_{\text{all}}(z)$, where $H_{\min}(z)$ is a minimum phase system and $H_{all}(z)$ is an allpass system. Any pole or zero of $H(z)$ that is inside the unit circle is also in $H_{min}(z)$. Any pole or zero of $H(z)$ that is outside the unit circle is in $H_{min}(z)$ in the conjugate reciprocal location. Therefore, a minimum phase system with the same magnitude can be formed from a non-minimum phase system by reflecting inside the unit circle all those zeros that were outside the unit circle. In like manner, a non-minimum phase system can be formed from a minimum phase system by reflecting some or all of the inside zeros to outside the unit circle.

2.7.1 Phase Response from Frequency Magnitude Response

Loudspeakers are typically minimum phase systems but their transfer functions are too complex to accurately model analytically. Nevertheless their phase characteristics can be computed directly from the magnitude of the amplitude response. The phase response of a minimum phase system $S(e^{j\omega T_s})$ can be reconstructed from the two-side magnitude response $|S(e^{j\omega T_s})|$ using the Hilbert Transform. This is commonly done in loudspeaker testing where it is relatively easy to measure the magnitude response but not the phase response, due to the uncertainty of the testing signal delay through the air. The phase response $\arg S(e^{j\omega T_s})$ in radians is found by taking the real part of the Hilbert transform $H\{\}$ of $\ln(|S(e^{j\omega T_s})|)$ as defined by Equation 2.33.

$$\angle S(e^{j\omega T_s}) = real\{H\{\ln(|S(e^{j\omega T_s})|)\}\} \tag{2.33}$$

If a Discrete Fourier Transform (DFT) is used to compute the Hilbert transform (Poularikas 1996) then the number of sample points N should be chosen such that NT_s is at least twice the length of the impulse response $s(nT_s)$. The phase in radians is found in the real part of the final result. The imaginary part is a constant, which varies in relation to the scale of the original signal to preserve the energy relation between the Fourier, transform pairs. Assume that a discrete Hilbert transform pair $u[n]$ and $v[n]$ is defined by $u[n] \overset{H}{\longleftrightarrow} v[n]$, and a discrete Fourier transform pair $v[n]$ and $V(e^{j\omega T_s})$ is defined by $v[n] \overset{DFT}{\longleftrightarrow} V(e^{j\omega T_s})$.
If

$$u[n] \overset{DFT}{\longleftrightarrow} U(e^{j\omega T_s}) \text{ and } v[n] \overset{DFT}{\longleftrightarrow} V(e^{j\omega T_s}),$$

then,

$$v[n] \overset{DFT}{\longleftrightarrow} V(e^{j\omega T_s}) = -j\,sgn(\omega)U(e^{j\omega T_s}),$$

where: $\mathrm{sgn}(\omega) = \begin{cases} +1 & 0 < \omega < \dfrac{\omega_2}{2} \\ 0 & \omega = \dfrac{\omega_2}{2} \\ -1 & -\dfrac{\omega_2}{2} < \omega < 0 \end{cases}$,

it follows that, $u[n] \overset{\mathrm{DFT}}{\longleftrightarrow} U(e^{j\omega T_s}) \rightarrow V(e^{j\omega T_s}) = -j\,\mathrm{sgn}(\omega)U(e^{j\omega T_s}) \overset{\mathrm{DFT}^{-1}}{\longrightarrow} v[n]$.

2.8 Linear Algebra Summary

Vector and matrix notation is very useful because it simplifies mathematical expressions and many useful results from linear algebra can be employed to solve them. For present purposes it is only necessary to summarise some of the more significant tools of vector and matrix analysis. Refer to (Hayes 1996) for more comprehensive details. The convention has been adopted, that bold lowercase letters represent vectors, bold uppercase letters represent matrices, and normal upper and lower case letters are scalars, unless otherwise stated or implied.

2.8.1 Vectors

As already mentioned vectors are written as column vectors and their transposes as row vectors. Vectors can be real or complex vectors. A vector having p elements is said to be a p-dimensional vector. The Hermitian transpose, H, of a complex vector $\mathbf{x}$ is defined by Equation 2.34 as $\mathbf{x}^H$.

$$\mathbf{x}^H = (\mathbf{x}^*)^T = (\mathbf{x}^T)^* = [x_1^* \quad x_2^* \quad .. \quad x_p^*] \tag{2.34}$$

where:

$$\mathbf{x} = \begin{bmatrix} x_1 \\ x_2 \\ \vdots \\ x_p \end{bmatrix} = [x_1 \quad x_2 \quad .. \quad x_p]^T$$

The magnitude of a vector is commonly defined according to the Euclidean or L_2 norm defined by Equation 2.35, where p is the dimension or number of vector coefficients x_i. This most commonly used norm is usually represented by just $\|\mathbf{x}\|$.

$$\|\mathbf{x}\|_2 = \sqrt{\sum_{i=1}^{p} |x_i|^2} = \|\mathbf{x}\| \tag{2.35}$$

Other useful norms are the L_1 and L_∞ norms defined by Equations 2.36 and 2.37 respectively. The L_∞ norm represents the maximum vector coefficient value.

$$\|\mathbf{x}\|_1 = \sum_{i=1}^{p} |x_i| \tag{2.36}$$

$$\|\mathbf{x}\|_\infty = \lim_{m \to \infty} \left[\sum_{i=1}^{p} |x_i|^m \right]^{\frac{1}{m}} = \max_i |x_i| \tag{2.37}$$

A vector $\mathbf{x}$, if $\|\mathbf{x}\| \neq 0$, can be normalised to a unit magnitude vector $\mathbf{v}_x$ by simply dividing by its norm $\|\mathbf{x}\|$ as defined by Equation 2.38.

$$\mathbf{v}_x = \frac{\mathbf{x}}{\|\mathbf{x}\|} \tag{2.38}$$

The dot product or inner product of two complex vectors $\mathbf{a} = [a_1,...., a_p]^T$ and $\mathbf{b} = [b_1,...., b_p]^T$ is scalar and defined by Equation 2.39.

$$\mathbf{a}.\mathbf{b} = \langle \mathbf{a}, \mathbf{b} \rangle = \mathbf{a}^H \mathbf{b} = \sum_{i=1}^{p} a_i^* b_i \tag{2.39}$$

Two vectors are said to be orthogonal if their inner product is zero and orthonormal if they also have unit norms. One use of the dot product is to represent the output of a LSI FIR filter, e.g.,

$$y[n] = \sum_{i=0}^{N-1} h[i]x[n-i] = \mathbf{h}^T \mathbf{x}[n],$$

where:

$$\mathbf{h} = \begin{bmatrix} h[0] \\ h[1] \\ \vdots \\ h[N-1] \end{bmatrix}, \text{ and } \mathbf{x}[n] = \begin{bmatrix} x[n] \\ x[n-1] \\ \vdots \\ x[n-N+1] \end{bmatrix}$$

2.8.2 Linear Independence, Vector spaces, and Basis Vectors

A set of n vectors $\{\mathbf{v}_1, \mathbf{v}_2,....., \mathbf{v}_n\}$ are linearly independent if $\alpha_1 \mathbf{v}_1 + \alpha_2 \mathbf{v}_2 + ...+ \alpha_n \mathbf{v}_n = 0$, with $\alpha_i \neq 0$ for all $i = 1,..., n$. If a set of nonzero α_i can be found that will make the equation hold then the vectors are linearly dependent. If the vectors are linearly dependent then at least one of the vectors, say $\mathbf{v}_1$, can be expressed as a linear combination of the others, i.e., $\mathbf{v}_1 = \beta_2 \mathbf{v}_2 + \beta_3 \mathbf{v}_3 + ...+ \beta_n \mathbf{v}_n$, for some set of scalars β_i. For p-dimensional vectors no more than p of them can be linearly independent.

For a set of n vectors, $\{\mathbf{v}_1, \mathbf{v}_2,..., \mathbf{v}_n\}$, the set of all vectors S that may be formed from a linear combination of the vectors $\mathbf{v}_i$, $\mathbf{v} = \sum\limits_{i=1}^{n} \alpha_i \mathbf{v}_i$, forms a linear vector space and the vectors $\mathbf{v}_i$ are said to span the space S. If the vectors $\mathbf{v}_i$ are linearly independent they form a basis for the space S and the number of vectors in the basis, n, is the dimension of the space.

2.8.3 Matrices

A $n \times m$ matrix $\mathbf{A}$ is a real or complex array of numbers or functions formed into n rows and m columns as defined by Equation 2.40.

$$\mathbf{A} = \{a_{ij}\} = \begin{bmatrix} a_{11} & a_{12} & .. & a_{1m} \\ a_{21} & a_{22} & .. & a_{2m} \\ : & : & : & : \\ a_{n1} & a_{n2} & .. & a_{nm} \end{bmatrix} \tag{2.40}$$

If $n = m$ the matrix is called a square matrix. A $n \times m$ matrix is sometimes represented by a set of m column vectors, or a set of n row vectors or a partition of submatrices as defined by Equation 2.41.

$$\mathbf{A} = \begin{bmatrix} \mathbf{c}_1 & \mathbf{c}_2 & .. & \mathbf{c}_m \end{bmatrix} = \begin{bmatrix} \mathbf{r}_1^T \\ \mathbf{r}_2^T \\ : \\ \mathbf{r}_n^T \end{bmatrix} = \begin{bmatrix} \mathbf{A}_{11} & \mathbf{A}_{12} \\ \mathbf{A}_{21} & \mathbf{A}_{22} \end{bmatrix} \tag{2.41}$$

The transpose of a $n \times m$ matrix $\mathbf{A}$, denoted by $\mathbf{A}^T$, is a $m \times n$ matrix formed by simply interchanging the rows and columns of $\mathbf{A}$ as defined by Equations 2.42 and 2.43.

$$\mathbf{A} = \{a_{ij}\} = \begin{bmatrix} a_{11} & a_{12} & .. & a_{1m} \\ a_{21} & a_{22} & .. & a_{2m} \\ : & : & : & : \\ a_{n1} & a_{n2} & .. & a_{nm} \end{bmatrix} \tag{2.42}$$

$$\mathbf{A}^T = \{a_{ji}\} = \begin{bmatrix} a_{11} & a_{12} & .. & a_{1n} \\ a_{21} & a_{22} & .. & a_{2n} \\ : & : & : & : \\ a_{m1} & a_{m2} & .. & a_{mn} \end{bmatrix} \tag{2.43}$$

If a square matrix is equal to its transpose it is said to be a symmetric matrix. For complex matrices the Hermitian transpose is the complex conjugate of the transpose, i.e., $\mathbf{A}^H = (\mathbf{A}^*)^T = (\mathbf{A}^T)^*$. The rank $\rho(\mathbf{A})$ of a matrix $\mathbf{A}$ is equal to the number of linearly independent rows and columns, therefore $\rho(\mathbf{A}) \leq \min(m, n)$. If $\rho(\mathbf{A}) = \min(m, n)$, then $\mathbf{A}$ is said to be of full rank. If $\mathbf{A}$ is a square matrix of full rank then there exists a unique matrix $\mathbf{A}^{-1}$, called the inverse of $\mathbf{A}$ such that $\mathbf{A}^{-1}\mathbf{A} = \mathbf{A}\mathbf{A}^{-1} = \mathbf{I}$, where $\mathbf{I}$ is the square identity matrix defined by Equation 2.44.

$$\mathbf{I} = \begin{bmatrix} 1 & 0 & .. & 0 \\ 0 & 1 & .. & 0 \\ : & : & : & : \\ 0 & 0 & .. & 1 \end{bmatrix} \tag{2.44}$$

In this case $\mathbf{A}$ is said to be invertible or nonsingular. If $\mathbf{A}$ is not of full rank it is said to be noninvertible or singular.

Some properties of matrices $\mathbf{A}$ $(n \times n)$, $\mathbf{B}$ $(n \times m)$ and $\mathbf{C}$ $(m \times m)$, where matrices $\mathbf{A}$ and $\mathbf{B}$ are nonsingular, are as follows,

$$(\mathbf{AB})^{-1} = \mathbf{B}^{-1}\mathbf{A}^{-1}$$
$$(\mathbf{A}^H)^{-1} = (\mathbf{A}^{-1})^H$$

A formula that is useful for efficiently inverting matrices, especially in adaptive filtering algorithms is defined by Equation 2.45.

$$(\mathbf{A} + \mathbf{BCD})^{-1} = \mathbf{A}^{-1} - \mathbf{A}^{-1}\mathbf{B}(\mathbf{C}^{-1} + \mathbf{DA}^{-1}\mathbf{B})^{-1}\mathbf{DA}^{-1} \tag{2.45}$$

In the special case when $\mathbf{C} = 1$, $\mathbf{B} = \mathbf{u}$, and $\mathbf{D} = \mathbf{v}^H$, and $\mathbf{u}$ and $\mathbf{v}$ are n-dimensional vectors Equation 2.45 is referred to as Woodbury's identify and it is expressed as follows, $(\mathbf{A} + \mathbf{uv}^H)^{-1} = \mathbf{A}^{-1} - \dfrac{\mathbf{A}^{-1}\mathbf{uv}^H\mathbf{A}^{-1}}{1 + \mathbf{v}^H\mathbf{A}^{-1}\mathbf{u}}$.

A special case of Woodbury's Identity occurs when $\mathbf{A} = \mathbf{I}$ as follows,

$(\mathbf{I} + \mathbf{uv}^H)^{-1} = \mathbf{I} - \dfrac{1}{1 + \mathbf{v}^H\mathbf{u}}\mathbf{uv}^H$

The determinant of a $n \times n$ matrix $\mathbf{A}$, $\det(\mathbf{A})$, is defined recursively in terms of the determinants of $(n - 1) \times (n - 1)$ matrices as follows,

For any j, $\det(\mathbf{A}) = \sum_{i=1}^{n}(-1)^{i+j}a_{ij}\det(\mathbf{A}_{ij})$,

where : $\mathbf{A}_{ij}$ is the $(n-1) \times (n-1)$ matrix formed

by deleting the ith row and jth column of $\mathbf{A}$.

If $\mathbf{A} = a_{11}$, $\det(\mathbf{A}) = a_{11}$.

If $\mathbf{A} = \begin{bmatrix} a_{11} & a_{12} \\ a_{21} & a_{22} \end{bmatrix}$, $\det(\mathbf{A}) = a_{11}a_{22} - a_{12}a_{21}$.

A $n \times n$ matrix $\mathbf{A}$ is invertible if and only if $\det(\mathbf{A}) \neq 0$. Some properties of the determinant involving $n \times n$ matrices $\mathbf{A}$ and $\mathbf{B}$ are as follows,

$$\det(\mathbf{AB}) = \det(\mathbf{A})\det(\mathbf{B})$$

$$\det(\mathbf{A}^T) = \det(\mathbf{A})$$

$$\det(\alpha\mathbf{A}) = \alpha^n \det(\mathbf{A}), \quad \text{where} : \alpha \text{ is a constant.}$$

$$\det(\mathbf{A}^{-1}) = \frac{1}{\det(\mathbf{A})}, \quad \text{if } \mathbf{A} \text{ is invertible.}$$

The trace of a $n \times n$ matrix $\mathbf{A}$, $\text{tr}(\mathbf{A})$, is the sum of the terms along the diagonal.

2.8.4 Linear Equations

The solution of linear equations is an important part of signal modelling, Wiener filtering and spectrum estimation. Consider the set of linear Equations 2.46 in m unknowns $x_1, x_2,..., x_m$.

$$a_{11}x_1 + a_{12}x_2 + ... + a_{1m}x_m = b_1$$
$$a_{21}x_1 + a_{22}x_2 + ... + a_{2m}x_m = b_2$$
$$\vdots \qquad\qquad\qquad\qquad\qquad (2.46)$$
$$a_{1n}x_1 + a_{n2}x_2 + ... + a_{nm}x_m = b_n$$

These equations can be written more efficiently in matrix form as defined by Equation 2.47.

$$\mathbf{Ax} = \mathbf{b} \qquad\qquad (2.47)$$

For a square matrix where $n = m$ the solution is $\mathbf{x} = \mathbf{A}^{-1}\mathbf{b}$, if $\mathbf{A}$ is nonsingular. If $\mathbf{A}$ is singular, then there may be no solution if the equations are inconsistent or there may be many possible solutions.

If $n < m$, then there are fewer equations than unknowns, and provided that the equations are not inconsistent there may be many solution vectors (this is an undetermined or incompletely specified solution). One way to find a unique solution for this case is to satisfy the equation that has the minimum norm, i.e., min‖x‖ such that $\mathbf{Ax} = \mathbf{b}$. If the rows of $\mathbf{A}$ are linearly independent ($\rho(\mathbf{A}) = n$), then the $n \times n$ matrix $\mathbf{AA}^H$ is invertible and the minimum norm solution is defined by the Equation 2.48.

$$\mathbf{x}_0 = \mathbf{A}^H(\mathbf{AA}^H)^{-1}\mathbf{b} \qquad\qquad (2.48)$$

The matrix $\mathbf{A}^+ = \mathbf{A}^H(\mathbf{AA}^H)^{-1}$ is known as the pseudoinverse of the matrix $\mathbf{A}$ for the underdetermined problem.

If $n > m$, then there are more equations than unknowns and the equations are inconsistent and the solution is said to be overdetermined, which in general means that no solution exists. In this case the least squares solution can be sought producing a vector $\mathbf{x}$ that minimises the norm of error e defined by Equation 2.49.

$$\left\| e^2 \right\| = \left\| \mathbf{b} - \mathbf{Ax} \right\|^2 \tag{2.49}$$

To solve this equation the so called normal Equations 2.50 are constructed.

$$\mathbf{A}^H \mathbf{Ax} = \mathbf{A}^H \mathbf{b} \tag{2.50}$$

If the columns of $\mathbf{A}$ are linearly independent ($\mathbf{A}$ has full rank), then the matrix $\mathbf{A}^H\mathbf{A}$ is invertible and the least squares solution is defined by Equation 2.51.

$$\begin{aligned} \mathbf{x}_0 &= (\mathbf{A}^H \mathbf{A})^{-1} \mathbf{A}^H \mathbf{b} \\ &= \mathbf{A}^+ \mathbf{b} \end{aligned} \tag{2.51}$$

where :

$\mathbf{A}^+ = (\mathbf{A}^H \mathbf{A})^{-1} \mathbf{A}^H$ is the pseudo - inverse for the overdetermined problem.

2.8.5 Special Matrices

A diagonal matrix is a square matrix that has the form defined by Equation 2.52.

$$\mathbf{A} = \begin{bmatrix} a_{11} & 0 & .. & 0 \\ 0 & a_{22} & .. & 0 \\ : & : & : & : \\ 0 & 0 & .. & a_{nn} \end{bmatrix} = \text{diag}\{a_{11}, a_{22}, .. a_{nn}\} \tag{2.52}$$

The identity matrix is a diagonal matrix. If its entries along the diagonal are replaced with matrices then $\mathbf{A}$ is said to be a block diagonal matrix as defined by Equation 2.53.

$$\mathbf{A} = \begin{bmatrix} \mathbf{A}_{11} & 0 & .. & 0 \\ 0 & \mathbf{A}_{22} & .. & 0 \\ : & : & : & : \\ 0 & 0 & .. & \mathbf{A}_{nn} \end{bmatrix} \tag{2.53}$$

The exchange matrix $\mathbf{J}$ is defined by Equation 2.54, like the identity matrix but with the cross-diagonal (diagonal which is perpendicular to the main diagonal) populated by 1s.

$$\mathbf{J} = \begin{bmatrix} 0 & .. & 0 & 1 \\ 0 & .. & 1 & 0 \\ : & : & : & : \\ 1 & .. & 0 & 0 \end{bmatrix} \tag{2.54}$$

Since $J^2 = I$ therefore J is its own inverse. If a matrix A is multiplied on the left by the exchange matrix the order of each column vector is reversed. If it is multiplied on the right then the order of the entries in each row are reversed. The effect of the product $J^T A J$ is to reverse the order of each row and column. These operations are demonstrated below as follows,

$$\text{if } A = \begin{bmatrix} a_{11} & a_{12} & \cdot\cdot & a_{1m} \\ a_{21} & a_{22} & \cdot\cdot & a_{2m} \\ \vdots & \vdots & \vdots & \vdots \\ a_{n1} & a_{n2} & \cdot\cdot & a_{nm} \end{bmatrix}, J^T A^T = \begin{bmatrix} a_{n1} & a_{n2} & \cdot\cdot & a_{mn} \\ \vdots & \vdots & \vdots & \vdots \\ a_{21} & a_{22} & \cdot\cdot & a_{2m} \\ a_{11} & a_{12} & \cdot\cdot & a_{1m} \end{bmatrix}$$

$$AJ = \begin{bmatrix} a_{1m} & \cdot\cdot & a_{12} & a_{11} \\ a_{2m} & \cdot\cdot & a_{22} & a_{21} \\ \vdots & \vdots & \vdots & \vdots \\ a_{nm} & \cdot\cdot & a_{n2} & a_{n1} \end{bmatrix}, J^T A J = \begin{bmatrix} a_{nm} & \cdot\cdot & a_{n2} & a_{n1} \\ \vdots & \vdots & \vdots & \vdots \\ a_{2m} & \cdot\cdot & a_{22} & a_{21} \\ a_{1m} & \cdot\cdot & a_{12} & a_{11} \end{bmatrix}$$

An upper triangular matrix is one in which all the terms below the diagonal are zero and a lower triangle matrix is one in which all the terms above the diagonal are zero. The transpose of an upper triangle is a lower triangle matrix. The determinant of a lower or upper triangle matrix is equal to the products along the diagonal. The inverse of an upper (lower) triangle matrix is an upper (lower) triangle matrix. The product of two upper (lower) triangle matrices is an upper (lower) triangle matrix.

A $n \times n$ matrix A is said to be Toeplitz if all the elements along each of the diagonals have the same value, i.e., $a_{ij} = a_{i+1,j+1}$, for all $i < n$ and $j < n$. An example of a 4 x 4 Toeplitz matrix is,

$$A = \begin{bmatrix} 1 & 2 & 3 & 4 \\ 5 & 1 & 2 & 3 \\ 6 & 5 & 1 & 2 \\ 7 & 6 & 5 & 1 \end{bmatrix}$$

A Toeplitz matrix is fully defined by the terms in the first row and first column.

The Hankel matrix has equal elements along the diagonals which are perpendicular to the main diagonal (cross-diagonals), i.e., $a_{ij} = a_{i+1,j-1}$, for all $i < n$ and $j \leq n$. An example of a 4 x 4 Hankel matrix is,

$$A = \begin{bmatrix} 4 & 3 & 2 & 1 \\ 3 & 2 & 1 & 5 \\ 2 & 1 & 5 & 6 \\ 1 & 5 & 6 & 7 \end{bmatrix}$$

The exchange matrix J is a Hankel matrix.

Toeplitz matrices are a special case of persymmetric matrices, i.e., symmetrical about the cross-diagonal (diagonal from bottom left to top right corners). If a Toeplitz is symmetric, or Hermitian in the case of a complex matrix, then all the elements are determined by either the first row or first column. Symmetric Toeplitz and Hermitian Toeplitz matrices **A** can be represented as follows,

$$\mathbf{A} = \text{Toep}\{a(0), a(1),.., a(p)\}$$

where:

$a(i)$ are the elements in the first column.

Symmetric Toeplitz matrices are centrosymmetric matrices, which are both symmetric and persymmetric. An example of a 4 x 4 Symmetric Toeplitz matrix is,

$$\mathbf{A} = \begin{bmatrix} 1 & 2 & 3 & 4 \\ 2 & 1 & 2 & 3 \\ 3 & 2 & 1 & 2 \\ 4 & 3 & 2 & 1 \end{bmatrix}$$

If **A** is a symmetric Toeplitz matrix then,

$$\mathbf{J}^T \mathbf{A} \mathbf{J} = \mathbf{A}$$

If **A** is a Hermitian Toeplitz matrix, then,

$$\mathbf{J}^T \mathbf{A} \mathbf{J} = \mathbf{A}^*$$

A summary of relationships between various matrices and their inverses is listed below,

Matrix	Inverse
Symmetric	*Symmetric*
Hermitian	*Hermitian*
Persymmetric	*Persymmetric*
Centrosymmetric	*Centrosymmetric*
Toeplitz	*Persymmetric*
Hankel	*Symmetric*
Triangular	*Triangular*

A real n x n matrix is said to be orthogonal if the columns (and rows) are orthogonal. If $\mathbf{A}^T \mathbf{A} = \mathbf{I}$ then **A** is said to be orthonormal (this implies that its rows and columns are orthonormal), and the inverse of **A** is equal to its transpose, i.e., $\mathbf{A}^{-1} = \mathbf{A}^T$. The exchange matrix **J** is an example of this type of orthogonal matrix since $\mathbf{J}^T\mathbf{J} = \mathbf{J}^2 = \mathbf{I}$. A complex $n \times n$ matrix said to be unitary if the columns (and rows) are orthogonal, therefore $\mathbf{A}^H\mathbf{A} = \mathbf{I}$ and $\mathbf{A}^{-1} = \mathbf{A}^H$.

2.8.6 Quadratic and Hermitian Forms

The quadratic form of a real symmetric $n \times n$ matrix $\mathbf{A}$ is the scalar defined by Equation 2.55.

$$Q_A(\mathbf{x}) = \mathbf{x}^T\mathbf{A}\mathbf{x} = \sum_{i=1}^{n}\sum_{j=1}^{n} x_i a_{ij} x_j \tag{2.55}$$

where:
   $\mathbf{x}$ is an n-dimensional real variable vector.

The Hermitian form of a $n \times n$ Hermitian matrix $\mathbf{A}$ is the scalar defined by Equation 2.56

$$Q_A(\mathbf{x}) = \mathbf{x}^H\mathbf{A}\mathbf{x} = \sum_{i=1}^{n}\sum_{j=1}^{n} x_i^* a_{ij} x_j \tag{2.56}$$

where:
   $\mathbf{x}$ is an n-dimensional complex variable vector.

If the quadratic form of matrix $\mathbf{A}$ is positive for all nonzero vectors $\mathbf{x}$, then $\mathbf{A}$ is said to be positive definite, i.e., $\mathbf{A} > 0$. If the quadratic form is nonnegative for all nonzero vectors it is said to be positive semidefinite. Likewise for the negative and nonpositive cases $\mathbf{A}$ is said to be negative definite and negative semidefinite respectively.

2.8.7 Eigenvalues and Eigenvectors

Eigenvalues can be used to determine if a matrix is positive definite, invertible as well as indicate how sensitive the determination of the inverse will be to numerical errors.   Given the set of linear Equations 2.57 for a $n \times n$ matrix $\mathbf{A}$, the matrix $(\mathbf{A}-\lambda\mathbf{I})$ must be singular, i.e., $\det(\mathbf{A}-\lambda\mathbf{I}) = 0$, in order for a nonzero vector to be the solution to the set of linear equations represented by Equation 2.57.

$$\mathbf{A}\mathbf{v} = \lambda\mathbf{v}, \text{ i.e., } (\mathbf{A}-\lambda\mathbf{I})\mathbf{v} = 0 \tag{2.57}$$

The characteristic nth order polynomial of the matrix $\mathbf{A}$ is defined by Equation 2.58.

$$p(\lambda) = \det(\mathbf{A}-\lambda\mathbf{I}) = 0 \tag{2.58}$$

The n roots, λ_i, of the characteristic polynomial are known as the eigenvalues of $\mathbf{A}$. For each eigenvalue the matrix $(\mathbf{A}-\lambda_i\mathbf{I})$ will be singular and there will be at least one nonzero vector, $\mathbf{v}_i$, that solves the equation $\mathbf{A}\mathbf{v} = \lambda\mathbf{v}$, i.e., $\mathbf{A}\mathbf{v}_i = \lambda_i\mathbf{v}_i$. These vectors $\mathbf{v}_i$ are called the eigenvectors. For any eigenvector $\mathbf{v}_i$, $\alpha\mathbf{v}_i$ will, for any constant α, also be an eigenvector, therefore eigenvectors are often normalised, i.e., $\| \mathbf{v}_i \| = 1$. The nonzero eigenvectors, $\mathbf{v}_1$, $\mathbf{v}_2$,..., $\mathbf{v}_n$, corresponding to distinct eigenvalues, λ_1, λ_2,..., λ_n, are linearly independent.

If $\mathbf{A}$ is a $n \times n$ singular matrix, then there are nonzero solutions to the homogeneous Equations 2.59, and it follows that $\lambda = 0$ is an eigenvalue of $\mathbf{A}$. Then $\mathbf{A}$ will have $\rho(\mathbf{A})$ nonzero eigenvalues and $(n - \rho(\mathbf{A}))$ eigenvalues equal to zero.

$$\mathbf{A}\mathbf{v}_i = 0 \tag{2.59}$$

The eigenvalues of a Hermitian matrix are real. A Hermitian matrix is positive definite if and only if the eigenvalues of the matrix are positive. Similar properties hold for positive semidefinite, negative definite, and negative semidefinite matrices. The determinant of a $n \times n$ matrix $\mathbf{A}$ is related to its eigenvalues by the following relationship,

$$\det(A) = \prod_{i=1}^{n} \lambda_i$$

Consequently, a matrix is invertible if and only if all of its eigenvalues are nonzero and any positive definite matrix is nonsingular. The eigenvectors of a Hermitian matrix corresponding to distinct eigenvalues are orthogonal, i.e., if $\lambda_i \neq \lambda_j$, then $\langle \mathbf{v}_i, \mathbf{v}_j \rangle = 0$.

For any $n \times n$ matrix $\mathbf{A}$ having a set of linearly independent eigenvectors it is possible to perform the following eigenvalue decomposition,

$$\mathbf{A} = \mathbf{V}\mathbf{\Lambda}\mathbf{V}^{-1} = [\mathbf{v}_1, \mathbf{v}_2,..., \mathbf{v}_n]\, \mathrm{diag}[\lambda_1, \lambda_2,..., \lambda_n][\mathbf{v}_1, \mathbf{v}_2,..., \mathbf{v}_n]^{-1}$$

where:

$\mathbf{V}$ is a matrix that contains the eigenvectors of $\mathbf{A}$.

$\mathbf{\Lambda}$ is a diagonal matrix that contains the eigenvalues.

For a Hermitian matrix $\mathbf{A}$, $\mathbf{V}$ is unitary and the eigenvalue decomposition becomes Equation 2.60.

$$\mathbf{A} = \mathbf{V}\mathbf{\Lambda}\mathbf{V}^H = \sum_{i=1}^{n} \lambda_i \mathbf{v}_i \mathbf{v}_i^H \tag{2.60}$$

This result is known as the Spectral Theorem, which states that: Any Hermitian matrix $\mathbf{A}$ may be decomposed as defined by Equation 2.61.

$$\mathbf{A} = \mathbf{V}\mathbf{\Lambda}\mathbf{\Lambda}^{-1} = \lambda_1 \mathbf{v}_1 \mathbf{v}_1^H, \lambda_2 \mathbf{v}_2 \mathbf{v}_2^H,...., \lambda_n \mathbf{v}_n \mathbf{v}_n^H \tag{2.61}$$

An application of the Spectral Theorem is to find the inverse of a nonsingular Hermitian matrix $\mathbf{A}$ as defined by Equation 2.62.

$$\mathbf{A}^{-1} = (\mathbf{V}\mathbf{\Lambda}\mathbf{V}^H)^{-1} = (\mathbf{V}^H)^{-1}\mathbf{\Lambda}^{-1}\mathbf{V}^{-1} = \mathbf{V}\mathbf{\Lambda}^{-1}\mathbf{V}^H = \sum_{i=1}^{n} \frac{1}{\lambda_i} \mathbf{v}_i \mathbf{v}_i^H \tag{2.62}$$

If $\mathbf{B}$ is a $n \times n$ matrix with eigenvalues λ_i and $\mathbf{A}$ is a matrix related to $\mathbf{B}$ by $\mathbf{A} = \mathbf{B} + \alpha\,\mathbf{I}$, then $\mathbf{A}$ and $\mathbf{B}$ have the same eigenvectors and the eigenvalues are $\lambda_i + \alpha$. This property can be used to stabilise a problem solution in some signal processing applications if $\mathbf{B}$ is singular or ill-conditioned (one or more eigenvalues are close to zero). By applying this remedy to $\mathbf{B}$ it does not change the eigenvectors but it conditions the eigenvalues in such a way as to stabilise the solution.

For a symmetrical positive definite matrix $\mathbf{A}$, the equation $\mathbf{x}^T\mathbf{A}\mathbf{x} = 1$ defines an ellipse in n dimensions whose axes are in the direction of the eigenvectors $\mathbf{v}_j$ of $\mathbf{A}$ with the half-length of these axes equal to $1/\sqrt{\lambda_j}$. A loose upper bound for the largest eigenvalue $\lambda_{\max}$ is as defined by Equation 2.63.

$$\lambda_{\max} \le \sum_{i=1}^{n} \lambda_i = \operatorname{tr}(\mathbf{A}) \tag{2.63}$$

A tighter upper bound is given by,

$$\lambda_{\max} \le \max_i \sum_{j=1}^{n} a_{ij}$$

also,

$$\lambda_{\max} \le \max_j \sum_{i=1}^{n} a_{ij}$$

where:

$$\mathbf{A} = \{a_{ij}\}$$

2.9 Introduction to Stochastic Processes

A stochastic process most commonly, but not always, represents the noise or unknown and unwanted part of a measurement made on a physical system. Stochastic or random processes are those that may only be described probabilistically or in terms of their expectation or average behaviour. The most common statistical averages are the mean, variance, and autocorrelation. These averages are strictly the ensemble averages over all possible process outputs over all times and situations typical of that process. A single infinite time history of a random process is called a sample function (or a sample record when recorded over a finite time interval) (Bendat and Piersol 1971). The collection of all sample functions, called the ensemble, is defined as the random process, since it encompasses all the possible process information. Random processes may be categorised as either stationary or nonstationary. Stationary processes may be further categorised as ergodic or nonergodic as shown in Figure 2.12.

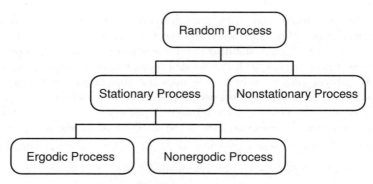

Figure 2.12. Basic Classifications of Random Processes

The notion of stationarity and the concept of ergodicity are important to understand in the context of systems engineering. Stationarity refers to statistical-time-invariance, that is, all the probability density functions associated with the process are invariant with respect to translation in time. An ergodic process is one whose statistical properties can be estimated from the time averages of a single sample function or realisation. This means that the same statistical information about a process can be determined by averaging outputs measured at different times (possible to do) as by averaging different outputs obtained at the same time (strictly impossible to do). Fortunately, in practice, many processes of engineering interest are in fact ergodic and therefore amenable to mathematical analysis.

A process is called wide-sense or weakly stationary if both its mean and variance are finite and constant, and its autocorrelation function depends only on the difference between the times of occurrence of the samples, or lag time. Wide-sense stationarity is a common assumption made about stochastic processes. This is an adequate enough assumption for many signals of interest if a short enough time interval is taken, although it may not be strictly true in practice. However, for many practical applications an assumption of strong stationarity can often be made if weak stationarity is verified.

A nonstationary random process is one that does not meet the requirements for stationarity. Therefore nonstationary processes are those whose statistics are time-variant. In practice it is necessary to impose some restrictions on them to be able to accurately estimate their statistics from a single sample record. Without restrictions, such as an assumption that the nonstationary process can be modelled by a stationary process multiplied by a deterministic function of time, it is often not feasible to obtain a sufficient number of sample records to achieve accurate ensemble averaging. Each type of nonstationarity must be analysed and dealt with separately since there are so many possible types with their own special characteristics.

2.10 Random Signals

A Signal is defined as the output from a process of some type, mathematical or physical, and it can be either deterministic or random. A deterministic signal is one that can be reproduced exactly by repeating the process. For example, the impulse response of a linear time-invariant continuous-time filter or linear shift-invariant discrete-time filter is a deterministic signal. A random signal, on the other hand, is one that is not repeatable in a predictable manner. An example of a random signal is the type of static noise that adds to a speech or music signal during an analogue radio transmission process. Depending on the signal processing that will be done on the signal the same type of signal can be considered to be either deterministic or random. Speech, for example, can be defined as deterministic if it is seen as a set of specific waveform signals, and defined as random if it is seen as all possible signals emanating from a general speech process.

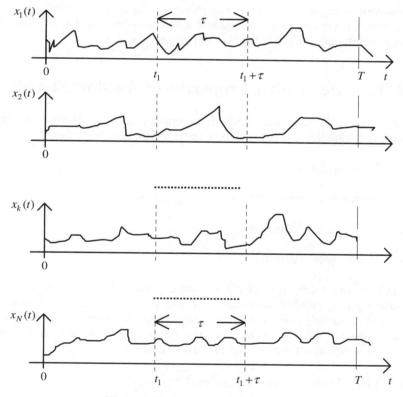

Figure 2.13. Sample Function Ensemble

A signal is termed random if it is not possible to specify deterministically what its amplitude value will be at any given time. Instead, a probability distribution is associated with each instant of time, which describes the likelihood of occurrence within some amplitude range. A stochastic process can also be defined as a collection of random variables, one for each instant of time. In the most general sense a stochastic or random signal can be seen as a continuous or discrete R^n valued function of time. If the process is represented by a vector $\mathbf{x} = [x_1, x_2,..., x_n]^T$ then each random real variable x_i represents a sample function associated with that instant of time. Refer to Figure 2.13.

Wide-sense stationary random processes, denoted by $\{x(t)\}$ or $\{y(t)\}$, are the most typically encountered types (the brackets "{ }" indicate a set or ensemble of sample functions). The statistics of stationary processes don't change with time, which effectively means that the process can be fully described by the moments of the process, i.e., the mean, variance etc. For example, a Gaussian, or second order random process, is fully described by its mean and variance alone. Many practical problems assume a Gaussian noise process. For least squares estimation problems Gaussian noise statistics will ensure that the optimum solution is achieved, since it attempts to minimise a second order or quadratic error function. The processes of main interest to engineering are ergodic processes. These are stationary processes whose means, covariance and other statistical functions can be computed from time averages on arbitrary sample functions from the process.

2.11 Basic Descriptive Properties of Random Signals

The main types of statistical functions commonly used to describe the basic properties of stationary random signals (Bendat and Piersol 1971) are the,

1. Mean square value.

2. Probability density function.

3. Autocorrelation function.

4. Power spectral density.

The mean square value provides a rough description of the signal intensity. The probability density provides statistical information about the signal amplitude. The power spectral density is the Fourier transform of the autocorrelation function and the power spectral density provides the same information in the frequency domain that the autocorrelation function does in the time domain.

2.11.1 The Mean Square Value and Variance

For a random real valued process $\{x(t)\}$ the estimates of the ensemble mean, mean square and variance are computed by taking the instantaneous value for each of say N sample functions of the ensemble at some arbitrary value t_i as follows,

$$\text{Mean value of } \{x(t_i)\} \text{ is} \qquad \bar{x}(t_i) = \lim_{N \to \infty} \frac{1}{N} \sum_{k=1}^{N} x_k(t_i)$$

$$\text{Mean square value of } \{x(t_i)\} \text{ is} \qquad \Psi_x^2(t_i) = \lim_{N \to \infty} \frac{1}{N} \sum_{k=1}^{N} x_k^2(t_i)$$

$$\text{Variance of } \{x(t_i)\} \text{ is} \qquad \sigma_x^2(t_i) = \lim_{N \to \infty} \frac{1}{N} \sum_{k=1}^{N} (x_k(t_i) - \bar{x}(t_i))^2$$

$$\text{where}: \sigma_x^2(t_i) = \Psi_x^2(t_i) - \bar{x}^2(t_i)$$

The time-averaged mean, mean square and variance for an arbitrary kth sample function of the real valued process $\{x(t)\}$ are as follows,

$$\text{Mean value of } x_k(t) \text{ is} \qquad \bar{x}_k = \lim_{T \to \infty} \frac{1}{T} \int_0^T x_k(t) \, dt$$

$$\text{Mean square value of } x_k(t) \text{ is} \qquad \Psi_{x_k}^2 = \lim_{T \to \infty} \frac{1}{T} \int_0^T x_k^2(t) \, dt$$

$$\text{Variance of } x_k(t) \text{ is} \qquad \sigma_{x_k}^2 = \lim_{T \to \infty} \frac{1}{T} \int_0^T (x_k(t) - \bar{x}_k)^2 \, dt$$

$$\text{where}: \sigma_{x_k}^2 = \Psi_{x_k}^2 - \bar{x}_k^2$$

If the ensemble averages of a random process are equal to the time averages of any sample function the process is said to be ergodic. Only stationary processes can be ergodic. If the mean of x is zero the mean square value of x is equal to the variance of x. This can be very advantageous in some filter algorithm computations therefore the mean is often removed before processing.

2.11.2 The Probability Density Function

The Probability Density Function (PDF) of a signal describes the probability that the signal will assume a particular amplitude value within some defined range at any instant in time.

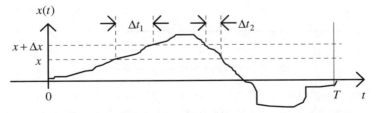

Figure 2.14. Probability Density Function Measurement

Consider the sample record of $x(t)$ in Figure 2.14. The probability that the signal $x(t)$ assumes an amplitude value between x and $x + \Delta x$ can be computed by taking the ratio of time that the signal is in the interval $T_x = \Sigma \, \Delta t_i = \Delta t_1 + \Delta t_2$ compared to the total time of the record T. For an ergodic signal the ratio T_x/T will approach the exact probability density function $p(x)$ as T approaches infinity,

$$p(x) = \lim_{\Delta x \to 0} \frac{\text{Probability}[x < x(t) < x + \Delta x]}{\Delta x} = \lim_{\Delta x \to 0} \frac{1}{\Delta x}\left[\lim_{T \to \infty} \frac{T_x}{T}\right]$$

The probability density function is always real-valued and nonnegative. The probability that the instantaneous value $x(t)$ is less than or equal to some value x is defined by $P(x)$, which is also know as the cumulative probability distribution function,

$$P(x) = \text{Probability}[x(t) \le x] = \int_{-\infty}^{x} p(\xi)d\xi$$

Conversely, the probability density function can be computed as the differential of the probability distribution function as defined by Equation 2.64.

$$p(x) = \frac{dP(x)}{dx} \tag{2.64}$$

$P(x)$ is also a nonnegative function since it is bounded by zero at $-\infty$, i.e., $P(-\infty) = 0$, and one at $+\infty$, i.e., $P(\infty) = 1$. The probability that $x(t)$ lies between the open interval range (x_1, x_2) is defined by Equation 2.65.

$$P(x_2) - P(x_1) = \text{Probability}[x_1 < x(t) \le x_2] = \int_{x_1}^{x_2} p(\xi)d\xi \tag{2.65}$$

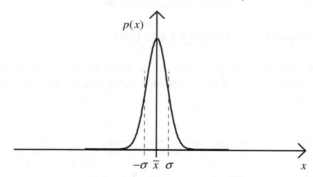

Figure 2.15. The Univariate Gaussian PDF

The Gaussian PDF is a very common distribution associated with noise statistics, especially those related to the optimum filtering problem. The univariate Gaussian probability density function $p(x)$ of a random variable x, shown in Figure 2.15, is defined by Equation 2.66.

$$p(x) = \frac{1}{\sqrt{2\pi\sigma_x^2}} e^{\frac{-[(x-\bar{x})(\sigma_x^2)^{-1}(x-\bar{x})]}{2}} \tag{2.66}$$

A measure of the linear association between the variables x_{ij} and x_{kj} of N random vectors $\mathbf{x}_j = [x_{1j}, x_{2j}, ..., x_{pj}]^T$ is provided by the sample covariance or the average product of deviations from their respective means as defined by Equation 2.67.

$$\sigma_{ik} = \frac{1}{N} \sum_{j=1}^{N} (x_{ij} - \bar{x}_{ij})(x_{kj} - \bar{x}_{kj}), \ i = 1,2,....p, \ k = 1,2,....p \tag{2.67}$$

The notation defined in Equation 2.68 is often used to represent the covariance matrix of the matrix $\mathbf{X}$ containing N samples of vector $\mathbf{x}_j$,

$$\Sigma = \text{Covariance } (\mathbf{X}) = \begin{bmatrix} \sigma_{11} & \sigma_{12} & .. & \sigma_{1p} \\ \sigma_{21} & \sigma_{22} & .. & \sigma_{2p} \\ \vdots & \vdots & \ddots & \vdots \\ \sigma_{p1} & \sigma_{p2} & .. & \sigma_{pp} \end{bmatrix} \tag{2.68}$$

The multivariate Gaussian probability density function $p(\mathbf{x})$ of a p-dimensional random vector $\mathbf{x}$ is defined by Equation 2.69.

$$p(\mathbf{x}) = \frac{1}{(2\pi)^{\frac{p}{2}} |\Sigma|^{\frac{1}{2}}} e^{\frac{-[(\mathbf{x}-\bar{\mathbf{x}})^T (\Sigma)^{-1}(\mathbf{x}-\bar{\mathbf{x}})]}{2}} \tag{2.69}$$

where:
Σ is the covariance matrix of the vector process $\{\mathbf{x}\}$.

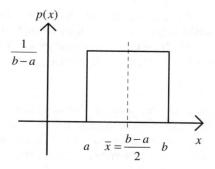

Figure 2.16. The Tophat PDF

Another important PDF is the tophat PDF shown in Figure 2.16, also known as a uniform or rectangular PDF and is defined by Equation 2.70.

$$p(x) = \frac{1}{b-a}, \quad \text{for } a \le x \le b$$

$$p(x) = 0, \qquad \text{for } x \text{ otherwise}$$

(2.70)

2.11.3 Jointly Distributed Random Variables

It may be possible that there is some statistical dependence between two random variables. The statistical dependence between random variables is measured by the joint probability density function. The joint probability density function $p(x, y)$ of two random variables or sample records, defined by Equation 2.71, describes the probability they will simultaneously assume values within a defined pair of ranges at any instant in time. The probability that the signal $x(t)$ assumes an amplitude value between x and $x+\Delta x$ while a signal $y(t)$ simultaneously assumes an amplitude value between y and $y+\Delta y$ can be computed by taking the ratio of time that the signals are both in their respective intervals, T_{xy} compared to the total time of the record T. For an ergodic signal the ratio T_{xy}/T will approach the exact probability function as T approaches infinity.

$$p(x, y) = \lim_{\substack{\Delta x \to 0 \\ \Delta y \to 0}} \frac{1}{\Delta x \, \Delta y} \left[\lim_{T \to \infty} \frac{T_{x,y}}{T} \right]$$

(2.71)

The joint probability density function is always real-valued and nonnegative. The probability that the instantaneous values $x(t)$ and $y(t)$ are less than or equal to some values x and y, $P(x,y)$, is defined by Equation 2.71, the joint probability distribution function.

$$P(x, y) = \text{Probability}[x(t) \le x, y(t) \le y] = \int_{-\infty}^{x} \int_{-\infty}^{x} p(\xi, \eta) d\xi d\eta$$

(2.71)

If two processes $\{x(t)\}$ and $\{y(t)\}$ are statistically independent, the joint probability density function is the product of their individual probability functions as defined by Equation 2.72. If two random variables are statistically independent then they are also linearly independent. The converse is not necessarily true.

$$p(x, y) = p(x)p(y)$$

(2.72)

The main application of the joint probability function is to establish a probabilistic description for an event associated with two sets of correlated random data. For example, to determine the probability of detecting two similar signals being transmitted randomly, but with some correlation, from separate sources to the same receiver.

2.11.4 The Expectation Operator

In the most general terms, if $f(\mathbf{x})$ is a (deterministic) m-dimensional vector function of a n-dimensional continuous vector variable $\mathbf{x}$, the expected value or expectation of $f(\mathbf{x})$ is defined by Equation 2.73. The expected value of the function is effectively the function's statistical average.

$$E\{f(\mathbf{x})\} = \int_{-\infty}^{+\infty} f(\mathbf{x})p(\mathbf{x})d\mathbf{x} \qquad (2.73)$$

where:
    $E\{.\}$ is the linear expectation operator.
    $p(\mathbf{x})$ is the probability density function of $\mathbf{x}$.

For the case of a discrete m-dimensional vector function $f(\mathbf{x}(k))$ of a n-dimensional discrete vector variable $\mathbf{x}(k)$, the expectation of $f(\mathbf{x}(k))$ is defined by Equation 2.74.

$$E\{f(\mathbf{x}(k))\} = \lim_{N \to \infty} \frac{1}{N} \sum_{k=-\infty}^{+\infty} f(\mathbf{x}(k)) \qquad (2.74)$$

The statistical moments of a signal can be defined in terms of the expectation operator and probability density function $p(x)$ of x as follows,

$\bar{x} = E\{x\} = \int_{-\infty}^{+\infty} xp(x)dx,$	Mean = first statistical moment.
$\Psi_x^2 = E\{x^2\} = \int_{-\infty}^{+\infty} x^2 p(x)dx,$	Mean square = second statistical moment.
$\sigma_x^2 = E\{(x-\bar{x})^2\} = \int_{-\infty}^{+\infty} (x-\bar{x})^2 p(x)\, dx,$	Variance = second central moment.
$M_x^n = E\{x^n\} = \int_{-\infty}^{+\infty} x^n p(x)dx,$	nth statistical moment.
$CM_x^n = E\{(x-\bar{x})^n\} = \int_{-\infty}^{+\infty} (x-\bar{x})^n p(x)dx,$	nth central moment.

2.11.5 The Autocorrelation and Related Functions

The autocorrelation function of a stationary real valued signal $x(t)$ describes the general dependence of its values at one time on its values at another time. An estimate of the autocorrelation between the values of $x(t)$ at the times t and $t + \tau$ may be made by taking the product of the two values and averaging over an observation time T. The exact autocorrelation function $r_x(\tau)$, defined by Equation 2.75, is approached as T approaches infinity,

$$r_x(\tau) = E\{x(t)x(t+\tau)\} = \lim_{T \to \infty} \frac{1}{T} \int_0^T x(t)x(t+\tau)dt \qquad (2.75)$$

Some properties of the autocorrelation function are as follows,

$$r_x(-\tau) = r_x(\tau)$$

$$\bar{x} = \sqrt{r_x(\infty)}$$

$$\Psi_x^2 = r_x(0) = \text{Average signal power.}$$

The autocorrelation function is always a real valued even function with its maximum at $\tau = 0$. A lowpassed or narrow band white noise example of the autocorrelation function is shown in Figure 2.17. For infinite band white noise the autocorrelation function is an impulse function at $\tau = 0$ and zero elsewhere.

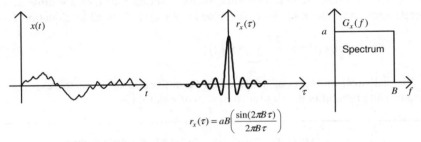

$$r_x(\tau) = aB\left(\frac{\sin(2\pi B \tau)}{2\pi B \tau}\right)$$

Figure 2.17. Band-limited White Noise

The cross-correlation function between $x(t)$ and $y(t)$ describes the general dependence of the values of $x(t)$ at one time on the values of $y(t)$ at another time and is defined by Equation 2.76.

$$r_{xy}(\tau) = E\{x(t)y(t+\tau)\} = \lim_{T\to\infty}\frac{1}{T}\int_0^T x(t)y(t+\tau)dt \qquad (2.76)$$

Some useful properties of the cross-correlation function are as follows,

$$r_{xy}(-\tau) = r_{yx}(\tau)$$

$$\left|r_{xy}(\tau)\right|^2 \le r_x(0)\,r_y(0)$$

$$\left|r_{xy}(\tau)\right| \le \frac{1}{2}[r_x(0) + r_y(0)]$$

When $r_{xy} = 0$ then $x(t)$ and $y(t)$ are uncorrelated. If $x(t)$ and $y(t)$ are statistically independent then $r_{xy}(\tau) = 0$ if either $\bar{x}$ or $\bar{y} = 0$, and if neither mean is zero then $r_{xy} = \bar{x}\,\bar{y}$. The cross-correlation function is often used for measurement of time delays, determination of transmission paths and detection and recovery of signals in noise.

The autocovariance function is defined by Equation 2.77.

$$c_x(\tau) = E\{(x(t)-\bar{x})(x(t+\tau)-\bar{x})\}$$

$$c_x(\tau) = \lim_{T\to\infty} \frac{1}{T} \int_0^T [x(t)-\bar{x}][x(t+\tau)-\bar{x}]dt$$

$$c_x(\tau) = r_x(\tau) - \bar{x}^2 \tag{2.77}$$

Covariance = Correlation when the mean is zero.

As for the mean square value and the variance, the covariance and correlation functions are equal when the mean is zero, which can be advantageous for some filter processing algorithms.

The cross-covariance function is defined by Equation 2.78.

$$c_{xy}(\tau) = \lim_{T\to\infty} \frac{1}{T} \int_0^T [x(t)-\bar{x}][y(t+\tau)-\bar{y}]dt$$

$$c_{xy}(\tau) = r_{xy}(\tau) - \bar{x}\,\bar{y} \tag{2.78}$$

Cross‑covariance = Cross‑correlation when either mean is zero.

The correlation function coefficient (normalised cross-variance function) is defined by Equation 2.79.

$$\rho_{xy} = \frac{c_{xy}(\tau)}{\sqrt{c_x(0)\,c_y(0)}}, \quad -1 \le \rho_{xy} \le 1 \tag{2.79}$$

where:

$$c_x(0) = \sigma_x$$
$$c_y(0) = \sigma_y$$

This correlation function coefficient ρ_{xy} measures the degree of linear dependence between x and y for a displacement of τ in y relative to x.

A measure of the correlation between the variables x_{ij} and x_{kj} of N random p-dimensional vectors $\mathbf{x}_j = [x_{1j}, x_{2j},..., x_{pj}]^T$ is provided by the sample correlation defined by Equation 2.80.

$$r_{ik} = \frac{1}{N} \sum_{j=1}^N (x_{ij})(x_{kj}), \quad i = 1,2,....p, \quad k = 1,2,....p \tag{2.80}$$

The notation defined by Equation 2.81 is often used to represent the correlation matrix of the matrix $\mathbf{X}$ containing N samples of vector $\mathbf{x}_j$,

$$\text{Correlation}(\mathbf{X}) = \begin{bmatrix} r_{11} & r_{12} & \cdots & r_{1p} \\ r_{21} & r_{22} & \cdots & r_{2p} \\ \vdots & \vdots & \ddots & \vdots \\ r_{p1} & r_{p2} & \cdots & r_{pp} \end{bmatrix} \tag{2.81}$$

2.11.6 Power Spectral Density Functions

The Power Spectral Density (PSD) function describes the general frequency composition of a signal in terms of the spectral density of its mean square value. It gives the distribution of signal power per unit frequency. The PSD can be defined as either a one-sided or two-sided function. The one-sided PSD $G_x(f)$ of a signal $x(t)$ at a particular frequency f is found by taking the limit as defined by Equation 2.82.

$$G_x(f) = \lim_{\Delta f \to 0} \frac{1}{\Delta f} \Psi_x^2(f, \Delta f), \quad \text{only for } f \geq 0 \tag{2.82}$$

where:

$$\Psi_x^2(f, \Delta f) = \lim_{T \to \infty} \frac{1}{T} \int_0^T x^2(t, f, \Delta f) dt$$

T is the observation time.

$x(t, f, \Delta f)$ is the portion of $x(t)$ in the frequency range from f to $f + \Delta f$.

$$\Psi_x^2 = \int_0^\infty G_x(f) df$$

$G_x(f)$ is derived by first postulating the filtering of a signal sample $x(t)$ with a bandpass filter having sharp cut off characteristics between f and $f + \Delta f$ and then computing the average of the squared output from the filter (Bendat and Piersol 1971). This average squared value approaches an exact square value as the observation time T approaches infinity. The single sided power spectral density $G_x(f)$ is then defined by Equation 2.82 as the filter bandwidth Δf approaches zero.

$G_x(f)$ is termed one-sided as it is only defined for $f \geq 0$. The two-sided PSD is defined over all f for stationary signals. According to the Wiener-Khinchine theorem the power spectral density is equal to the Fourier transform of the autocorrelation function as defined by Equation 2.83.

$$P_x(f) = \int_{-\infty}^{+\infty} r_x(\tau) e^{-j2\pi f \tau} d\tau, \tag{2.83}$$

The relation between these two PSDs is defined by Equation 2.84.

$$G_x(f) = 2P_x(f) = 2P_x(-f) \tag{2.84}$$

The two cross-spectral density functions for two stationary processes $\{x(t)\}$ and $\{y(t)\}$ are defined as the Fourier transforms of the respective cross-correlation functions as defined by Equations 2.85 and 2.86 respectively.

$$P_{xy}(f) = FT[r_{xy}] = \int_{-\infty}^{+\infty} r_{xy}(\tau) e^{-j2\pi f\tau} d\tau \tag{2.85}$$

$$P_{yx}(f) = FT[r_{yx}] = \int_{-\infty}^{+\infty} r_{yx}(\tau) e^{-j2\pi f\tau} d\tau \tag{2.86}$$

P_{xy} and P_{yx} are complex conjugates of each other and their sum is real. Furthermore, if $\{x(t)\}$ and $\{y(t)\}$ are zero mean then,

$$P_{x+y}(f) = P_x(f) + P_{xy}(f) + P_{yx}(f) + P_y(f)$$

2.11.7 Coherence Function

The coherence function $\gamma_{xy}(f)$, defined by Equation 2.87, between two wide-sense stationary random processes $\{x(t)\}$ and $\{y(t)\}$ is equal to the cross-power spectrum $P_{xy}(f)$ divided by the square root of the product of the two auto-power spectra (Chen 1988). This coherence function is a normalised cross-spectral density function. The Magnitude-Squared Coherence (MSC) function is defined Equation 2.88.

$$\gamma_{xy}(f) = \frac{P_{xy}(f)}{\sqrt{P_x(f)P_y(f)}}, \tag{2.87}$$

Analogous to the correlation function coefficient $\rho_{xy}(\tau)$.

$$S_{xy}(f) = \int_{-\infty}^{+\infty} r_{xy}(\tau) e^{-j2\pi f\tau} d\tau,$$

Fourier Transform of the cross-correlation function.

$$MSC_{xy}(f) = \left|\gamma_{xy}(f)\right|^2, \quad 0 \le MSC_{xy}(f) \le 1 \tag{2.88}$$

When the MSC = 0 for all frequencies then $x(t)$ and $y(t)$ are statistically independent and when the MSC = 1 for all frequencies they are said to be fully coherent. Since the coherence function, and particularly the MSC function, is a measure of the relative linearity of two processes, or correlation between them, it can be used for numerous purposes, including,

1.   System identification.

2.   Measurement of Signal to Noise Ratio (SNR).

3.   Determination of time delay.

However, the use of the coherence function is valid only when it can be estimated accurately.

2.11.8 Discrete Ergodic Random Signal Statistics

The statistical properties discussed above were primarily for continuous-time real valued random signals. Discrete-time real valued signals have similar properties. Some of the main statistical properties for discrete-time wide-sense stationary random signals generated by ergodic discrete-time random real valued processes $\{x[k]\}$ and $\{y[k]\}$ are,

Mean value of $\{x[k]\}$,	$\bar{x} = E\{x[k]\} = \lim\limits_{N \to \infty} \dfrac{1}{N} \sum\limits_{k=1}^{N} x[k]$
Mean square value of $\{x[k]\}$,	$\Psi_x^2 = E\{(x[k])^2\} = \lim\limits_{N \to \infty} \dfrac{1}{N} \sum\limits_{k=1}^{N} x^2[k]$
Variance of $\{x[k]\}$,	$\sigma_x^2 = E\{(x[k] - \bar{x})^2\} = \lim\limits_{N \to \infty} \dfrac{1}{N} \sum\limits_{k=1}^{N} (x^2[k] - \bar{x}]$
	$\sigma_x^2 = E\{(x[k])^2\} - \bar{x}^2 = \Psi_x^2 - \bar{x}^2$
Autocorrelation of $\{x[k]\}$,	$r_x(m) = E\{x[k]x[k+m]\} = \lim\limits_{N \to \infty} \dfrac{1}{N} \sum\limits_{k=1}^{N} x[k]x[k+m]$
Autocovariance of $\{x[k]\}$,	$c_x(m) = E\{(x[k] - \bar{x})(x[k+m] - \bar{x})\}$
Cross - correlation of $\{x[k]\}$ & $\{y[k]\}$,	$r_{xy}(m) = E\{x[k]y[k+m]\}$
Cross - covariance of $\{x[k]\}$ & $\{y[k]\}$,	$c_{xy}(m) = E\{(x[k] - \bar{x})(y[k+m] - \bar{x})\}$
	where : m is the lag factor

Some relational properties are as follows,

$$r_x(-m) = r_x(m)$$
$$\bar{x} = \sqrt{r_x(\infty)}$$
$$\Psi_x^2 = r_x(0) = \text{Average signal power.}$$

For discrete-time wide-sense stationary random signals generated by ergodic discrete-time random complex valued processes $\{x[k]=x_r[k]+jx_i[k]\}$ and $\{y[k]=y_r[k]+jy_i[k]\}$ the autocorrelation, autocovariance, cross-correlation and cross-covariance functions are defined and related as follows,

Autocorrelation of complex $\{x[k]\}$,	$r_x(m) = E\{x[k](x[k+m])^*\}$
	$= [r_{x_r}(m) + r_{x_i}(m)] + j[-r_{x_r x_i}(m) + r_{x_i x_r}(m)]$
Autocovariance of complex $\{x[k]\}$,	$c_x(m) = E\{(x[k+m] - \bar{x})(x[k] - \bar{x})^*\}$
	$c_x(m) = r_x(m) - \bar{x}[k]\bar{x}^*[m]$
Cross-correlation of complex $\{x[k]\} \& \{y[k]\}$,	$r_{xy}(m) = E\{x[k](y[k+m])^*\}$
Cross-covariance of complex $\{x[k]\} \& \{y[k]\}$,	$c_{xy}(m) = E\{(x[k] - \bar{x})(y[k+m] - \bar{y})^*\}$
	$c_{xy}(m) = r_{xy}(m) - \bar{x}[k]\bar{x}^*[m]$
	where : m is the lag factor

The autocorrelation function of complex wide-sense stationary processes is a very important function which is used extensively. Consequently, it deserves closer attention. The autocorrelation between the random variables $x[k]$ and $x[i]$ depends only on the difference, $k - i$, separating the two random variables in time, i.e.,

$$r_x(k,i) = r_x(k-i,0) \equiv r_x(k-i)$$

The difference, $m = k - i$, is called the lag. The autocorrelation sequence of a wide-sense stationary process is a conjugate symmetric function of k, $r_x(m) = r_x^*(-m)$. This property is evident from the definition Equation 2.89,

$$r_x(m) = E\{x(k+m)x^*(k)\} = E\{x(k)x^*(k+m)\} = r_x(-m) \tag{2.89}$$

where:
   m is the lag.

2.11.9 Autocovariance and Autocorrelation Matrices

The autocovariance and autocorrelation sequences are important second-order moments of discrete-time random processes that are often represented in a matrix form. For a $(p+1)$-dimensional vector $\mathbf{x} = [x[0], x[1], x[2],...., x[p]]^T$ of a wide-sense stationary process $\{x[n]\}$ its outer product is a $(p+1) \times (p+1)$ matrix defined by Equation 2.90.

$$\mathbf{xx}^H = \begin{bmatrix} x[0]x^*[0] & x[0]x^*[1] & .. & x[0]x^*[p] \\ x[1]x^*[0] & x[1]x^*[1] & .. & x[1]x^*[p] \\ \vdots & \vdots & .. & \vdots \\ x[p]x^*[0] & x[p]x^*[1] & .. & x[p]x^*[p] \end{bmatrix} \tag{2.90}$$

The expectation of this matrix is the $(p+1) \times (p+1)$ autocorrelation matrix, $\mathbf{R}_x$, as defined by Equation 2.91.

$$\mathbf{R}_x = E\{\mathbf{x}\mathbf{x}^H\} = \begin{bmatrix} r_x(0) & r_x^*(1) & .. & r_x^*(p) \\ r_x(1) & r_x(0) & .. & r_x^*(p-1) \\ \vdots & \vdots & \vdots & \vdots \\ r_x(p) & r_x(p-1) & : & r_x(0) \end{bmatrix} \tag{2.91}$$

where:

$r_x(m) = r_x^*(-m)$, according to Hermitian symmetry.

By the same process the expectation of the outer product of the vector $\mathbf{x}$ minus the mean vector of the process, i.e., $(\mathbf{x} - \overline{\mathbf{x}})$, produces the autocovariance matrix, $\mathbf{C}_x$, which for a zero mean process is equal to the autocorrelation matrix. The autocovariance matrix is defined by Equation 2.92.

$$\mathbf{C}_x = E\{(\mathbf{x} - \overline{\mathbf{x}})(\mathbf{x} - \overline{\mathbf{x}})^H\} = \mathbf{R}_x - \overline{\mathbf{x}}\overline{\mathbf{x}}^H \tag{2.92}$$

The autocorrelation matrix of a wide-sense stationary process is a Hermitian matrix with all the diagonal values real and equal. For a real valued random process it is a symmetric Toeplitz matrix.

A wide-sense stationary Gaussian process with covariance $c_x(m)$ is referred to as autocorrelation ergodic if,

$$\lim_{N \to \infty} \frac{1}{N} \sum_{m=0}^{N-1} c_x^2(m) = 0$$

In most applications it is not practical to determine whether a given process is ergodic. Therefore, often, time averages are simply used to estimate the ensemble averages and the validity of the assumption is tested by the performance of the algorithm requiring the information.

2.11.10 Spectrum of a Random Process

The power Spectrum of a discrete-time wide-sense stationary random process, $\{x[n]\}$, is the Fourier transform of its autocorrelation sequence $r_x(k)$ as defined by Equation 2.93.

$$P_x(e^{j\theta}) = \sum_{k=-\infty}^{\infty} r_x(k)e^{-jk\theta} \qquad P_x(z) = \sum_{k=-\infty}^{\infty} r_x(k)z^{-k} \tag{2.93}$$

where:

$$\theta = \omega T = \frac{2\pi f}{F_s}$$

The autocorrelation sequence may be computed by taking the inverse Fourier transform of $P_x(e^{j\theta})$ as defined by Equation 2.94.

$$r_x(k) = \frac{1}{2\pi} \int_{-\pi}^{+\pi} P_x(e^{j\theta}) e^{jk\theta} d\theta \tag{2.94}$$

The power spectrum of a wide-sense random process $x[n]$ is nonnegative and real-valued, i.e., $P_x(e^{j\theta}) = P_x^*(e^{j\theta})$, and $P_x(z)$ satisfies the symmetry condition, $P_x(z) = P_x^*(1/z)$. If $x[n]$ is real then the power spectrum is even, i.e., $P_x(e^{j\theta}) = P_x(e^{-j\theta})$, which implies that $P_x(z) = P_x^*(z)$. The total power in a zero mean wide-sense stationary process is proportional to the area under the power spectral curve as defined by Equation 2.95.

$$E\{|x[n]|^2\} = \frac{1}{2\pi} \int_{-\pi}^{+\pi} P_x(e^{j\theta}) d\theta \tag{2.95}$$

The eigenvalues of the $n \times n$ autocorrelation matrix of a zero mean wide-sense stationary random process are upper and lower bounded by the maximum and minimum values of the power spectrum as defined by Equation 2.96.

$$\min_\theta P_x(e^{j\theta}) \le \lambda_i \le \max_\theta P_x(e^{j\theta}) \tag{2.96}$$

The power spectrum can also be seen as the expected value of the squared Fourier magnitude, $P_N(e^{j\theta})$, in the limit as $N \to \infty$ for $2N + 1$ samples of a given realisation of the random process, i.e., refer to Equation 2.97.

$$P_x(e^{j\theta}) = E\{P_N(e^{j\theta})\} = \sum_{k=-\infty}^{\infty} r_x(k) e^{-jk\theta}$$

$$= \lim_{N \to \infty} \frac{1}{2N+1} E\left\{ \left| \sum_{n=-N}^{N} x(n) e^{-jn\theta} \right|^2 \right\} \tag{2.97}$$

where:

$$P_N(e^{j\theta}) = \frac{1}{2N+1} \left| \sum_{n=-N}^{N} x(n) e^{-jn\theta} \right|^2$$

If the power spectrum $P_x(e^{j\theta})$ of a wide-sense stationary process is a continuous function of θ, then $P_x(z)$ may be factored into a product of a form known as the "spectral factorisation" of $P_x(z)$ as defined by Equation 2.98.

$$P_x(z) = \sigma_0^2 Q(z) Q^*\left(\frac{1}{z^*}\right) \tag{2.98}$$

where:

$$\sigma_0^2 = \exp\left\{\frac{1}{2\pi}\int_{-\pi}^{\pi}\ln P_x(e^{j\theta})d\theta\right\}$$

For a real-valued process the spectral factorisation is defined by Equation 2.99.

$$P_x(e^z) = \sigma_0^2 Q(z)Q(z^{-1}) \qquad\qquad (2.99)$$

Any process that can be factored in this way is called a regular process and has the following properties,

1. The process $x[n]$ can be realised as the output of a causal and stable filter $H(z)$ that is driven by white noise having a variance of σ_0^2.

2. If the process $x[n]$ is filtered with the inverse filter $\dfrac{1}{H(z)}$ the output $v[n]$ is white noise having a variance of σ_0^2. In this case $\dfrac{1}{H(z)}$ is known as a whitening filter.

3. Since $v[n]$ and $x[n]$ are related by an invertible transformation then they both contain the same information and may be derived from each other.

For the special case when $P_x(z) = \dfrac{N(z)}{D(z)}$, a rational function, then according to the spectral factorisation $P_x(z)$ may be factored in the following form,

$$P_x(z) = \sigma_0^2 Q(z)Q^*\left(\frac{1}{z^*}\right) = \sigma_0^2 \left[\frac{B(z)}{A(z)}\right]\left[\frac{B^*\left(\frac{1}{z^*}\right)}{A^*\left(\frac{1}{z^*}\right)}\right]$$

where:

σ_o^2 is a constant.

$B(z) = 1 + b[1]z^{-1} + .. + b[q]z^{-q}$, is a monic polynomial having all its roots inside the unit circle.

$A(z) = 1 + a[1]z^{-1} + .. + a[p]z^{-p}$, is monic polynomial having all its roots inside the unit circle.

2.11.11 Filtering of Random Processes

The output $y[n]$ of a stable LSI filter, $h[n]$, driven by $x[n]$, a wide-sense stationary process, is defined by Equation 2.100.

$$y[n] = x[n] * h[n] = \sum_{k=-\infty}^{\infty} h[k]x[n-k]$$ (2.100)

The mean of $y[n]$ is defined by Equation 2.101.

$$E\{y[n]\} = E\left\{ \sum_{k=-\infty}^{\infty} h[k]x[n-k] \right\} = \sum_{k=-\infty}^{\infty} h[k]E\{x[n-k]\}$$

$$= \bar{x}[n] \sum_{k=-\infty}^{\infty} h[k] = \bar{x}[n]H(e^{j0})$$ (2.101)

The crosscorrelation between $y[n]$ and $x[n]$, $r_{yx}(n+k,n)$, depends only on the difference between $n + k$ and n and is defined by Equation 2.102.

$$r_{yx}(k) = E\{y[n+k]x^*[n]\} = r_x(k) * h[k]$$ (2.102)

where:
   k is the difference between $n + k$ and n .

The autocorrelation of $y[n]$ is defined by Equation 2.103.

$$r_y(k) = r_{yx}(k) * h^*[-k]$$

$$= \sum_{l=-\infty, m=-\infty}^{\infty} h[l]r_x(m-l+k)h^*[m]$$ (2.103)

$$= r_x(k) * h[k] * h^*[-k]$$

It can also be said that,

$$r_h(k) = h[k] * h^*[-k] = \sum_{n=-\infty}^{\infty} h[n]h^*[n+k]$$

therefore,

$$r_y(k) = r_x(k) * r_h(k)$$

The variance of the output process is defined by Equation 2.104.

$$E\{|y[n]|^2\} = \sigma_y^2 = r_y(0) = \sum_{l=-\infty}^{\infty} \sum_{m=-\infty}^{\infty} h[l]r_x(m-l)h^*[m]$$ (2.104)

In the special case when $h[n]$ is finite with a length of N then the variance, or power, of $y[n]$ may be expressed in terms of the autocorrelation matrix $\mathbf{R}_x$ of $x[n]$ and the vector filter coefficients $\mathbf{h}$ as defined by Equation 2.105.

$$\sigma_y^2 = E\{|y[n]|^2\} = \mathbf{h}^H \mathbf{R}_x \mathbf{h}$$ (2.105)

The power spectrum of $x[n]$ and $y[n]$ are related as follows,

$$P_y(e^{j\theta}) = P_x(e^{j\theta})\left|H(e^{j\theta})\right|^2$$

and,

$$P_y(z) = P_x(z)H(z)H^*(\frac{1}{z^*})$$

If $h[n]$ is real, then $H(z) = H^*(z^*)$ and $P_y(z) = P_x(z)H(z)H(1/z)$.

2.11.12 Important Examples of Random Processes

Some random processes that are typically found and used in relation to signal processing algorithms are Gaussian, white noise and sequences, Gauss-Markov, and random telegraph wave processes.

2.11.12.1 Gaussian Process
A Gaussian stochastic process is one for which all the probability density functions are Gaussian distributions, including all the joint probability density functions as well as the distribution functions at individual instants. Stationary Gaussian processes have the property that they are completely determined by their autocorrelation functions.

2.11.12.2 White Noise
Any stationary stochastic process $\{x(t)\}$ having a constant spectral density function is described as white noise. Its power spectral density function is $P_x(f) = a$ for some constant a, and the autocorrelation function is $r_x(\tau) = a\delta(\tau)$, where $\delta(\tau)$ is the Dirac delta-function (impulse function). These relations strictly only apply to the ideal case of a process having an infinite bandwidth.

Band-limited white noise is a more realistic process model that has a density function defined by Equation 2.106.

$$P_x(f) = \begin{cases} a, & W_1 \le f \le W_2 \\ 0, & |f| < W_1, |f| > W_2 \end{cases} \tag{2.106}$$

where:
 W_1 and W_2 are the lower and upper limits of the passband in Hertz.

The autocorrelation function of band-limited white noise is defined by Equation 2.107.

$$r_x(\tau) = 2a\left[W_2\frac{\sin(2\pi W_2\tau)}{2\pi W_2\tau} - W_1\frac{\sin(2\pi W_1\tau)}{2\pi W_1\tau}\right]$$

$$= 2a\Delta W\frac{\sin(\pi\Delta W\tau)}{\pi\Delta W\tau}\cos(2\pi W_0\tau), \tag{2.107}$$

where:

$$\Delta W = W_2 - W_1$$

$$W_0 = \frac{(W_1 + W_2)}{2}$$

2.11.12.3 White Sequences

The discrete equivalent of white noise is a white sequence, which is defined to be a sequence of zero mean, uncorrelated random variables $x[n]$ each having a variance of σ_x^2. If the random variables $x[n]$ are normally distributed, the sequence is called a Gaussian white sequence. A wide-sense stationary process is said to be white if the autocovariance function $c_x(m)$ is zero for all $m \neq 0$ and it is defined by Equation 2.108.

$$c_x(m) = \sigma_x^2 \delta(m) \tag{2.108}$$

Since white noise is defined only in terms of the second-order moment there are an infinite variety of white noise random processes that are possible.

2.11.12.4 Gauss-Markov Processes

A zero mean stationary Gaussian process $\{x(t)\}$ with an exponential autocorrelation function is called a Gauss-Markov process. The autocorrelation function is defined by Equation 2.109.

$$r_x(\tau) = \sigma_x^2 e^{-\beta|\tau|} \tag{2.109}$$

where:

σ_x^2 is the mean square value of the process and $1/\beta$ is its time constant.

The power spectral density function of the Gauss-Markov process is defined by Equation 2.110.

$$P_x(j\omega) = \frac{2\sigma^2 \beta}{\omega^2 + \beta^2} \tag{2.110}$$

2.11.12.5 The Random Telegraph Wave

A random telegraph wave is a voltage waveform with the following properties,

1. The voltage is either +1 or -1 volts.

2. The voltage at time $t = 0$ may be either +1 or -1 with equal likelihood.

3. The distribution of changes of voltage is a Poisson distribution.

The third property means that the probability of k number of voltage changes in a time interval T is defined by the Poisson distribution Equation 2.111.

$$P(k) = (aT)^k e^{-aT} k!$$

(2.111)

where:
  a is the average number of voltage changes per unit time.

The autocorrelation function of the random telegraph wave is defined by Equation 2.112, which is the same as the autocorrelation of the Gauss-Markov process. However, the random telegraph wave is not a Gaussian process and the two waveforms do look different.

$$r_x(\tau) = e^{-2a|\tau|}$$

(2.112)

2.12 Exercises

The following Exercises identify some of the basic ideas presented in this Chapter.

2.12.1 Problems

2.1.   Does the equation of a straight line $y = \alpha x + \beta$, where α and β are constants, represent a linear system? Show the proof.

2.2.   Show how the z-Transform can become the DFT.

2.3.   Which of the FIR filters defined by the following impulse responses are linear phase filters?

   a. $h[n] = \{0.2, 0.3, 0.3, 0.2\}$
   b. $h[n] = \{0.1, 0.2, 0.2, 0.1, 0.2, 0.2\}$
   c. $h[n] = \{0.2, 0.2, 0.1, 0.1, 0.2, 0.2\}$
   d. $h[n] = \{0.05, 0.15, 0.3, -0.15, -0.15\}$
   e. $h[n] = \{0.05, 0.3, 0.0, -0.3, -0.05\}$

2.4.   Given the following FIR filter impulse responses what are their $H(z)$ and $H(z^{-1})$? What are the zeros of the filters? Express the transfer functions in terms of zeros and poles? Prove that these filters have linear phases.

   a. $h[n] = \{0.5, 0.5\}$
   b. $h[n] = \{0.5, 0.0, -0.5\}$
   c. $h[n] = \{0.5, 0.0, 0.5\}$
   d. $h[n] = \{0.25, -0.5, 0.25\}$

2.5.   Which of the following vector pairs are orthogonal or orthonormal?

   a. $[1, -3, 5]^T$ and $[-1, -2, -1]^T$

b. $[0.6, 0.8]^T$ and $[4, -3]^T$
c. $[0.8, 0.6]^T$ and $[0.6, -0.8]^T$
d. $[1, 2, 3]^T$ and $[4, 5, 6]^T$

2.6.   Which of the following matrices are Toeplitz?

a. $\begin{bmatrix} 3 & 2 & 1 \\ 4 & 3 & 2 \\ 5 & 4 & 3 \end{bmatrix}$, b. $\begin{bmatrix} 1 & 2 & 3 \\ 1 & 2 & 3 \\ 1 & 2 & 3 \end{bmatrix}$, c. $\begin{bmatrix} 1 & 1 & 1 \\ 1 & 1 & 1 \\ 1 & 1 & 1 \end{bmatrix}$, d. $\begin{bmatrix} 3 & 2 & 1 \\ 2 & 1 & 2 \\ 1 & 2 & 3 \end{bmatrix}$, e. $\begin{bmatrix} 1 & 2 & 3 \\ 2 & 1 & 2 \\ 3 & 2 & 1 \end{bmatrix}$

f. $\begin{bmatrix} 1 & (1+j) & (1-j) \\ (1-j) & 1 & (1+j) \\ (1+j) & (1-j) & 1 \end{bmatrix}$

What is special about matrices c, d, e, and f?

2.7.   Which of the following matrices are orthogonal? Compute the matrix inverses of those that are orthogonal.

a. $\begin{bmatrix} 0 & 1 & 0 \\ 0 & 0 & 1 \\ 1 & 0 & 0 \end{bmatrix}$, b. $\begin{bmatrix} 0 & 0 & 1 \\ 0 & 1 & 0 \\ 1 & 0 & 0 \end{bmatrix}$, c. $\begin{bmatrix} 1 & 1 & 1 \\ 1 & 1 & 1 \\ 1 & 1 & 1 \end{bmatrix}$, d. $\begin{bmatrix} 1 & 0 & 0 \\ 0 & 2 & 0 \\ 0 & 0 & 3 \end{bmatrix}$, e. $\begin{bmatrix} 1 & 2 & 3 \\ 2 & 1 & 2 \\ 3 & 2 & 1 \end{bmatrix}$

2.8.   Why are ergodic processes important?

2.9.   Find the eigenvalues of the following 2 x 2 Toeplitz matrix,

$$\mathbf{A} = \begin{bmatrix} a & b \\ b & a \end{bmatrix}$$

Find the eigenvectors for $a = 4$ and $b = 1$.

2.10.   Compute the rounding quantization error variance for an Analogue to Digital Converter (ADC) with a quantization interval equal to Δ. Assume that the signal distribution is uniform and that the noise is stationary white noise.

2.11.   What is the mean and autocorrelation of the random phase sinusoid defined by, $x[n] = A\sin(n\omega_0 + \phi)$, given that A and ω_0 are fixed constants and ϕ is a random variable that is uniformly distributed over the interval $-\pi$ to π. The probability density function for ϕ is,

$$p_\phi(\alpha) = \begin{cases} \dfrac{1}{2\pi} &, -\pi \le \alpha < \pi \\ 0 &, \text{elsewhere} \end{cases}$$

Repeat the computations for the harmonic process, $x[n] = Ae^{j(n\omega_0 + \phi)}$.

2.12.   Given the autocorrelation function for the random phase sinusoid in the previous Problem 2.11 compute the 2 x 2 autocorrelation matrix.

2.13.   The autocorrelation sequence of a zero mean white noise process is $r_v(k) = \sigma_v^2 \delta(k)$ and the power spectrum is $P_v(e^{j\theta}) = \sigma_v^2$, where σ_v^2 is the variance of the process. For the random phase sinusoid the autocorrelation sequence is,

$$r_x(m) = \frac{1}{2} A^2 \cos(m\omega_0)$$

and the power spectrum is,

$$P_x(e^{j\theta}) = \frac{1}{2} \pi A^2 [u_0(\omega - \omega_0) + u_0(\omega + \omega_0)]$$

where:
   $u_0(\omega - \omega_0)$ represents an impulse at frequency ω_0.

What is the power spectrum of the first-order autoregressive process that has an autocorrelation sequence of,

$$r_x(m) = \alpha^{|m|}$$

where:
   $|\alpha| < 1$

2.14.   Let $x[n]$ be a random process that is generated by filtering white noise $w[n]$ with a first-order LSI filter having a system transfer function of,

$$H(z) = \frac{1}{1 - 0.25z^{-1}}$$

If the variance of the white noise is $\sigma_w^2 = 1$ what is the power spectrum of $x[n]$, $P_x(z)$? Find the autocorrelation of $x[n]$ from $P_x(z)$.

2.15.   If $x[n]$ is a zero mean wide-sense stationary white noise process and $y[n]$ is formed by filtering $x[n]$ with a stable LSI filter $h[n]$ then is it true that,

$$\sigma_y^2 = \sigma_x^2 \sum_{n=-\infty}^{\infty} |h[n]|^2$$

where :

   σ_y^2 and σ_x^2 are the variances of $x[n]$ and $y[n]$ respectively.

PART II. MODELLING

"The study of modelling is inseparable from the practice of modelling" (Gershenfeld 1999). According to Gershenfeld there are no rigorous ways to make choices about mathematical modelling issues, but once they have been made there are rigorous ways to use and apply them. Although it is easy to say that the "best" model is the smallest model (Occam's Razor), unfortunately, there is no analytic way to find such a model or to determine the definitive metric to judge it by. In the end, the "best model" is the one that does the best job according to the designer's satisfaction. This Part II comprising two chapters presents some of these rigorous ways that can be applied to signal and system modelling, keeping in mind that adequacy of the model is the business of the designer and his/her preconceptions. Optimisation is an important tool employed in the modelling process.

Optimisation involves finding the best possible solution to a problem given the constraints of model choice and noise statistics. It is usually associated with the differential equations of a suitable model formulated in such a way as to identify the extrema of the model equations, where an optimal result is expected to be found. Least Squares Estimation (LSE) is a solution method that is fundamental to classical linear and nonlinear signals and systems model optimisation and is actually a general solution method for simultaneous linear equations. It has been used successfully for over two hundred years in a very wide range of modelling applications and is especially useful where the measurement error statistics associate with the observations are Gaussian or at least symmetrical.

Parametric signal and system modelling is generally concerned with the efficient mathematical representation of signals and systems by choosing an appropriate parametric form for the model and then finding the parameters that provide "the best" approximation to the signal or system according to some suitable metric. In this signals and systems context the metric is very often chosen and configured as the Mean Square Error (MSE) between desired and observed quantities as a function of the unknown parameters. The optimisation proceeds by minimising the MSE through differentiation and ultimately by the solution of sets of simultaneous equations using LSE methods.

Parametric signal modelling can be applied to both deterministic and stochastic signals. Key methods of deterministic signal modelling that are covered include the least squares method, the Padé approximation method, Prony's method and the autocorrelation and covariance methods. Stochastic signal modelling can be performed by autoregressive moving average modelling, autoregressive models and moving average models, amongst others. Within these approaches a common special linear equation form is seen repeatedly, which can be solved efficiently using the Levinson-Durbin recursion. This solution method leads to a number of interesting results including lattice filters and efficient Toeplitz matrix inversion.

3. Optimisation and Least Squares Estimation

Optimisation theory is fundamental to signal processing and adaptive systems in so far as it involves finding the best possible solution to a problem given the constraints of system structure and noise statistics. The theory is used in the design of filters given a desired response specification and it is also incorporated in the design of some adaptive filter algorithms to ensure proper convergence to the best solution. It forms the basis of gradient descent approaches that rely on moving consistently down hill of an error function in order to achieve the lowest error and therefore the best solution.

Least Squares Estimation (LSE) is an optimisation method that aims to find the best solution to a set of linear equations where the data error statistics are Gaussian or at least symmetrical. LSE is fundamental to classical linear and nonlinear filter theory, where the model's dependence on its parameters is linear and measurement errors are Gaussian or symmetrical with constant variance. In this Chapter it is shown how LSE may be applied to arbitrary problems by first developing solution equations for simple geometric problems related to navigation position calculations given noisy navigation measurements. From these specific examples it is easy to see how LSE can be adapted and used to also solve problems from other domains.

3.1 Optimisation Theory

Optimisation theory is generally involved with finding minima or maxima of functions of one or more variables. The simplest and most common application of optimisation theory is to find the global minimum of a scalar function $f(x)$ of a single variable x. If the function is differentiable then all the stationary points of the function, including the local minima and global minimum must satisfy the following conditions,

$$\frac{df(x)}{dx} = 0, \quad \text{and,} \quad \frac{d^2 f(x)}{dx^2} > 0$$

If the function $f(x)$ is strictly convex then there is only one solution to the equation $\frac{df(x)}{dx} = 0$, otherwise each stationary point must be checked to see if it is the global minimum. A function is strictly convex over a closed interval $[a, b]$ if, for any two pairs x_1 and x_2 in $[a, b]$, and for any scalar α such that $0 \le \alpha \le 1$, then $f(\alpha x_1 + (1-\alpha) x_2) < \alpha f(x_1) + (1-\alpha) f(x_2)$.

For a scalar function of n real variables, $f(\mathbf{x}) = f(x_1, x_2, ..., x_n)$, the minima are found by first computing the gradient, i.e., a vector of partial derivatives as defined by Equation 3.1.

$$\nabla_x f(x) = \frac{d}{d\mathbf{x}} f(\mathbf{x}) = \begin{bmatrix} \dfrac{\partial}{\partial x_1} f(\mathbf{x}) \\ : \\ \dfrac{\partial}{\partial x_n} f(\mathbf{x}) \end{bmatrix} \qquad (3.1)$$

The gradient vector points in the direction of the greatest rate of the function's change and is equal to zero at the stationary points of the function. A stationary point $\mathbf{x}$ is the minimum if the Hessian matrix $\mathbf{H}_x$ is positive definite, i.e., $\mathbf{H}_x > 0$. The Hessian matrix $\mathbf{H}_x$ is a $n \times n$ matrix of second-order partial derivatives with respect to the (i, j)th element, and is defined by Equation 3.2.

$$\{\mathbf{H}_x\}_{i,j} = \frac{\partial^2}{\partial x_i \partial x_j} f(\mathbf{x}) \qquad (3.2)$$

If $f(\mathbf{x})$ is strictly convex, then the solution to the equation $\nabla_x f(\mathbf{x}) = 0$ is unique and is equal to the global minimum of $f(\mathbf{x})$.

When the function is a real-valued function of complex vectors $\mathbf{z}$ and $\mathbf{z}^*$, finding the minimum of $f(\mathbf{z}, \mathbf{z}^*)$ is complicated by the fact that the function is not differentiable. If $f(\mathbf{z}, \mathbf{z}^*)$ is a real-valued function of the complex vectors $\mathbf{z}$ and $\mathbf{z}^*$ then the vector pointing in the direction of the maximum rate of change of the function is $\nabla_{z^*} f(\mathbf{z}, \mathbf{z}^*)$, which is the derivative of $f(\mathbf{z}, \mathbf{z}^*)$ with respect to $\mathbf{z}^*$. Therefore, the stationary points of the function $f(\mathbf{z}, \mathbf{z}^*)$ are solutions to Equation 3.3.

$$\nabla_{z^*} f(\mathbf{z}, \mathbf{z}^*) = 0 \qquad (3.3)$$

An example of a minimisation problem found in array processing is to minimise the quadratic form, $\mathbf{z}^H \mathbf{R} \mathbf{z}$, with the constraint that, $\mathbf{z}^H \mathbf{a} = 1$, where $\mathbf{z} = [z_1, z_2, ..., z_n]^T$ is the complex solution vector, $\mathbf{R}$ is a positive definite Hermitian matrix, and $\mathbf{a}$ is a given complex vector. One way to solve this problem for $\mathbf{z}$ is to introduce a Lagrange multiplier λ and minimise the unconstrained objective function as defined by Equation 3.4.

$$Q_R(\mathbf{z}, \lambda) = \frac{1}{2} \mathbf{z}^H \mathbf{R} \mathbf{z} + \lambda(1 - \mathbf{z}^H \mathbf{a}) \qquad (3.4)$$

By setting the gradient to zero the result is $\nabla_{z^*} Q_R(z, \lambda) = Rz + \lambda a = 0$, then $z = \lambda R^{-1} a$. To solve for λ it is also necessary to compute the derivative of $Q_R(z, \lambda)$ with respect to λ and set it to zero as follows, $\dfrac{\partial Q_R(z, \lambda)}{\partial \lambda} = 1 - z^H a = 0$.

The Lagrange multiplier λ can then be computed by combining these two results as defined by Equation 3.5.

$$\lambda = \frac{1}{a^H R^{-1} a} \tag{3.5}$$

The solution for z is now defined by Equation 3.6.

$$z = \frac{R^{-1} a}{a^H R^{-1} a} \tag{3.6}$$

The minimum value of $z^H R z$ is achieved by substituting the solution z into the quadratic form $Q_R(z, \lambda) = z^H R z$ to arrive at the solution defined by Equation 3.7

$$\min_{z}\{ z^H R z \} = \frac{z^H a}{a^H R^{-1} a} = \frac{1}{a^H R^{-1} a} \tag{3.7}$$

3.2 Optimisation Methods in Digital Filter Design

The digital filter design problem can be formulated as an approximation problem utilising optimisation methods. The form of the filter's transfer function is first assumed and then an error function is formulated on the basis of some desired amplitude and/or phase response. A norm of the error function is minimised with respect to the transfer function coefficients to achieve some arbitrary amplitude or phase response. The optimisation methods used to achieve this are iterative, requiring extensive computations and they are not guaranteed to converge. The same methods may be used for analogue filter design, as well as for any modelling problem given the form of the transfer function.

Consider the design of a digital filter whose transfer function $H(z)$ is to approximate a prescribed transfer function $\hat{H}(z)$ over some frequency interval at discretely chosen frequency points. Suppose that the type of filter has already been specified in relation to form, i.e., recursive, nonrecursive, stability, sensitivity etc. The design problem is then stated as follows,

Obtain the parameter (coefficients) values of the transfer function so that it is as close as possible to the ideal transfer function at the prescribed frequency points.

To give a more precise definition to the term "as close as possible" an error norm must be chosen. Let $M(\omega)$ be the specified or required filter amplitude response and $M(\mathbf{x}, \omega)$ be the actual filter's amplitude response, where $\mathbf{x}$ is a vector of the filter's coefficients. The approximation error $e(\mathbf{x}, \omega)$ can be defined as the difference between the filter's response and the desired response, i.e., $e(\mathbf{x}, \omega) = M(\mathbf{x}, \omega) - M(\omega)$. By sampling $e(\mathbf{x}, \omega)$ at K discrete frequencies ω_1, $\omega_2, \ldots, \omega_K$, the error vector can then be formed as defined by Equation 3.8.

$$\mathbf{e}(\mathbf{x}) = \begin{bmatrix} e_1(\mathbf{x}) \\ e_2(\mathbf{x}) \\ \vdots \\ e_K(\mathbf{x}) \end{bmatrix} \tag{3.8}$$

where:

$$e_i(\mathbf{x}) = e(\mathbf{x}, \omega_i)$$

If a solution exists then an error or objective function $\Psi(\mathbf{x})$ is needed that is a scalar differentiable function whose minimisation with respect to $\mathbf{x}$ should lead to the minimisation of all the elements of $\mathbf{e}(\mathbf{x})$ in some sense. The L_p norm of $\mathbf{e}(\mathbf{x})$ is commonly used as the objective function. It is defined by Equation 3.9.

$$\Psi(\mathbf{x}) = \left\| \mathbf{e}(\mathbf{x}) \right\|_p = L_p = \left[\sum_{i=1}^{K} \left| e_i(\mathbf{x}) \right|^p \right]^{\frac{1}{p}} \tag{3.9}$$

L_p norms of possible use are the L_1, L_2 (Euclidean) and L_∞ (minimax) norms defined respectively as follows,

$$\left\| \mathbf{e}(\mathbf{x}) \right\|_1 = \left[\sum_{i=1}^{K} \left| e_i(\mathbf{x}) \right| \right]$$

$$\left\| \mathbf{e}(\mathbf{x}) \right\|_2 = \left[\sum_{i=1}^{K} \left| e_i(\mathbf{x}) \right|^2 \right]^{\frac{1}{2}}$$

$$\left\| \mathbf{e}(\mathbf{x}) \right\|_\infty = \lim_{m \to \infty} \left[\sum_{i=1}^{K} \left| e_i(\mathbf{x}) \right|^m \right]^{\frac{1}{m}} = \max_i \left| e_i(\mathbf{x}) \right|$$

The optimisation problem then becomes a matter of minimising one of these norms. However, L_1 and L_∞ are not differentiable and cannot be used for the proposed optimisation. The Euclidean norm is the sum of squares and is therefore often used to compute the mean square error function. Most optimisation algorithms operate in essentially the same iterative manner. Given a small minimum error tolerance of say E, a typical iteration process for optimisation is as follows,

1. Make a reasonable guess on the coefficient values given any known constraints.

2. Use the coefficient values of Step 1 to compute the objective function $\Psi(\mathbf{x})$. If $\Psi(\mathbf{x}) > E$ then go to Step 3 else stop.

3. Introduce small changes to the coefficient values, i.e., $\mathbf{x}$ is replaced with $\mathbf{x} + \delta\mathbf{x}$. Using these adjusted coefficient values compute the corresponding transfer function and objective function values. $\delta\mathbf{x}$ is chosen such that the following two conditions are satisfied,

   i. $\Psi(\mathbf{x} + \delta\mathbf{x}) < \Psi(\mathbf{x})$
   ii. $|\Psi(\mathbf{x} + \delta\mathbf{x}) - \Psi(\mathbf{x})|$ is maximised.

   The first condition i guarantees that the adjustment of the element values is in the right direction in the sense that the new coefficient values are one step closer to the desired one. The second condition ii assures that $\delta\mathbf{x}$ is chosen in an optimal fashion, in the sense that the difference between the original objective function and the adjusted one is as large as possible. Ideally the two conditions should be satisfied simultaneously.

4. If $|\Psi(\mathbf{x} + \delta\mathbf{x})| < E$, then stop the iterations and take $\mathbf{x} + \delta\mathbf{x}$ as the optimal coefficient vector.
   If $|\Psi(\mathbf{x} + \delta\mathbf{x})| > E$, then use $\mathbf{x} + \delta\mathbf{x}$ as the new coefficient vector and go back to Step 3 and repeat the process until one of the following three conditions are satisfied,

   i. $|\Psi(\mathbf{x} + \delta\mathbf{x})| < E$
   ii. $|\Psi_{i+1} - \Psi_i|$ for $i < n$, a preassigned positive integer.
   iii. The number of loops back to Step 1 exceeds some specified maximum number implying that the problem cannot be solved to the required error tolerance.

   The second condition above implies that the objective function has reached a minimum point or a saddle point and further iterations will not reduce it any further. If this happens then try going back to Step 1 and start again.

Although this technique will lead to a minimum it will not necessarily be the global minimum. If the minimum reached is suspected not to be the global minimum the process can be repeated with new initial coefficient guesses until a more satisfactory solution is achieved. The lack of certainty in reaching the global minimum is the main drawback of this type of optimisation process. In most practical problems however, the advantages heavily outweigh the disadvantages.

The heart of this iterative optimisation technique is in the method of computation of the optimal $\delta\mathbf{x}$. This is commonly done using the method of steepest descent. The parameter vector $\mathbf{x}$ must be adjusted to minimise the objective function as stated above. Firstly, the change in the current vector $\mathbf{x}_i$, $\Delta\mathbf{x}_i$ is defined as the difference between the new vector value, $\mathbf{x}_{i+1}$, and the current value $\mathbf{x}_i$ as defined by Equation 3.10.

$$\Delta\mathbf{x}_i = \mathbf{x}_{i+1} - \mathbf{x}_i \tag{3.10}$$

The change vector $\Delta\mathbf{x}_i$ indicates both the direction and the magnitude of the difference between vectors $\mathbf{x}_{i+1}$ and $\mathbf{x}_i$ in the parameter vector space. The vector $\Delta\mathbf{x}_i$ needs to be found such that Equation 3.11 is satisfied.

$$\Psi(\mathbf{x}_{i+1}) = \Psi(\mathbf{x}_i + \Delta\mathbf{x}_i) < \Psi(\mathbf{x}_i) \tag{3.11}$$

To do this, the multi-dimensional Taylor series is taken of $\Psi(\mathbf{x}_i + \Delta\mathbf{x}_i)$ about $\mathbf{x}_i$ and only the linear part is retained, i.e., the first two terms, as defined by Equation 3.12.

$$\Psi(\mathbf{x}_i + \Delta\mathbf{x}_i) \approx \Psi(\mathbf{x}_i) + \left[\nabla\Psi(\mathbf{x}_i)\right]\Delta\mathbf{x}_i < \Psi(\mathbf{x}_i) \tag{3.12}$$

The gradient vector $\nabla\Psi(\mathbf{x}_i)$ in Equation 3.12 is a row vector defined by Equation 3.13.

$$\nabla\Psi(\mathbf{x}_i) = \left[\frac{d\Psi}{dx_1}\bigg|_{\mathbf{x}=\mathbf{x}_i}, \frac{d\Psi}{dx_2}\bigg|_{\mathbf{x}=\mathbf{x}_i}, \ldots, \frac{d\Psi}{dx_n}\bigg|_{\mathbf{x}=\mathbf{x}_i}\right] \tag{3.13}$$

In order to decrease $\Psi(\mathbf{x}_i)$ it is necessary that $\left[\nabla\Psi(\mathbf{x}_i)\right]\Delta\mathbf{x}_i < 0$. If a unit vector $\mathbf{s}_i = \Delta\mathbf{x}_i / |\Delta\mathbf{x}_i|$ is defined to have the same direction as the change vector $\Delta\mathbf{x}_i$, then Equation 3.14 can be formulated to define $\Delta\mathbf{x}_i$ in terms of $\mathbf{s}_i$.

$$\Delta\mathbf{x}_i = \mathbf{x}_{i+1} - \mathbf{x}_i = \alpha_i\mathbf{s}_i \tag{3.14}$$

where:
   α_i is the step size, a real number representing the magnitude of $\Delta\mathbf{x}_i$.

The step size α_i and direction, represented by the unit vector $\mathbf{s}_i$, must be chosen such that $\alpha_i\left[\nabla\Psi(\mathbf{x}_i)\right]\mathbf{s}_i < 0$, for $\alpha_i > 0$. The equation $\left[\nabla\Psi(\mathbf{x}_i)\right]\mathbf{s}_i$ is the dot product between two vectors as defined by Equation 3.15.

$$\left[\nabla\Psi(\mathbf{x}_i)\right]\mathbf{s}_i = |\nabla\Psi(\mathbf{x}_i)||\mathbf{s}_i|\cos\theta$$
$$\tag{3.15}$$

i.e., it is most negative when $\theta = 180°$

To ensure that $\Psi(\mathbf{x}_i + \Delta\mathbf{x}_i)$ decreases $\mathbf{x}_i$ must be changed along a direction that is opposite to the gradient $\nabla\Psi(\mathbf{x}_i)$. Since the gradient represents the direction of the greatest increase in a function, the parameter vector can be adjusted along the direction of the steepest descent, i.e., opposite to the gradient. The next step is to find a suitable value for α_i along the direction of the steepest descent that will minimise the function $\Psi(\mathbf{x}_i + \alpha_i\mathbf{s}_i)$. The search for α_i is a one-dimensional search for which there are a number of methods. The simplest method is to start with a very small estimate for α_i and progressively double it, i.e., $\alpha_{\text{estimate}} = 2^j \alpha_{\text{small}}$, $j = 0,1,....$, until the value of $\Psi(\mathbf{x}_i + 2^j \alpha_{\text{small}}\mathbf{s}_i)$ achieves its lowest value at $j = k$. Since the error function is unimodal the minimum of $\Psi(\mathbf{x}_i + \alpha_i\mathbf{s}_i)$ can be assumed to be between $2^{k-1}\alpha_{\text{small}} < \alpha_{\text{min}} < 2^{k+1}\alpha_{\text{small}}$ and then the bisection method can be used to find $\alpha_i = \alpha_{\text{min}}$. This approach has the disadvantage that a small step size is needed, therefore it may only find a local minimum and it also has slow convergence. The other simple gradient descent methods also suffer limitations, nevertheless they still find use in many areas.

In addition to the steepest descent methods a number of other optimisation methods exist that do help to overcome some of the limitations of steepest descent. These include,

1.   The conjugate gradient method.

2.   Simulated annealing.

3.   The simplex method.

4.   The minimax method.

3.3 Least Squares Estimation

Least Squares Estimation (LSE) is an extremely important and widely used statistical technique for solving both linear and nonlinear equation sets. In one way or another LSE is fundamental to general signal processing, pattern recognition and neural network theory through its links to probability theory (maximum likelihood estimation). It is fundamental to stochastic approximation, linear and nonlinear regression, Backpropagation-of-error and other artificial neural network learning and in optimum and nonlinear filter design.

Given a set of determined or over determined linear or nonlinear equations, that model a system or a process, it is possible to solve for the dependent variables (model parameters) given values for the independent variables (measured values) using LSE. Provided that the model's dependence on its parameters is linear and measurement errors are Gaussian with constant variance, the least squares solution is a maximum likelihood estimation (Press *et al* 1986). Serious problems can arise with LSE if the measurement errors are not close to Gaussian, especially if there are large outliers in the measurements. Therefore measures must be taken to correct these problems before LSE can be used with good effect. The basic LSE problem can be stated as follows,

Given a real $m \times n$ matrix $\mathbf{A}$ of rank $k \leq \min(m, n)$, and given a real m-dimensional vector $\mathbf{y}$, find a real n-dimensional vector $\mathbf{x}^1$ that minimises the Euclidean length of vector $\mathbf{Ax}\text{-}\mathbf{y}$, (Lawson and Hanson 1974).

The matrix $\mathbf{A}$ is the design matrix that relates the unknowns $\mathbf{x}$ to the measured quantities $\mathbf{y}$ by linear equations. The equation $\mathbf{Ax} = \mathbf{y}$ can be solved for the unknown $\mathbf{x}$ most economically by Gaussian elimination, but Gaussian elimination can suffer from rounding error when solving large matrices. In practical situations, orthogonalization methods such as Householder Orthogonalization, Modified Gram-Schmidt, Bidiagonalization or Singular Valued Decomposition (SVD) are favoured. Orthogonalization methods have guaranteed stability and they do not suffer badly from rounding errors. In cases of ill-conditioning or over determination, the SVD method is unsurpassed (Golub and Van Loan 1983). It is also possible to use relaxation techniques like the Gauss-Seidel method (Miller 1981). These start with an initial guess to the solution and improve it iteratively until it relaxes to the true solution. Their advantages are that they do not suffer from rounding problems, they can solve very large matrices and the equations do not need to be linear. However the disadvantage is that they may not always converge.

If equation $\mathbf{Ax} = \mathbf{y}$ is multiplied by $\mathbf{A}$ transpose $(\mathbf{A}^T)$ the so called normal equations of the LSE problem result, as defined by Equation 3.16.

$$\mathbf{A}^T \mathbf{A}\, \mathbf{x} = \mathbf{A}^T \mathbf{y} \qquad\qquad (3.16)$$

These normal equations can be solved using similar techniques to those described above. The major disadvantage of the normal equations is that in many practical cases they can be close to singular if two equations or two different combinations of the same equation in the $\mathbf{A}$ matrix happen to fit the measured data either equally well or equally badly, i.e., ill-conditioned. However, this can be avoided by good design in the first place. The matrix $\mathbf{A}^T\mathbf{A}$ is now a square $n \times n$ matrix for which it is at least possible to compute an inverse.

In cases where the error variances of the measured quantities are not equal, it is desirable to add a weight matrix $\mathbf{W}$, which is the inverse of the variance-covariance matrix of the measured quantities. The normal equations then become as defined by Equation 3.17 (Cross 1981).

$$\mathbf{A}^T \mathbf{W} \mathbf{A}\, \mathbf{x} = \mathbf{A}^T \mathbf{W} \mathbf{y} \qquad\qquad (3.17)$$

The weight matrix ensures that more significance is given to more accurately measured quantities in the final solution. If the measurements are uncorrelated then the $\mathbf{W}$ matrix simply becomes a diagonal matrix with each diagonal element being the inverse of the respective measurement variance. If the measurements are correlated then it contains values in the off-diagonal and it can be more difficult to compute or estimate.

For nonlinear problems the design equations are partially differentiated so that now $\mathbf{B} = \partial \mathbf{y}/\partial \mathbf{x}$ becomes the matrix that relates the unknown differentials $\partial \mathbf{x}$ to the differences between the measured quantities $\mathbf{y}$ and the calculated ones using the provisional solution $\mathbf{x}_0$. Given $\mathbf{B}\partial \mathbf{x} = \partial \mathbf{y} = \Delta \mathbf{y}$ (difference between observed and calculated values) + $\mathbf{r}$ (residuals), the estimate of the solution becomes $\hat{\mathbf{x}} = \mathbf{x}_0 + \partial \hat{\mathbf{x}}$ and the process is iterated until a satisfactory solution is reached when $\partial \hat{\mathbf{x}}$ converges to a very small value. The design equation $\mathbf{B}\,\partial \hat{\mathbf{x}} = \Delta \mathbf{y}$ can be solved, for each iteration, in a similar manner to equation $\mathbf{A}\mathbf{x} = \mathbf{y}$ as described above.

The LSE method finds the best solution for a set of observations or measurements that include noise. To achieve the best benefit from the LSE a design should include more measurement equations than solution variables. The more measurement equations the more precision is achieved through error variance reduction. This can be achieved in two ways, or by a combination of both ways. A range of different equations related to the solution variables can be developed. Otherwise, extra measurements for the same equations can be taken and integrated into the LSE design.

3.4 Least Squares Maximum Likelihood Estimator

Least squares is a maximum likelihood estimator. Suppose that it is desired to fit N data points (x_i, y_i), $i = 1,\ldots, N$, to a model that has M adjustable parameters a_j, $j = 1,\ldots, M$. The model predicts a functional relationship between the measured independent variables and the desired dependent variables,

$$y(x) = y(x ; a_1,\ldots, a_M)$$

The least squares fit minimises the error over the variables $a_1,\ldots, a_M$, as defined by Equation 3.18

$$\text{Error} = \sum_{i=1}^{N} \left[y_i - y(x_i : a_1,\ldots,a_M) \right]^2 \tag{3.18}$$

Suppose that each data point y_i has a measurement error that is independently random and a Gaussian distribution around the "true" model $y(x)$. Also assume, as is often the case, that the error standard deviations of the measured data are the same for all points and represented by σ. Then the probability P that a given set of fitted parameters $a_j,\ldots, a_M$ is correct is the product of the probabilities of each point as defined by Equation 3.19.

$$P = \prod_{i=1}^{N}\{\exp[-\frac{1}{2}(\frac{y_i - y(x_i)}{\sigma})^2]\Delta y\} \tag{3.19}$$

Maximising P (or its log) is like minimising the negative of its log, namely,

$$[\sum_{i-1}^{N} \frac{(y_i - y(x_i))^2}{2\sigma^2}] - N \log \Delta y$$

Since N, σ and Δy are constants, minimising this equation is equivalent to minimising,

$$\sum_{i=1}^{N} [y_i - y(x_i)]^2$$

On the other hand if each data point (x_i, y_i) has its own measurement error σ_i then the Chi-square metric χ^2, as defined by Equation 3.20, can be used.

$$\chi^2 = \sum_{i=1}^{N} \left(\frac{y_i - y(x_i; a_1 ..... a_M)}{\sigma_i^2} \right)^2 \tag{3.20}$$

Here the derivative of χ^2 with respect to each parameter a_k is set equal to zero as defined by Equation 3.21.

$$\sum_{i=1}^{N} (\frac{y_i - y(x_i)}{\sigma_i^2})(\frac{\partial y(x_i; ..... a_k ....)}{\partial a_k}) = 0, \quad k = 1,....M \tag{3.21}$$

Equation 3.21 is a set of M nonlinear simultaneous equations that need to be solved for the M unknowns a_k to find the least squares fit.

3.5 Linear Regression - Fitting Data to a Line

The LSE linear regression solution, or the fitting of a straight line to a two-dimensional data set, is defined by Equation 3.22. The model parameters of a line in two dimensions are the y intercept a and the line's slope b.

$$y(x) = y(x; a, b) = a + bx \tag{3.22}$$

To solve for these parameters a and b Equation 3.20 is applied as defined by Equation 3.23.

$$\chi^2(a,b) = \sum_{i=1}^{N} \left(\frac{y_i - a - bx_i}{\sigma_i} \right)^2 \tag{3.23}$$

The solution to Equation 3.23 is in accordance with Equation 3.21 and results in the set of solution Equations 3.24 to 3.27.

$$\Delta = S S_{xx} - S_x^2 \tag{3.24}$$

$$a = \frac{S_{xx}S_y - S_x S_{xy}}{\Delta}$$

$$b = \frac{SS_{xy} - S_x S_y}{\Delta}$$

(3.25)

where:

$$S \equiv \sum_{i=1}^{N} \frac{1}{\sigma_i^2}, \quad S_x \equiv \sum_{i=1}^{N} \frac{x_i}{\sigma_i^2}, \quad S_y \equiv \sum_{i=1}^{N} \frac{y_i}{\sigma_i^2}$$

(3.26)

$$S_{xx} \equiv \sum_{i=1}^{N} \frac{x_i^2}{\sigma_i^2}, \quad S_{xy} \equiv \sum_{i=1}^{N} \frac{x_i y_i}{\sigma_i^2}$$

(3.27)

The variances of the a and b estimates are defined by Equations 3.28.

$$\sigma_a^2 = \frac{S_{xx}}{\Delta}$$

$$\sigma_b^2 = \frac{S}{\Delta}$$

(3.28)

3.6 General Linear Least Squares

The general linear least squares problem is involved with fitting a set of data points (x_i, y_i) to a linear combination of any M specified functions of x. Equation 3.29 defines a specific polynomial example using basis functions 1, x, x^2,...., x^{M-1}, whereas Equation 3.30 defines the more general case.

$$y(x) = a_1 + a_2 x + a_3 x^2 + ...+a_M x^{M-1}$$

(3.29)

$$y(x) = \sum_{k=1}^{M} a_k X_k(x)$$

(3.30)

where:

$X_1(x),...., X_M(x)$ are arbitrary functions of x, i.e., the basis functions.

Linear, in the context of least squares refers to the model's dependence on its parameters a_k not the function itself. Consequently, LSE can be used for either linear or nonlinear function estimation so long as the problem can be formulated in terms of an equation like Equation 3.30.

3.7 A Ship Positioning Example of LSE

The workings of LSE is best demonstrated with a practical application problem such as ship position at sea. Such a problem requires that a set of nonlinear geometric equations be solved. Therefore, a general introduction to the LSE solution for nonlinear problems is given first before showing its application to ship positioning using range information gathered from fixed navigation radio beacons.

Take the case of a nonlinear differentiable vector function model $\mathbf{y} = F(\mathbf{x})$. To compute a LSE solution for the model parameters $\mathbf{x}$ it is first necessary to take the partial differentials of the measured (observed) variables $\mathbf{y}$ with respect to the model parameters $\mathbf{x}$, i.e., $\mathbf{B} = \partial\mathbf{y}/\partial\mathbf{x}$. This can be reformulated in terms of Equation 3.31, which shows that the function differential can be estimated by an observed minus a calculated term ($\Delta\mathbf{y} = [\mathbf{O} - \mathbf{C}]$) plus a residual vector $\mathbf{r}$ quantity.

$$\partial\mathbf{y} = \mathbf{B}\partial\mathbf{x} = \Delta\mathbf{y} + \mathbf{r} = [\mathbf{O} - \mathbf{C}] + \mathbf{r} \tag{3.31}$$

The estimates of the parameter differentials $\partial\hat{\mathbf{x}}$ and residuals $\mathbf{r}$ are defined by Equations 3.32 and 3.33 respectively.

$$\partial\hat{\mathbf{x}} = [\mathbf{B}^T \mathbf{W} \mathbf{B}]^{-1} \mathbf{B}^T \mathbf{W} \Delta\mathbf{y} \tag{3.32}$$

$$\hat{\mathbf{r}} = \mathbf{B}\partial\hat{\mathbf{x}} - \Delta\mathbf{y} \tag{3.33}$$

The normal LSE equations for this problem configuration are defined by Equation 3.34, which also include a weight matrix $\mathbf{W}$.

$$\mathbf{B}^T \mathbf{W} \mathbf{B} \partial\hat{\mathbf{x}} = \mathbf{B}^T \mathbf{W} \Delta\mathbf{y} \tag{3.34}$$

The required normal equation solution is then defined by Equation 3.35.

$$\hat{\mathbf{x}} = \mathbf{x}_0 + \partial\hat{\mathbf{y}} = \mathbf{x}_0 + [\mathbf{B}^T \mathbf{W} \mathbf{B}]^{-1} \mathbf{B}^T \mathbf{W} \Delta\mathbf{y} \tag{3.35}$$

Starting with an initial provisional parameter estimate $\mathbf{x}_0$ the solution Equation 3.35 is iterated whereby each estimate $\hat{\mathbf{x}}$ becomes the next provisional estimate $\mathbf{x}_0$ until $\partial\hat{\mathbf{x}}$ approaches zero or is less than some very small value. This is a gradient descent adaptation algorithm, where the variables are defined as follows,

$\partial\mathbf{y}$ = differential observed or measures patterns. (Vector size m).

$\mathbf{B}$ = matrix of partial differentials - function of $\mathbf{y}$. (Matrix size $m \times n$).

$\partial\mathbf{x}$ = differentials of unknowns. (Vector size n).

$\mathbf{W}$ = weight matrix. Inverse of covariance matrix of observations or measurements. If observations are independent then $W_{ij} = \dfrac{1}{\sigma_i^2}$.

(Matrix size $m \times m$).

$\Delta\mathbf{y}$ = difference between observed and calculated values ($\mathbf{O} - \mathbf{C}$).

(Vector size m).

r   = residuals. (Vector size m).

$\mathbf{x}_0$ = solution estimate, starts with a first guess then becomes the solution estimate of the previous iteration and also the provisional estimate. (Vector size n).

x   = solution vector. (Vector size n).

The problem Equation 3.31, $\mathbf{B}\partial\,\mathbf{x} = \Delta\mathbf{y}+\mathbf{r} = (\mathbf{O}\text{-}\mathbf{C})+\mathbf{r} = \partial\,\mathbf{y}$, is approximately linear near the solution and can therefore be expressed as set of linear equations defined by Equation 3.36. These set of equations are repeatedly solved for each iteration of the process.

$$
\begin{bmatrix}
\partial y_1/\partial x_1 & \partial y_1/\partial x_2 & \cdots & \cdots & \partial y_1/\partial x_n \\
\partial y_2/\partial x_1 & \partial y_2/\partial x_2 & \cdots & \cdots & \partial y_1/\partial x_n \\
\vdots & \vdots & \vdots & \vdots & \vdots \\
\vdots & \vdots & \vdots & \vdots & \vdots \\
\partial y_m/\partial x_1 & \partial y_m/\partial x_2 & \cdots & \cdots & \partial y_m/\partial x_n
\end{bmatrix}
\begin{bmatrix}
\partial x_1 \\ \partial x_2 \\ ..... \\ ..... \\ \partial x_n
\end{bmatrix} =
$$

$$
\begin{bmatrix}
(O-C)_1 \\ (O-C)_2 \\ .... \\ .... \\ (O-C)_m
\end{bmatrix}
+
\begin{bmatrix}
r_1 \\ r_2 \\ ... \\ ... \\ r_m
\end{bmatrix}
=
\begin{bmatrix}
\partial y_1 \\ \partial y_2 \\ ...... \\ ...... \\ \partial y_m
\end{bmatrix}
$$

(3.36)

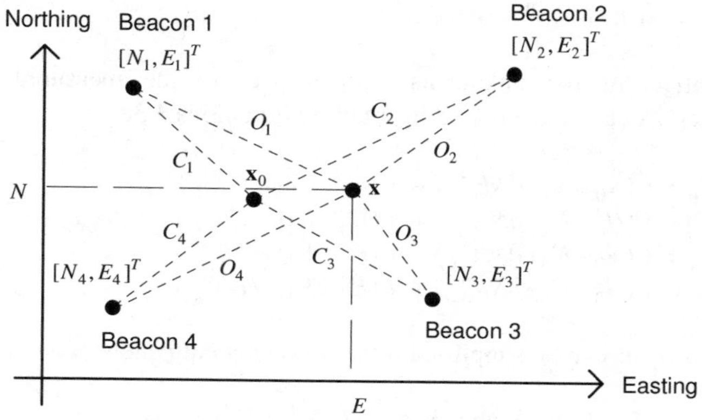

Figure 3.1. A Ship Positioning Example

The best way to illustrate LSE is with a detailed worked example. A typical problem might be to estimate the position of a ship by taking distance measurements to a number of fixed and known navigation beacons. Refer to Figure 3.1. The measurements to the beacons have different measurement accuracies but all errors may be assumed to be zero mean Gaussian. This examples is taken from a paper from the Hydrographic Journal (Cross 1981), which gives a complete working and analysis of the problem with numeric examples. The variables relevant to this example are defined as follows,

$\mathbf{x} = [N, E]^T$ the actual ship's northing and easting position.
$\mathbf{x}_0 = [N_0, E_0]^T$ the ship's provisional position.
$O_1 =$ observed, measured distance to beacon number 1 to the ship.
$O_2 =$ observed, measured distance to beacon number 2 to the ship.
$O_3 =$ observed, measured distance to beacon number 3 to the ship.
$O_4 =$ observed, measured distance to beacon number 4 to the ship.
$C_1 =$ computed distance from beacon number 1 to the ship's provisional position.
$C_2 =$ computed distance from beacon number 2 to the ship's provisional position.
$C_3 =$ computed distance from beacon number 3 to the ship's provisional position.
$C_4 =$ computed distance from beacon number 4 to the ship's provisional position.
$\partial \mathbf{x} = [\partial N, \partial E]^T$ differentials of the unknowns $\mathbf{x} = [N, E]^T$.

This problem's model equations $\mathbf{y} = F(\mathbf{x})$ are defined by the set of Equations 3.37.

$$C_1 = \sqrt{((N_1-N_0)^2 + (E_1-E_0)^2)}$$
$$C_2 = \sqrt{((N_2-N_0)^2 + (E_2-E_0)^2)}$$
$$C_3 = \sqrt{((N_3-N_0)^2 + (E_3-E_0)^2)} \tag{3.37}$$
$$C_4 = \sqrt{((N_4-N_0)^2 + (E_4-E_0)^2)}$$

The differentials of the unknowns with respect to the measured quantities $\partial \mathbf{y}/\partial \mathbf{x} = \partial F(\mathbf{x})/\partial \mathbf{x}$ can be expressed in detail by Equations 3.38.

$$\partial C_1 = ((N_0 - N_1)\, \partial N/C_1) + ((E_0 - E_1)\, \partial E/C_1)$$
$$\partial C_2 = ((N_0 - N_2)\, \partial N/C_2) + ((E_0 - E_2)\, \partial E/C_2)$$
$$\partial C_3 = ((N_0 - N_3)\, \partial N/C_3) + ((E_0 - E_3)\, \partial E/C_3) \tag{3.38}$$
$$\partial C_4 = ((N_0 - N_4)\, \partial N/C_4) + ((E_0 - E_4)\, \partial E/C_4)$$

Subsequent analysis can be simplified if the following equivalences are adopted,

$$L_i = ((N_0 - N_i)/C_i) \quad \text{and} \quad K_i = ((E_0 - E_i)/C_i)$$

Given these new symbols the general problem Equation 3.39 can then be more economically expressed as Equation 3.40.

$$\partial \mathbf{y} = \mathbf{B}\, \partial \mathbf{x} = \Delta \mathbf{y} + \mathbf{r} = (\mathbf{O} - \mathbf{C}) + \mathbf{r} \tag{3.39}$$

$$\partial C_1 = L_1\, \partial N + K_1\, \partial E = (O_1 - C_1) + r_1 \tag{3.40}$$

From Equation 3.40 the matrix **B** is,

$$\mathbf{B} = \begin{bmatrix} L_1 K_1 \\ L_2 K_2 \\ L_3 K_3 \\ L_4 K_4 \end{bmatrix}$$

In principle it is possible to add more equations to improve the estimate if there were more beacons. The more redundant equations there are the better the LSE solution. Although two beacons are theoretically enough to compute a two-dimensional position coordinate there would be two possible solutions because the two equations cannot distinguish solutions on either side of the baseline between the two beacons. Therefore in this problem it is necessary to have a minimum of three beacons to compute a unique solution.

The problem requires a solution for the position variables ∂E and ∂N, i.e., $\partial \mathbf{x}$ as defined by Equation 3.41, which are a set of simultaneous and hopefully linearly independent equations.

$$\partial \hat{\mathbf{x}} = [\mathbf{B}^T \mathbf{W} \mathbf{B}]^{-1} \mathbf{B}^T \mathbf{W} \Delta \mathbf{y} \tag{3.41}$$

W is the inverse covariance matrix of observations. In this case it can be assumed that the measurements are independent of each other. **W** can often be computed by theory or else estimated by measurement. The weight matrix ensures that the factors with the lowest variances carry more weight toward the solution. The weight matrix **W** is,

$$\mathbf{W} = \begin{bmatrix} \sigma_1^{-2} & & & \\ & \sigma_2^{-2} & & \\ & & \sigma_3^{-2} & \\ & & & \sigma_4^{-2} \end{bmatrix}$$

Equation 3.41 can be expanded, for clarity, as follows,

$$\begin{bmatrix} \partial \hat{N} \\ \partial \hat{E} \end{bmatrix} = \begin{bmatrix} L_1 L_2 L_3 L_4 \\ K_1 K_2 K_3 K_4 \end{bmatrix} \begin{bmatrix} \sigma_1^{-2} & & & \\ & \sigma_2^{-2} & & \\ & & \sigma_3^{-2} & \\ & & & \sigma_4^{-2} \end{bmatrix} \begin{bmatrix} L_1 K_1 \\ L_2 K_2 \\ L_3 K_3 \\ L_4 K_4 \end{bmatrix}^{-1} \begin{bmatrix} L_1 L_2 L_3 L_4 \\ K_1 K_2 K_3 K_4 \end{bmatrix} \begin{bmatrix} \sigma_1^{-2} & & & \\ & \sigma_2^{-2} & & \\ & & \sigma_3^{-2} & \\ & & & \sigma_4^{-2} \end{bmatrix} \begin{bmatrix} O_1 - C_1 \\ O_2 - C_2 \\ O_3 - C_3 \\ O_4 - C_4 \end{bmatrix}$$

| (2x1) | (2x4) | (4x4) | (4x2) | (2x4) | (4x4) | (4x1) |

To find a solution to Equation 3.35 it is necessary to first start with a reasonable guess for $\mathbf{x}_0$, possibly from a graphic solution. Equation 3.35 can be re-expressed as defined by Equation 3.42.

$$\hat{\mathbf{x}} = \begin{bmatrix} N \\ E \end{bmatrix} = \mathbf{x}_0 + \partial\hat{\mathbf{x}} = \begin{bmatrix} N_0 \\ E_0 \end{bmatrix} + \begin{bmatrix} \partial\hat{N} \\ \partial\hat{E} \end{bmatrix} \tag{3.42}$$

The solution is finally found by iterating Equation 3.42 until either $\left|E - E_0\right| < \varepsilon$ and $\left|N - N_0\right| < \varepsilon$ or $\left|\partial E\right| < \varepsilon$ and $\left|\partial N\right| < \varepsilon$, where ε is a small acceptable error margin. During each iteration the residuals are estimated by $\hat{\mathbf{r}} = \mathbf{B}\partial\hat{\mathbf{y}} - \Delta\mathbf{y}$. If the residuals are zero or very close to zero at convergence it means that the model chosen and Gaussian assumption for noise variance are probably correct. If not it indicates there is a bias, most probably due to an inadequate model.

If observation measurements were made at regular or known time intervals it is possible to introduce ship velocity into the equations without additional velocity sensors and thus make better ship position estimates for a moving ship. Refer to (Cross 1981) for details.

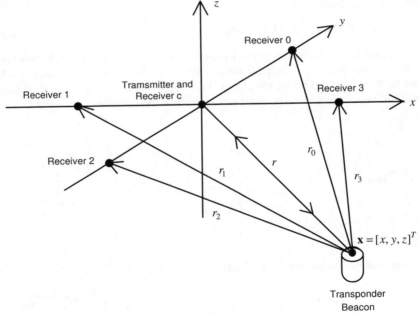

Figure 3.2. A Beacon Position Example

3.8 Acoustic Positioning System Example

Another instructional example of LSE involves a three-dimensional acoustic positioning system. The system can be designed a number of different ways but the following method has been chosen because it is an easy way of doing it. The problem is to compute the position of an underwater transponder beacon by using array of short base-line acoustic transducers as shown in Figure 3.2. In the transducer array there are four receiver transducers and one transmitter/receiver transducer c placed at the centre of the array. The centre transmitter transmits a signal to the beacon, which after detection the beacon immediately sends back another signal that is subsequently detected by all five transducers in the array. Allowing for the beacon detection delay it is then possible to measure the turn around times from transducer c to the beacon and back to each of the five array transducers, receiver c and receiver 0 to receiver 3. The array is in a flat plane and each transducer is placed on a rectangular axis location.

To compute the position, $\mathbf{x} = [x, y, z]^T$, (in relation to the array's orthogonal axes) of the underwater transponder beacon a geometric model of the flight times of a signal transmitted to the beacon and back again to each receiver on the array is required. The appropriate model is defined by the set of Equations 3.43 and 3.44.

$$r = \sqrt{x^2 + y^2 + z^2}$$

$$r_1 = \sqrt{(x + \frac{d_{13}}{2})^2 + y^2 + z^2}$$

$$r_3 = \sqrt{(x - \frac{d_{13}}{2})^2 + y^2 + z^2} \qquad (3.43)$$

$$r_2 = \sqrt{x^2 + (y + \frac{d_{20}}{2})^2 + z^2}$$

$$r_0 = \sqrt{x^2 + (y - \frac{d_{20}}{2})^2 + z^2}$$

where:

$d_{13} = 0.20$ m, fixed distance between receiver 1 and 3, receiver c in the middle.
$d_{20} = 0.20$ m, fixed distance between receiver 2 and 0, receiver c in the middle.

$$T_{cc} = \frac{2r}{V}$$

$$T_{c1} = (r + \sqrt{(x + \frac{d_{13}}{2})^2 + y^2 + z^2}) / V$$

$$T_{c3} = (r + \sqrt{(x - \frac{d_{13}}{2})^2 + y^2 + z^2}) / V \qquad (3.44)$$

$$T_{c2} = (r + \sqrt{x^2 + (y + \frac{d_{20}}{2})^2 + z^2}) / V$$

$$T_{c0} = (r + \sqrt{x^2 + (y - \frac{d_{20}}{2})^2 + z^2}) / V$$

where:
  r = range from c the centre of the array to the beacon.
  r_i = ranges from receiver i to the beacon.
  T_{cc} = flight time of a turn around signal from c to beacon back to c.
  T_{ci} = flight time of a turn around signal from c to beacon to i.
  V = 1507.9 m/s velocity of sound in sea water.

From these basic model equations it is necessary to construct some linearly independent observation equations. It is possible to use all the five T_{cc} and T_{ci} Equations 3.44 directly. However, this would result in a very poor LSE solution because the differences between flight times are small, especially over longer ranges, resulting in a set of ill-conditioned equations. To achieve a more robust solution it is much better to make appropriate combinations of them such that the average difference between measurements is as large as possible. It is possible to make three suitable equations as defined by Equations 3.45 to 3.47. Three equations are the minimum number required for a solution of three unknowns. Since the transducer array is in a flat plane the system cannot distinguish between solutions above or below the plane, but in this case it doesn't matter since the beacon can only be located below the plane. Otherwise, this ambiguity can easily be resolved by adding another receiver transducer on the z-axis (preferably on the negative side) and adding the appropriate model equation for that.

$$G_0 = V(T_{c1} - T_{c3}), \quad \text{i.e., } r_1 - r_3 \tag{3.45}$$

$$H_0 = V(T_{c2} - T_{c0}), \tag{3.46}$$

$$r = V(T_{cc/2}) \tag{3.47}$$

These are a set of equations that give measured observations in terms of array geometry and the measured return flight times of the acoustic signal. Taking the partial differentials of Equations 3.45 to 3.47 with respect to x, y, and z provides the **B** matrix as defined by Equations 3.48 to 3.50.

$$\mathbf{B} = \frac{\partial \mathbf{y}}{\partial \mathbf{x}} \tag{3.48}$$

$$\mathbf{B} = \begin{bmatrix} \dfrac{\partial G_0}{\partial x} & \dfrac{\partial G_0}{\partial y} & \dfrac{\partial G_0}{\partial z} \\[2mm] \dfrac{\partial H_0}{\partial x} & \dfrac{\partial H_0}{\partial y} & \dfrac{\partial H_0}{\partial z} \\[2mm] \dfrac{\partial r}{\partial x} & \dfrac{\partial r}{\partial y} & \dfrac{\partial r}{\partial z} \end{bmatrix} = \begin{bmatrix} b_{11} & b_{12} & b_{13} \\ b_{21} & b_{22} & b_{23} \\ b_{31} & b_{32} & b_{33} \end{bmatrix} \tag{3.49}$$

$$b_{11} = [(x + \frac{d_{13}}{2})/r_1] - [(x - \frac{d_{13}}{2})/r_3]$$

$$b_{12} = [y/r_1] - [y/r_3]$$

$$b_{13} = [z/r_1] - [z/r_3]$$

$$b_{21} = [x/r_2] - [x/r_0]$$

$$b_{22} = [(y + \frac{d_{20}}{2})/r_2] - [(y - \frac{d_{20}}{2})/r_0] \tag{3.50}$$

$$b_{23} = [z/r_2] - [z/r_0]$$

$$b_{31} = x/r$$

$$b_{32} = y/r$$

$$a_{33} = z/r$$

The weight matrix is defined by Equation 3.51 if it is assumed that all measurements are accurate to say 0.1%.

$$\mathbf{W} = \begin{bmatrix} \dfrac{1 \times 10^6}{d_{13}^2} & 0 & 0 \\ 0 & \dfrac{1 \times 10^6}{d_{20}^2} & 0 \\ 0 & 0 & \dfrac{1 \times 10^6}{r^2} \end{bmatrix} \tag{3.51}$$

$$W_{ij} = \frac{1}{\sigma_i^2} \text{ for } i = j, \text{ and } W_{ij} = 0 \text{ otherwise.}$$

It is fairly important to make the first guess of the unknowns as close as possible to the true values. There are regions of first guesses that will either not lead to convergence or converge to the wrong solution. In this, case since all the receivers are in the x-y plane there are two possible solutions (x, y, z) and $(x, y, -z)$. If the z guess is positive it will converge to the positive solution and vice versa.

The iteration process toward the solution is started by computing the estimate of $\partial \mathbf{x}$ and guessing the first estimate $\mathbf{x}_0$, i.e., $\hat{\mathbf{x}} = \mathbf{x}_0 + \partial \mathbf{x}$. To do this the following values must be computed,

B

$$\mathbf{c} = \begin{bmatrix} C_1 \\ C_2 \\ C_3 \end{bmatrix} = \begin{bmatrix} r_1 - r_3 \\ r_2 - r_0 \\ r \end{bmatrix}, \text{ given } \mathbf{x}_0 = \begin{bmatrix} x^1 \\ y^1 \\ z^1 \end{bmatrix}, \text{ the first guess.}$$

$$\Delta \mathbf{y} = \begin{bmatrix} O_1 - C_1 \\ O_2 - C_2 \\ O_3 - C_3 \end{bmatrix}, \text{ where: } \begin{matrix} O_1 = G_0 + \text{error} \\ O_2 = H_0 + \text{error} \\ O_3 = r + \text{error} \end{matrix}$$

$$\partial \mathbf{B}^T$$
$$\mathbf{B}^T \mathbf{W}$$
$$\mathbf{B}^T \mathbf{W} \Delta \mathbf{y}$$
$$\mathbf{B}^T \mathbf{W} \mathbf{B}$$

With these computations it is then possible to solve for $\partial \hat{\mathbf{x}}$ in the normal equations $[\mathbf{B}^T \mathbf{W} \mathbf{B}][\partial \mathbf{x}] = [\mathbf{B}^T \mathbf{W} \Delta \mathbf{y}]$. This can be done by using a number of possible techniques such as,

1. Gaussian elimination.

2. Householder orthogonalization.

3. Modified Gram-Schmidt.

4. Bidiagonalisation.

5. Singular Valued Decomposition (SVD) - (probably is the best one).

6. Gauss-Seidel.

Then the new provisional solution $\hat{\mathbf{x}}^1$ is defined by Equation 3.52.

$$\hat{\mathbf{x}}^1 = \mathbf{x}_0 + \partial \hat{\mathbf{x}} \tag{3.52}$$

Next, let the new provisional solution $\hat{\mathbf{x}}^1$ be the next estimate $\mathbf{x}_0$, i.e., $\mathbf{x}_0 = \hat{\mathbf{x}}^1$ and continue the iteration and stop when $\sqrt{\partial \hat{x}^2 + \partial \hat{y}^2 + \partial \hat{z}^2} < 1 \times 10^{-6}$.

3.9 Measure of LSE Precision

The LSE precision $\sigma(\partial \hat{\mathbf{x}})$ is determined from the variance - covariance matrix of the least squares solution and is defined by Equation 3.53.

$$\sigma(\partial \hat{\mathbf{x}}) = \sigma_0^2 [\mathbf{B}^T \mathbf{W} \mathbf{B}]^{-1} \tag{3.53}$$

Here, σ_0^2 is a unitless scaler quantity known as the unit variance (or often called the standard error of an observation of unit weight) and is defined by Equation 3.54.

$$\sigma_0^2 = \frac{(\hat{\mathbf{r}}^T \mathbf{W} \hat{\mathbf{r}})}{(m-j)} \tag{3.54}$$

where:

j = number of parameters in vector $\partial \mathbf{x}$ plus 2.

m = number of observations.

$m - j$ = number of degrees of freedom.

If the degrees of freedom $(m - j)$ is small σ_0^2 has a large uncertainty so it is best to compute it by taking an average of σ_0^2 over a large number of fixes. If $\mathbf{W}$ has been correctly estimated then σ_0^2 will be unity, therefore $\sigma(\partial \hat{\mathbf{y}}) = [\mathbf{B}^T \mathbf{W} \mathbf{B}]^{-1}$. If the average σ_0^2 turns out not to be unity it is best to make a new estimate of $\mathbf{W}$ from $\mathbf{W}_{\text{new}} = \dfrac{\mathbf{W}_{\text{old}}}{\sigma_0^2}$.

3.10 Measure of LSE Reliability

A reliable value is one that is known to not contain a gross error. Gross error detection in the ith observation can done by application of reliability Equation 3.55.

$$\text{rel}_i = \frac{\sigma_i^1}{\sigma_i} \tag{3.55}$$

where:

σ_i = standard error of ith observation.

σ_i^1 = a *posteriori* standard error of ith observation, determined by the least squares estimate $\partial \hat{\mathbf{y}}$.

A $\mathbf{R}$ matrix is defined by Equation 3.56, which is used to compute the least square reliability as defined by Equation 3.57.

$$\mathbf{R} = \mathbf{B}[\mathbf{B}^T \mathbf{W} \mathbf{B}]^{-1} \mathbf{B}^T \tag{3.56}$$

$$\text{rel}_i = \frac{\sqrt{R_i}}{\sigma_i} \tag{3.57}$$

where:

R_i = the diagonal elements of the reliability matrix $\mathbf{R}$.

In practice, if any of the reliabilities rel_i exceed 0.90 the fix may be unreliable and it should be investigated.

3.11 Limitations of LSE

LSE is used successfully in many applications but it does have some limitations that need to be considered. These are,

1. The accuracy of the solution along with its measure of precision and reliability is limited by the mathematical model used. If the model does not truly reflect the physical situation the results will be of limited value. However, it is not always easy to detect errors in the model.

2. The method depends on a proper choice of weight matrix **W**. Although errors in **W** are unlikely to seriously affect the final solution and its reliability, they will have a direct effect on the assessment of precision. In practice, observations are often considered to be uncorrelated when in fact they are unlikely to be so.

3. Precision and reliability analysis assumes that the constants in the model are perfectly known. If they are in error the solution will be thought to be of better quality than it actually is.

3.12 Advantages of LSE

Despite its limitations LSE does have significant advantages, as follows,

1. The LSE method has a sound statistical basis and on average it should give the solution with the minimum variance. If the observation errors are Gaussian then the least squares solution is the most probable, i.e., maximum likelihood estimate.

2. One general computer program can handle any kind of problem with any number of measurements. The measurements may be mixed. LSE is easy to program.

3. The LSE method is automatic, typically requiring no or at most limited human intervention.

4. The full precision and reliability analysis can be carried out before any observations are made because the precision and reliability equations do not depend on the observations.

3.13 The Singular Value Decomposition

LSE requires the computation of a matrix inversion, which is usually the most computationally intensive part of the solution. One way to achieve this is to use the Singular Value Decomposition (SVD). SVD is a generalisation of matrix diagonalisation, which can be used to define the pseudoinverse matrix, a one-sided generalisation of a square matrix. Once a square matrix has been diagonalised it is trivial to find its inverse by simply taking the reciprocals of all the diagonal elements. The SVD is one of the most powerful methods used for solving the LSE problem and it can also be used to perform spectral estimation.

To explain the meaning and interpretation of SVD let A be a $k \times m$ matrix. Then, there exists a $k \times k$ orthogonal matrix U and a $m \times m$ orthogonal matrix V and diagonal matrix Σ such that Equation 3.58 is true.

$$U^T A V = \begin{bmatrix} \Sigma & 0 \\ 0 & 0 \end{bmatrix}$$
(3.58)

where:

$$\Sigma = \begin{bmatrix} \sigma_1 & 0 & .. & 0 \\ 0 & \sigma_2 & .. & 0 \\ \vdots & \vdots & \vdots & \vdots \\ 0 & 0 & .. & \sigma_w \end{bmatrix}, \text{ with } \sigma_1 \geq \sigma_2 \geq ... \geq \sigma_w > 0, \text{ and } w \text{ is the rank of } A.$$

The $\sigma_1, \sigma_2, ..., \sigma_w$ are the singular values of A and the singular value decomposition of A is the factorisation defined by Equation 3.59.

$$A = U \begin{bmatrix} \Sigma & 0 \\ 0 & 0 \end{bmatrix} V^T$$
(3.59)

If A is a square and symmetric matrix the usual diagonalisation result can be obtained by making $U = V$. The columns of V can be considered to be a set of m real vectors $v_1, v_2, ..., v_m$, which are the right singular vectors of A. The columns of U can be considered to be a set of k real vectors $u_1, u_2, ..., u_m$, which are the left singular vectors of A. Since $UU^T = I$, then the following defining relationships hold,

$$AV = U \begin{bmatrix} \Sigma & 0 \\ 0 & 0 \end{bmatrix}, \quad A = \sum_{i=1}^{w} \sigma_i u_i v_i^T$$

and,

$$A^T U = V \begin{bmatrix} \Sigma & 0 \\ 0 & 0 \end{bmatrix}, \quad A^T = \sum_{i=1}^{w} \sigma_i v_i u_i^T$$

3.13.1 The Pseudoinverse

If $\mathbf{A}$ has the SVD defined by Equation 3.60 then the pseudoinverse, $\mathbf{A}^+$, of $\mathbf{A}$ is defined by Equation 3.61.

$$\mathbf{A} = \mathbf{U}\begin{bmatrix} \Sigma & 0 \\ 0 & 0 \end{bmatrix}\mathbf{V}^T \tag{3.60}$$

$$\mathbf{A}^+ = \mathbf{V}\begin{bmatrix} \Sigma^{-1} & 0 \\ 0 & 0 \end{bmatrix}\mathbf{U}^T \tag{3.61}$$

where:

$$\Sigma^{-1} = \begin{bmatrix} \sigma_1^{-1} & 0 & .. & 0 \\ 0 & \sigma_2^{-1} & .. & 0 \\ : & : & : & : \\ 0 & 0 & .. & \sigma_w^{-1} \end{bmatrix}$$

If $k > m$ and $w = m$, then it can be shown that $\mathbf{A}^+ = (\mathbf{A}^T\mathbf{A})^{-1}\mathbf{A}^T$, and therefore $\mathbf{A}^+\mathbf{A} = \mathbf{I}$. In this case it can also be shown that,

$$A^+ = \sum_{i=1}^{w}\frac{1}{\sigma_i}\mathbf{v}_i\mathbf{u}_i^T$$

On the other hand if $m > k$ and $w = k$, then $\mathbf{A}^+ = \mathbf{A}^T(\mathbf{A}\mathbf{A}^T)^{-1}$, and therefore $\mathbf{A}^+\mathbf{A} = \mathbf{I}$.

3.13.2 Computation of the SVD

Two common algorithms used to compute the SVD are the Jacobi algorithm and the QR algorithm.

3.13.2.1 The Jacobi Algorithm
The Jacobi algorithm for computing the SVD is an extension of the Jacobi algorithm for diagonalising a symmetric square matrix, which makes it suitable for dealing with matrices which are not symmetric and not square. The computation of the SVD for a 2 x 2 matrix $\mathbf{A}$ is as follows. Let,

$$\mathbf{A} = \begin{bmatrix} a_{11} & a_{12} \\ a_{21} & a_{22} \end{bmatrix}$$

and let a general rotation matrix $\mathbf{J}$ be defined as,

$$\mathbf{J} = \begin{bmatrix} c & s \\ -s & c \end{bmatrix} = \begin{bmatrix} \cos\theta & \sin\theta \\ -\sin\theta & \cos\theta \end{bmatrix}$$

where:

$$c^2 + s^2 = 1$$

To do the diagonalisation two rotation matrices $\mathbf{J}_1$ and $\mathbf{J}_2$ are needed such that $\mathbf{J}_1^T \mathbf{A} \mathbf{J}_2$ is a diagonal matrix, i.e.,

$$\mathbf{J}_1^T \mathbf{A} \mathbf{J}_2 = \begin{bmatrix} c_1 & s_1 \\ -s_1 & c_1 \end{bmatrix}^T \begin{bmatrix} a_{11} & a_{12} \\ a_{21} & a_{22} \end{bmatrix} \begin{bmatrix} c_2 & s_2 \\ -s_2 & c_2 \end{bmatrix} = \begin{bmatrix} d_1 & 0 \\ 0 & d_2 \end{bmatrix}$$

The diagonalisation is carried out in two steps. Firstly, a rotation matrix is found that makes $\mathbf{A}$ symmetrical and then another rotation matrix that annihilates the off-diagonal terms.

Step 1

To make $\mathbf{A}$ symmetrical a rotation $\mathbf{J}$ is needed such that $\mathbf{J}^T \mathbf{A}$ is symmetric. Since,

$$\mathbf{J}^T \mathbf{A} = \begin{bmatrix} c & s \\ -s & c \end{bmatrix}^T \begin{bmatrix} a_{11} & a_{12} \\ a_{21} & a_{22} \end{bmatrix} = \begin{bmatrix} ca_{11} - sa_{21} & ca_{12} - sa_{22} \\ sa_{11} + ca_{21} & sa_{12} + ca_{22} \end{bmatrix},$$

c and s must be found such that $ca_{12} - sa_{22} = sa_{11} + ca_{21}$ and $c^2 + s^2 = 1$. If ρ is defined such that $\rho = \dfrac{c}{s}$, the symmetric condition gives $\rho = \dfrac{a_{11} + a_{22}}{a_{12} - a_{21}}$ if $a_{12} \neq a_{21}$. This is true by hypothesis therefore, $s = \dfrac{1}{\sqrt{1+\rho^2}}$ and $c = s\rho$, thereby providing the required rotation matrix $\mathbf{J}$. Next, let,

$$\mathbf{B} = \mathbf{J}^T \mathbf{A} = \begin{bmatrix} b_{11} & b_{12} \\ b_{12} & b_{22} \end{bmatrix}, \text{ if } \mathbf{A} \text{ is not symmetric,}$$

and,

$$\mathbf{B} = \mathbf{A}, \text{ if } \mathbf{A} \text{ is symmetric.}$$

Step 2

A rotation matrix $\mathbf{J}_2$ is found to make $\mathbf{J}_2^T \mathbf{B} \mathbf{J}_2$ diagonal, by using the procedure for diagonalising a symmetric matrix as follows.

Set:

$$\xi = \frac{b_{22} - b_{11}}{2b_{12}},$$

and solve the following equation for t.

$$t^2 + 2\xi t - 1 = 0$$

i.e.,

$$t = -\xi \pm \sqrt{1 + \xi^2} \,.$$

Then put,

$$c_2 = \frac{1}{\sqrt{1 + t^2}} \quad \text{and} \quad s_2 = tc_2 \text{ to form } \mathbf{J}_2 \,.$$

Set:

$$\mathbf{J}_1 = \mathbf{J}\mathbf{J}_2,$$

then $\mathbf{J}_1^T \mathbf{A} \mathbf{J}_2$ will be a diagonal matrix.

There are two special cases that can be treated with a single rotation. If $a_{12} = 0 = a_{22}$, then $\mathbf{J}_2$ is the identity matrix. If $a_{21} = 0 = a_{22}$, then $\mathbf{J}_1$ is the identity matrix and $\mathbf{J}_2$ must be computed differently.

When $\mathbf{A}$ is a $n \times n$ matrix with $n > 2$, rotation matrices are used to annihilate symmetrically placed pairs of off-diagonal elements in turn. Each of these individual annihilations undoes previous annihilations, but still makes the resulting matrix closer to a diagonal matrix. The annihilation process is repeated until the off-diagonal elements are all smaller than some threshold. When $\mathbf{A}$ is not square it must be made square by the addition of zeros. If $\mathbf{A}$ is a $k \times m$ matrix with $k > m$ $\mathbf{A}$ is extended to a square matrix by adding columns of zeros to obtain $[\mathbf{A} \ 0]$. Diagonalising this matrix will provide a form defined by Equation 3.62.

$$\mathbf{U}^T [\mathbf{A} \ 0] \begin{bmatrix} \mathbf{V} & 0 \\ 0 & \mathbf{I} \end{bmatrix} = \begin{bmatrix} \sigma_1 & .. & 0 & 0 & .. & 0 \\ : & \ddots & : & : & \ddots & : \\ 0 & .. & \sigma_w & 0 & .. & 0 \\ 0 & .. & 0 & 0 & .. & 0 \\ : & \ddots & : & : & \ddots & : \\ 0 & .. & 0 & 0 & .. & 0 \end{bmatrix} \tag{3.62}$$

The diagonalisation of $\mathbf{A}$ is given by $\mathbf{U}^T \mathbf{A} \mathbf{V}$. If $\mathbf{A}$ is a $k \times m$ matrix with $m > k$ $\mathbf{A}$ is extended to a square matrix by adding rows of zeros and proceeding in a similar fashion as described above.

3.13.2.2 The QR Algorithm

The QR algorithm uses Householder matrices and rotation matrices to find the SVD. It is computed in two steps. Firstly, the matrix is reduced to bidiagonal form using Householder matrices and secondly, the off-diagonal terms are annihilated by an iterative process using rotation matrices.

If $\mathbf{v}$ is any nonzero vector, it determines a Householder matrix, $\mathbf{H}(\mathbf{v})$, as defined by Equation 3.63

$$\mathbf{H}(\mathbf{v}) = \mathbf{I} - \frac{2\mathbf{v}\mathbf{v}^T}{\|\mathbf{v}\|^2} \tag{3.63}$$

Householder matrices are useful for diagonalisation because of the following property.

If the first standard basis vector is,

$$\mathbf{e}_1 = \begin{bmatrix} 1 \\ 0 \\ \vdots \\ 0 \end{bmatrix},$$

and $\mathbf{x}$ is any nonzero vector and putting,

$$\mathbf{v} = \mathbf{x} - \|\mathbf{x}\|\mathbf{e}_1,$$

then if $\mathbf{x}$ is not a scalar multiple of $\mathbf{e}_1$, $\mathbf{H}(\mathbf{v})\mathbf{x}$ is a scalar multiple of $\mathbf{e}_1$, in fact $\mathbf{H}(\mathbf{v})\mathbf{x} = \|\mathbf{x}\|\mathbf{e}_1$.

The desired bidiagonal form has zeros everywhere except along the main diagonal from the top left corner and along the diagonal just above it.

To reduce a $k \times m$ matrix $\mathbf{A}$ to bidiagonal form, when $k \geq m$, a $k \times k$, Householder matrix $\mathbf{Q}_1$ is found such that $\mathbf{Q}_1^T \mathbf{A}$ has a first column consisting of zeros below the diagonal. Then a $m \times m$ Householder matrix $\mathbf{P}_1$ is found such that the first row of $\mathbf{Q}_1^T \mathbf{A} \mathbf{P}_1$ has a first row of zeros apart from the first two columns. The first column of $\mathbf{Q}_1^T \mathbf{A} \mathbf{P}_1$ will still have zeros below the diagonal. This process is repeated for the columns and rows in succession until the matrix is reduced to bidiagonal form.

The off-diagonal terms are annihilated by the Golub-Kahan algorithm, which uses rotations in a way similar to the Jacobi algorithm to reduce the absolute values of these terms (Haykin 1996).

3.14 Exercises

The following Exercises identify some of the basic ideas presented in this Chapter.

3.14.1 Problems

3.1.   Assume that you have a square piece of sheet steel which you wish to bend up into a square open tray. The sheet is 6 by 6 units in area and the bend lines are x units in from the edge as shown in the diagram below,

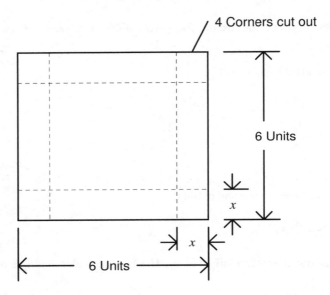

Solve for x to make a tray having the maximum volume, using differential calculus and a standard optimisation method.

3.2.   Show how you could use the LSE equations to solve Problem 3.1.

3.3.   Assume that you have three independent measurements, y_1, y_2 and y_3 with a measurement error variance of σ^2, of the volume of water in the tray of optimum volume (tray of Problem 3.1). Show how you would use LSE to solve for x, the tray depth.

3.4.   In the Acoustic Positioning System example why were the modified set of model equations,

$$G_0$$
$$H_0$$
$$T$$

used instead of the original model equations,

T_{cc}
T_{c1}
T_{c3}
T_{c2}
T_{c0}

to solve the LSE problem?

4. Parametric Signal and System Modelling

Parametric signal modelling involves the reduction of a complicated process with many variables to a simpler one with a smaller number of parameters. This is known as data compression and is common in speech and other areas including economic models, communication systems and efficient data storage systems. The reduction often requires approximation but even so, if the parameters of the model turn out to be physically meaningful, then one can gain insight into the behaviour of the overall process by understanding the influence of each parameter. Another important application for signal modelling is for signal prediction or extrapolation. In both data compression and prediction the signal is known over some interval of time or space and then the goal is to determine the signal over some other unknown interval.

Signal modelling is generally concerned with the efficient representation of signals. There are two main steps in this modelling process. The first is to choose an appropriate parametric form for the model and the second is to find the parameters that provide "the best" approximation to the signal. For example, if the class of signals is known to be sinusoidal then any signal in the class can be fully represented with only three sinusoid model parameters, the amplitude, frequency and phase. Finding the best approximations to the unknown signal's amplitude, frequency and phase allows the signal to be efficiently represented and fully reproduced given the sinusoidal model and parameter values.

Modelling of discrete-time signals can be done by the output of a Linear Time-Invariant (LTI) system driven by a fixed input signal such as a unit-sample signal or white Gaussian noise. Nonstationary signals can be modelled by modelling small sections of the signal with a model whose parameters can change from section to section, e.g., in speech where the fundamental block for analysis is one pitch period ($\approx$ 10ms). In adaptive filtering the coefficients are allowed to evolve over time according to an update strategy that is always working to minimise an error measure (e.g., LMS algorithm). The frequency domain model that is most commonly used is one that represents a signal as the output of a causal Linear Shift-Invariant (LSI) filter with the rational system function defined by Equation 4.1. Notice that the filter coefficients in Equation 4.1 have been defined by the parameter sets $\{a_k\}$ and $\{b_k\}$. They could have been alternatively represented by $\{a[k]\}$ and $\{b[k]\}$, but the first representation will be used throughout this Chapter to distinguish between filter coefficient values and discrete-time signal values.

Equation 4.1 can also be represented as the difference Equation 4.2 in the discrete-time domain.

$$\frac{B_q(z)}{A_p(z)} = \frac{b_0 + b_1 z^{-1} + ... + b_q z^{-q}}{a_0 + a_1 z^{-1} + ... + a_p z^{-p}},$$

$a_0 = 1$, typically, therefore,

$$\frac{B_q(z)}{A_p(z)} = \frac{\sum_{k=0}^{q} b_k z^{-k}}{1 + \sum_{k=1}^{p} a_k z^{-k}}$$

(4.1)

$$y[n] = \sum_{k=0}^{q} b_k x[n-k] - \sum_{k=1}^{p} a_k y[n-k]$$

(4.2)

where:
$p = 1,2,...$ and $q = 0,1,2,...$

Here, the linear stochastic model assumes that a discrete time series $y[n]$ is the result of applying a linear filtering operation to some unknown time series $x[n]$. Another way to state this is to say that the value of $y[n]$ can be predicted from the values of $y[m]$ for $m < n$ and the values of $x[m]$ for $m \le n$, by taking a linear combination of these values.

The roots of the numerator and denominator polynomials of Equation 4.1 are the zeros and poles of the model, respectively. For both Equations 4.1 and 4.2 if all the coefficients a_k, except for $a_0=1$, are zero the model is referred to as an all-zero or Moving Average (MA) model. If the all the coefficients b_k are zero for $k > 0$ the model is referred to as an all-pole or Autoregressive (AR) model. If at least one of each of the coefficients a_k and b_k for $k > 0$ are nonzero the model is referred to as a pole-zero or Autoregressive Moving-Average (ARMA) model.

4.1 The Estimation Problem

The estimation problem for linear stochastic models can be defined as follows,

Given a finite set of observations of the time series $\{x[n]\}$, determine the parameters $a_1, a_2,...,a_p, b_0, b_1,...,b_q$ of the model that generated the series $\{x[n]\}$.

This problem is not well posed. Firstly, the input to the filter is unknown. Secondly, there is no general *a priori* information regarding the number of unknown parameters. The first issue is adequately solved for many problems by assuming that input has the spectral characteristics of white noise. Finding the number of unknown parameters to be estimated is made difficult by a number of factors. It can be shown that any model with a finite number of zeros can be approximated arbitrarily closely by a model without any zeros, and conversely, that any model with a finite number of poles can be approximated arbitrarily closely by a model without any poles. Also, increasing the number of parameters does not lead to a consistent improvement in the fit. The typical behaviour is that the fit will improve as the number of parameters is increased until some point is reached where the fit will begin to slowly worsen. This effect tends to be followed by small variations in the fit in either direction as the number of parameters is increased.

In the absence of any objective criteria the choice of the number of poles and zeros to be used is generally based on considerations of efficiency and tractability. The most desirable model is one having the least parameters whose estimation requires the least amount of effort. The all-pole model is by far the simplest with respect to parameter estimation. The relationship between the pole coefficients and the autocorrelation function yields a set of simultaneous linear equations for the pole coefficients. The estimation of the parameters of an all-pole model can be performed by simply computing estimates of the autocorrelation terms, substituting the values into the normal equations, and solving the resulting set of simultaneous linear equations for the pole coefficients. These linear equations have special symmetries, which make it possible to devise efficient algorithms for their solution. Some procedures for the efficient solution of the equations arising from the all-pole estimation problem have been proposed by (Levinson 1947), (Durbin 1960) and (Robinson 1964). In 1977 Makhoul described a class of lattice methods for solving the problem (Makhoul 1977). In contrast, the estimation of parameters of a model having zeros requires the solution of non-linear equations and is correspondingly more difficult. These nonlinear equations are such that their solution requires iterative methods, which are both computationally expensive and numerically delicate. Box and Jenkins have developed suitable methods for solving these ARMA models as applied to stationary time series (Box and Jenkins 1970).

4.2 Deterministic Signal and System Modelling

There are a number of methods for deterministic signal modelling that have been developed over the years from diverse fields but no one of them stands out as the best. Some of the more important ones are,

1. The Least Squares (LS) Method.

2. The Padé Approximation Method (constrained LS).

3. Prony's Method (approximate matching - blend of LS and Padé).

4. Autocorrelation Method (finite data modification of Prony all-pole modelling).

5. Covariance Method (finite data modification of Prony all-pole modelling).

The first method is a direct method of signal modelling based on the method of least squares whereas the remaining four are indirect methods. Prony's method is the most important of the indirect methods. All these methods are based on the idea that the signal represents the impulse response of a linear system. Consequently, given the impulse response of a desired system or filter it is possible to use these same methods to model linear systems or filters.

The direct signal modelling approach is represented in Figure 4.1 (Lim and Oppenheim 1988).

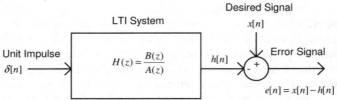

Figure 4.1. Direct Signal Modelling

The direct error, $e[n] = x[n] - h[n]$, is nonlinear in the coefficients of $A(z)$ and $B(z)$ so it is solved by solving nonlinear equations resulting from minimising the least squares error.

The indirect signal modelling approach as typified by Figure 4.2 (Lim and Oppenheim 1988) is a practical approach employed by most methods because it involves the solution of only linear equations.

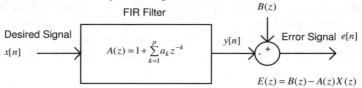

Figure 4.2. Indirect Signal Modelling

The indirect frequency domain error, $E(z) = B(z) - X(z) A(z)$, is linear in the coefficients of $A(z)$ and $B(z)$ so it can be solved by solving linear equations through minimising a least squares error generated by filtering the signal $x[n]$ with a FIR filter. The FIR filter attempts to remove the poles of $X(z)$, i.e., $A(z)$.

4.2.1 The Least Squares Method

Direct modelling is not widely used because it is difficult to solve the associated nonlinear equations. However, it does accurately state the goal of matching the impulse response of a rational linear shift invariant (LSI) system to an arbitrary deterministic signal. The direct method is also referred to as the least squares method. The least squares method of signal modelling attempts to solve the rational Equation set 4.1 by driving the LSI system with an impulse function as shown in Figure 4.1.

The desired signal $x[n]$ is assumed to be equal to zero for $n < 0$ and the filter $h[n]$ is assumed to be causal ($h[n] = 0$ for $n < 0$). The numerator, $B(z)$, and denominator, $A(z)$, polynomials of the rational LSI system are defined in terms of the unknown parameters, $\{a_k\}$ and $\{b_k\}$, by Equations 4.3 respectively.

$$B(z) = \sum_{k=0}^{q} b_k z^{-k}$$

$$A(z) = 1 + \sum_{k=1}^{p} a_k z^{-k} \tag{4.3}$$

The problem is to find the filter coefficients (parameters) $\{a_k\}$ and $\{b_k\}$, which make the error between the desired signal and the filter response, $e[n] = x[n] - h[n]$, as small as possible in the least squares sense, as defined by Equation 4.4.

$$\min_{A(z),B(z)} \varepsilon = \sum_{n=0}^{\infty} |e[n]|^2 \tag{4.4}$$

A necessary condition for parameters $\{a_k\}$ and $\{b_k\}$ to minimise the squared error ε is, the partial derivative of ε with respect to both a_k and b_k must vanish (Hayes 1996), i.e.,

$$\frac{\partial \varepsilon}{\partial a_k^*} = 0, \quad k = 1,2,.....,p \quad \text{and,} \quad \frac{\partial e}{\partial b_k^*} = 0, \quad k = 0,1,.....,q$$

Using Parseval's theorem, the error Equation 4.4 can be expressed in terms of the Fourier transform of $e[n]$, $E(e^{j\theta})$, as defined by Equation 4.5.

$$\varepsilon = \frac{1}{2\pi} \int_{-\pi}^{+\pi} \left| E(e^{j\theta}) \right|^2 d\theta \equiv \frac{1}{2\pi} \int_{-\pi}^{+\pi} \left| E(e^{-j\theta}) \right|^2 d\theta \tag{4.5}$$

where:

$$E(e^{-j\theta}) = X(e^{-j\theta}) - \frac{B(e^{-j\theta})}{A(e^{-j\theta})}$$

The Partial differentiation of Equation 4.5 with respect to variables a_k^* and b_k^* results in Equations 4.6 and 4.7 respectively.

$$\frac{\partial \varepsilon}{\partial a_k^*} = \frac{1}{2\pi} \int_{-\pi}^{+\pi} \left[X(e^{-j\theta}) - \frac{B(e^{-j\theta})}{A(e^{-j\theta})} \right] \frac{B^*(e^{-j\theta})}{\left[A^*(e^{-j\theta}) \right]^2} e^{-jk\theta} d\theta = 0,$$

for $k = 1,2,...,p$ \tag{4.6}

$$\frac{\partial \varepsilon}{\partial b_k^*} = -\frac{1}{2\pi} \int_{-\pi}^{+\pi} \left[X(e^{-j\theta}) - \frac{B(e^{-j\theta})}{A(e^{-j\theta})} \right] \frac{e^{-jk\theta}}{A^*(e^{-j\theta})} d\theta = 0,$$

for $k = 0,1,2,...,q$ \tag{4.7}

From Equations 4.6 and 4.7 it can be seen that optimum set of model parameters are defined explicitly in terms of a set of $p + q + 1$ nonlinear equations. These can be solved using iterative techniques such as the method of steepest descent or Newton's method or iterative prefiltering (Lim and Oppenheim 1988). However, these methods are not very suitable for real-time signal processing applications. It is for this reason that indirect methods of signal modelling are mostly used in practice, where the parameters can be solved much more easily.

4.2.2 The Padé Approximation Method

If the Padé approximation model, represented by Equation 4.1 and Figure 4.1, is forced to be exact over a fixed finite interval of the first $p + q + 1$ points of the signal, then this only requires the solution to a set of linear equations. However, the disadvantage is that there is no guarantee on how accurate the model will be for values outside this interval. In fact, the Padé approximation is not actually an approximation over the interval of the first $p + q + 1$ points of the signal, rather it is only likely to be approximate outside that interval. Also, unless the signal just happens to be exactly the impulse response of some low-order rational system the Padé method is not very practical. Nevertheless, the Padé method is an important lead into the more practical Prony method discussed in the next Section.

Given a causal signal $x[n]$ to be modelled with the rational model defined by Equation 4.8 $h[n]$ can be forced to equal $x[n]$ for $n = 0, 1,..., p + q$, (the interval [0, $p + q$]) by a suitable choice of parameters $\{a_k\}$ and $\{b_k\}$. Notice that the same method can also be used to design a filter given the filter's impulse response.

$$H(z) = \frac{B(z)}{A(z)} = \frac{\sum_{k=0}^{q} b_k z^{-k}}{1 + \sum_{k=1}^{p} a_k z^{-k}} = \sum_{n=0}^{\infty} h[n] z^{-n} \tag{4.8}$$

This can be achieved in a two-step process that conveniently requires the solution of only linear equations. Equation 4.8 can be rearranged to obtain $B(z) = A(z)H(z)$, which represents the fact that b_n is the convolution of $h[n]$ and a_n. The time-domain convolution is defined by Equation 4.9.

$$b_n = h[n] + \sum_{k=1}^{p} a_k h[n-k] \tag{4.9}$$

where:
  $h[n] = 0$ for $n < 0$ and $b_n = 0$ for $n < 0$ and $n > q$.

This general convolution equation can also be expressed in matrix form as defined by the matrix Equation 4.10.

$$
\begin{bmatrix}
h[0] & 0 & 0 & .. & 0 \\
h[1] & h[0] & 0 & .. & 0 \\
h[2] & h[1] & h[0] & .. & 0 \\
h[3] & h[2] & h[1] & .. & : \\
: & : & : & \ddots & h[0] \\
: & : & : & \ddots & h[1] \\
: & : & : & : & :
\end{bmatrix}
\begin{bmatrix}
1 \\
a_1 \\
a_2 \\
: \\
a_p
\end{bmatrix}
=
\begin{bmatrix}
b_0 \\
b_1 \\
: \\
b_q \\
0 \\
0 \\
:
\end{bmatrix}
\tag{4.10}
$$

After imposing the impulse matching restriction, $h[n] = x[n]$ for $n = 0, 1,...., p + q$, matrix Equation 4.10 then becomes matrix Equation 4.11.

$$
\begin{bmatrix}
x[0] & 0 & 0 & .. & 0 \\
x[1] & x[0] & 0 & .. & 0 \\
: & : & \ddots & \ddots & 0 \\
x[q] & x[q-1] & x[q-2] & .. & x[q-p] \\
x[q+1] & x[q] & x[q-1] & .. & x[q-p+1] \\
: & : & \ddots & \ddots & : \\
x[q+p] & x[q+p-1] & x[q+p-2] & .. & x[q]
\end{bmatrix}
\begin{bmatrix}
1 \\
a_1 \\
a_2 \\
: \\
a_p
\end{bmatrix}
=
\begin{bmatrix}
b_0 \\
b_1 \\
: \\
b_q \\
0 \\
: \\
0
\end{bmatrix}
\tag{4.11}
$$

$\mathbf{Xa} = \begin{bmatrix} \mathbf{b} \\ \mathbf{0}_p \end{bmatrix}$, where: $\mathbf{0}_p$ is a p-dimensional vector with p zeros.

Matrix Equation 4.11 represents a set of $p + q + 1$ linear equations in $p + q + 1$ unknowns as defined by Equation 4.12.

$$
x[n] + \sum_{k=1}^{p} a_k x[n-k] = \begin{cases} b_n & : & n = 0,1,..,q \\ 0 & : & n = q+1,..,q+p \end{cases}
\tag{4.12}
$$

The matrix Equation 4.11 can be partitioned into a top part and bottom part and written as matrix Equation 4.13.

$$
\begin{bmatrix} \mathbf{X}_1 \\ \mathbf{X}_2 \end{bmatrix} \mathbf{a} = \begin{bmatrix} \mathbf{b} \\ \mathbf{0}_p \end{bmatrix}
\tag{4.13}
$$

where:

$\mathbf{X}_1$ is the top $(q+1) \times (p+1)$ part of $\mathbf{X}$.

$\mathbf{X}_2$ is the bottom $p \times (p+1)$ part of $\mathbf{X}$.

$\mathbf{a}$ is a $p+1$ dimensional vector of 1 and $\{a_k\}$.

$\mathbf{b}$ is a $q+1$ dimensional vector of $\{b_k\}$.

The two-step approach used to solve for the parameters $\{a_k\}$ and $\{b_k\}$ proceeds by solving the denominator parameters $\{a_k\}$ first and then the denominator parameters $\{b_k\}$ after that. The bottom partition of Equation 4.13 is used to solve for $\{a_k\}$ since it is independent of $\{b_k\}$ and then the top partition for $\{b_k\}$. The first step involves solving Equation 4.14.

$$\mathbf{X}_2 \mathbf{a} = \mathbf{0}_p$$

i.e.,   (4.14)

$$\tilde{\mathbf{X}}_2 \tilde{\mathbf{a}} = -\tilde{\mathbf{x}}$$

where:

$\tilde{\mathbf{X}}_2$ is the matrix $\mathbf{X}_2$ less the first column, $p \times p$ matrix.

$\tilde{\mathbf{a}}$ is the vector $\mathbf{a}$ less the first element $= 1$, p - dimensional.

$\tilde{\mathbf{x}}$ is the first column of $\mathbf{X}_2$, p - dimensional.

The matrix equation $\tilde{\mathbf{X}}_2 \tilde{\mathbf{a}} = -\tilde{\mathbf{x}}$ represents a set of p linear equations in p unknowns as defined by Equation 4.15.

$$\begin{bmatrix} x[q] & x[q-1] & .. & x[q-p+1] \\ \vdots & \vdots & \ddots & \vdots \\ x[q+p-1] & x[q+p-2] & .. & x[q] \end{bmatrix} \begin{bmatrix} a_1 \\ \vdots \\ a_p \end{bmatrix} = - \begin{bmatrix} x[q+1] \\ \vdots \\ x[q+p] \end{bmatrix} \quad (4.15)$$

The solution of Equation 4.15 involves three cases where,

1. $\tilde{\mathbf{X}}_2$ is nonsingular and there is a unique solution, $\tilde{\mathbf{a}} = -\tilde{\mathbf{X}}_2^{-1}\tilde{\mathbf{x}}$.

2. $\tilde{\mathbf{X}}_2$ is singular but a nonunique solution exists to Equation 4.15. If r is the rank of $\tilde{\mathbf{X}}_2$ the solution with $(p - r)$ zero entries is preferred because it will produce the lowest order model. For any solution $\tilde{\mathbf{a}}$ to Equation 4.15, and solution $\mathbf{z}$ to equation $\tilde{\mathbf{X}}_2 \mathbf{z} = 0$ then $\tilde{\mathbf{a}} + \mathbf{z}$ is also a solution.

3. $\tilde{\mathbf{X}}_2$ is singular but no solution to Equation 4.15 exists. In this case the assumption that $a_0 = 1$ is incorrect. If a_0 is set to 0 a nonunique solution can be found to Equation 4.14 but it may not be able to match to $x[p+q]$.

Once suitable $\{a_k\}$ parameters are found then the $\{b_k\}$ parameters are found from the upper partition of Equation 4.13 by simple matrix multiplication. Refer to the Exercises Section of this Chapter for example computations.

As is evident from the approach taken there is no mechanism built into the Padé approximation method to ensure that the solution is stable. The method can be guaranteed to produce a stable solution and exact fit to the data over the interval [0, $p+q$] only if $\tilde{\mathbf{X}}_2$ is nonsingular. A second more serious problem with the method is that the order of the model is tied directly to the number of signal points being matched. In practice this inevitably requires a very high order model if a significant portion of the signal is to be matched. The only exception to this is the unlikely situation when it is possible to represent the signal with a low-order rational model. In practical applications having long signal records it is possible to derive lower order models using Prony's approximate matching method.

4.2.3 Prony's Method

Prony's method is a blend between the least squares approach and the Padé approximation method but it still only requires the solution to a set of linear equations. Even though the method is a true approximation technique it is on average still more accurate than the Padé method if taken over the whole signal. It is fundamentally based on signal approximation with a linear combination of adjustable exponentials.

The impulse matching problem for modelling an entire causal signal $x[n]$, $n = 0$, $1,...,\infty$, produces an infinite number of equations as defined by Equation 4.16. The orders of the model numerator and denominator coefficients are fixed and finite.

$$\begin{bmatrix} x[0] & 0 & .. & 0 \\ x[1] & x[0] & .. & 0 \\ x[2] & x[1] & .. & 0 \\ \vdots & \vdots & \ddots & \vdots \end{bmatrix} \begin{bmatrix} 1 \\ a_1 \\ a_2 \\ \vdots \\ a_p \end{bmatrix} \overset{?}{=} \begin{bmatrix} b_0 \\ b_1 \\ \vdots \\ b_q \\ 0 \\ 0 \\ \vdots \end{bmatrix}$$

$$\mathbf{X}\mathbf{a} \overset{?}{=} \begin{bmatrix} \mathbf{b} \\ \mathbf{0} \end{bmatrix}$$

(4.16)

Equation 4.16 can be partitioned into a top and bottom part to decouple the effects of the numerator and denominator coefficients as in the Padé method. The first $q + 1$ rows of matrix $\mathbf{X}$ are separated from the rest of the matrix and the vector $\mathbf{a}$ is separated as defined by Equation 4.17.

$$\begin{bmatrix} \mathbf{X}_1 \\ \mathbf{X}_2 \end{bmatrix} \mathbf{a} \overset{?}{=} \begin{bmatrix} \mathbf{b} \\ \mathbf{0} \end{bmatrix}$$

$$\begin{bmatrix} \mathbf{X}_1 \\ \mathbf{X}_2 \end{bmatrix} \begin{bmatrix} 1 \\ \tilde{\mathbf{a}} \end{bmatrix} \overset{?}{=} \begin{bmatrix} \mathbf{b} \\ \mathbf{0} \end{bmatrix}$$

$$\begin{bmatrix} \mathbf{X}_1 \\ \tilde{\mathbf{x}} & \tilde{\mathbf{X}}_2 \end{bmatrix} \begin{bmatrix} 1 \\ \tilde{\mathbf{a}} \end{bmatrix} \overset{?}{=} \begin{bmatrix} \mathbf{b} \\ \mathbf{0} \end{bmatrix}$$

(4.17)

where:

$\mathbf{X}_1$ is the top $(q+1) \times (p+1)$ part of $\mathbf{X}$.

X_2 is the bottom part of X.

$\tilde{a}$ is a p dimensional vector of $\{a_k\}$.

b is a $q+1$ dimensional vector of $\{b_k\}$.

$\tilde{X}_2$ is the matrix X_2 less the first column.

$\tilde{x}$ is the first column of X_2.

The lower partition contains an infinite number of equations to be solved for $\tilde{a}$ as defined by Equation 4.18, which is expanded in detail as Equation 4.18a.

$$X_2 a \overset{?}{=} 0$$

$$\begin{bmatrix} \tilde{x} & \tilde{X}_2 \end{bmatrix} a \overset{?}{=} 0$$

$$\begin{bmatrix} \tilde{x} & \tilde{X}_2 \end{bmatrix} \begin{bmatrix} 1 \\ \tilde{a} \end{bmatrix} \overset{?}{=} 0 \tag{4.18}$$

$$\tilde{x} + \tilde{X}_2 \tilde{a} \overset{?}{=} 0$$

$$\begin{bmatrix} x[q+1] & x[q] & x[q-1] & .. & x[q-p+1] \\ x[q+2] & x[q+1] & x[q] & \ddots & x[q-p+2] \\ \vdots & \vdots & \vdots & \ddots & \vdots \end{bmatrix} \begin{bmatrix} 1 \\ a_1 \\ \vdots \\ a_p \end{bmatrix} \overset{?}{=} \begin{bmatrix} 0 \\ 0 \\ \vdots \end{bmatrix}$$

$$\begin{bmatrix} x[q+1] \\ x[q+2] \\ \vdots \end{bmatrix} + \begin{bmatrix} x[q] & x[q-1] & .. & x[q-p+1] \\ x[q+1] & x[q] & \ddots & x[q-p+2] \\ \vdots & \vdots & \ddots & \vdots \end{bmatrix} \begin{bmatrix} a_1 \\ \vdots \\ a_p \end{bmatrix} \overset{?}{=} \begin{bmatrix} 0 \\ 0 \\ \vdots \end{bmatrix} \tag{4.18a}$$

In general Equation 4.18 describes a system of over determined linear equations that need not have an exact solution, even if the columns of $\tilde{X}_2$ are linearly independent vectors. This means that since the vector $\tilde{x}$ can only be approximated by the columns of matrix $\tilde{X}_2$, it is necessary to choose $\tilde{a}$ to minimise the equation error e defined by Equation 4.19. This error is not the same error as used before in the Padé approximation method.

$$e = X_2 a = \tilde{x} + \tilde{X}_2 \tilde{a} \tag{4.19}$$

The least squares error norm ε (min $e^T e$), as defined by Equation 4.20, is chosen to be minimised because of its mathematical simplicity.

$$\varepsilon = e^T e = \sum_{n=q+1}^{\infty} e^2[n] = \sum_{n=q+1}^{\infty} \left[x[n] + \sum_{k=1}^{p} a_k x[n-k] \right]^2 \tag{4.20}$$

Minimising the error Equation 4.20 by partial differentiation with respect to coefficients $\{a_k\}$ leads to the normal equations defined by Equation 4.21 (Lim and Oppenheim 1988).

$$\mathbf{X}_2^T\mathbf{X}_2\mathbf{a} = \begin{bmatrix} \{\varepsilon\}_{\min} \\ \mathbf{0} \end{bmatrix} \tag{4.21}$$

There are actually two equations in Equation 4.21. One is due to the orthogonality condition, $\mathbf{e}_{\min}^T\tilde{\mathbf{X}}_2 = \mathbf{0}$, used in the least squares minimisation and is defined by Equation 4.22.

$$\tilde{\mathbf{X}}_2^T\tilde{\mathbf{X}}_2\tilde{\mathbf{a}} = -\tilde{\mathbf{X}}_2^T\tilde{\mathbf{x}} \tag{4.22}$$

The other is an expression for the minimal value of the least squares error as defined by Equation 4.23.

$$\begin{aligned} \{\varepsilon\}_{\min} &= \mathbf{e}_{\min}^T(\tilde{\mathbf{X}}_2\tilde{\mathbf{a}} + \tilde{\mathbf{x}}) \\ &= \tilde{\mathbf{x}}^T\mathbf{X}\mathbf{a} \end{aligned} \tag{4.23}$$

In principle Equation 4.22 provides a solution for the optimum vector $\tilde{\mathbf{a}}$, which can then be used to find the solution to vector $\mathbf{b}$ by simple matrix multiplication in Equation 4.17. The solution is facilitated by the fact that the $p \times p$ matrix $\tilde{\mathbf{X}}_2^T\tilde{\mathbf{X}}_2$ is always symmetric, or Hermitian for complex data, and also positive semidefinite. This means that the matrix is always invertible if and only if the columns of $\tilde{\mathbf{X}}_2$ are linearly independent. Equation 4.21 has a special structure for the case of all-pole modelling that allows it to be solved using fast solution algorithms such as the Levinson-Durbin recursion.

The form of matrix $\mathbf{X}_2^T\mathbf{X}_2$ shows it to be an unscaled autocorrelation matrix, $\mathbf{R}_x$, defined by the matrix coefficient Equation 4.24.

$$r_x(i,j) = \sum_{m=q+1}^{\infty} x[m-i]x^*[m-j] = r_x^*(j,i), \quad i,j = 0,1,2,..,p \tag{4.24}$$

A summary of the Prony equations is as follows,

1.   The Prony normal equations are,

$$\sum_{m=1}^{p} a_m r_x(k,m) = -r_x(k,0); \quad k = 1,2,....,p$$

$$r_x(k,m) = \sum_{n=q+1}^{\infty} x[n-m]x^*[n-k]; \quad k,m \geq 0$$

2.   The numerator is,

$$b_n = x[n] + \sum_{k=1}^{p} a_k x[n-k]; \quad n = 0,1,..,q$$

3.   The minimum error is,

$$\{\varepsilon\}_{min} = r_x(0,0) + \sum_{k=1}^{p} a_k r_x(0,k)$$

Refer to the Exercises Section of this Chapter for example computations.

4.2.3.1 All-pole Modelling Using Prony's Method

All-pole models are important for various reasons. They can accurately model many physical processes. For example, in speech modelling the tube model for speech production is an all-pole model. Even if an all-pole model is not physically justified it has been found that an all-pole model may still be accurate in many practical signal modelling problems. Probably the most important reason is that there is a special structure in all-pole models that allows for the development of fast and efficient algorithms for solving the pole parameters.

An all-pole model for a signal $x[n] = 0$ for $n < 0$ is defined by Equation 4.25.

$$H(z) = \frac{b_0}{A(z)} = \frac{b_0}{1 + \sum_{k=1}^{p} a_k z^{-k}} = \sum_{n=0}^{\infty} h[n] z^{-n} \tag{4.25}$$

In Prony's method the pole parameters $\{a_k\}$ are found by minimising the error Equation 4.20 for $q = 0$. The coefficients $\{a_k\}$ that minimise the error Equation 4.20 also minimise $\varepsilon_p = \sum_{n=0}^{\infty} e^2[n]$, since the error at $n = 0$ is defined to be $x(0)^2$ and is not dependent on $\{a_k\}$. The all-pole normal equations that minimise $\varepsilon_p = \sum_{n=0}^{\infty} e^2[n]$ can be shown to be as defined by Equation 4.26 (Hayes 1996).

$$\sum_{m=1}^{p} a_m r_x(k,m) = -r_x(k,0), \quad k = 1,2,..,p \tag{4.26}$$

For a complex signal and the new error ε_p the autocorrelation function $r_x(k,m)$ is defined by Equation 4.27.

$$r_x(k,m) = \sum_{n=0}^{\infty} x[n-m] x^*[n-k] \tag{4.27}$$

There is an underlying structure in $r_x(k,m)$ that provides the simplification defined by Equation 4.28.

$$r_x(k+1,m+1) = r_x(k,m), \quad k \geq 0, m \geq 0$$

$$\therefore \tag{4.28}$$

$$r_x(k-m) \equiv r_x(k,m) = \sum_{n=0}^{\infty} x[n-m] x^*[n-k] = r_x^*(m-k)$$

The all-pole normal equations can therefore be defined by Equation 4.29.

$$\sum_{m=1}^{p} a_m r_x(k-m) = -r_x(k), \qquad k=1,....,p$$

$$\begin{bmatrix} r_x(0) & r_x^*(1) & r_x^*(2) & .. & r_x^*(p-1) \\ r_x(1) & r_x(0) & r_x^*(1) & .. & r_x^*(p-2) \\ r_x(2) & r_x(1) & r_x(0) & .. & r_x^*(p-3) \\ \vdots & \vdots & \vdots & \ddots & \vdots \\ r_x(p-1) & r_x(p-2) & r_x(p-3) & .. & r_x(0) \end{bmatrix} \begin{bmatrix} a_1 \\ a_2 \\ a_3 \\ \vdots \\ a_p \end{bmatrix} = - \begin{bmatrix} r_x(1) \\ r_x(2) \\ r_x(3) \\ \vdots \\ r_x(p) \end{bmatrix} \qquad (4.29)$$

The autocorrelation matrix in Equation 4.29 is Hermitian Toeplitz, which allows the Levinson-Durbin recursion to be used to solve the equation efficiently. The $\{a_k\}$ can also be determined by finding the least squares solution to set of overdetermined equations defined by Equation 4.18 for $q = 0$.

The minimum all-pole modelling error $\{\varepsilon_p\}_{min}$ is defined by Equation 4.30.

$$\{\varepsilon_p\}_{min} = r_x(0) + \sum_{k=1}^{p} a_k r^*(k) \qquad (4.30)$$

The remaining b_0 coefficient can be determined in two different ways. The first way is to simply make it equal to $x[0]$ as in the Prony method. However, this may be problematic if $x[0]$ takes on an anomalous or a bad data value. The other, preferred, way is to choose b_0 such that the energy in $x[n]$ is equal to the energy in $\hat{x}[n] = h[n]$, i.e., $r_x(0) = r_h(0)$. In this case the solution for b_0 is defined by Equation 4.31.

$$b_0 = \sqrt{\{\varepsilon_p\}_{min}} \qquad (4.31)$$

4.2.3.2 Linear Prediction
There is an equivalence between all-pole signal modelling and the linear prediction problem. In Prony's method the error is defined by Equation 4.32.

$$e[n] = x[n] + \sum_{k=1}^{p} a_k x[n-k] \qquad (4.32)$$

If this error is expressed as $e[n] = x[n] - \hat{x}[n]$, where $\hat{x}[n] = -\sum_{k=1}^{p} a_k x[n-k]$, then $\hat{x}[n]$ is a linear combination of the values of $x[n]$ over the interval $[n-p, n-1]$. Because $e[n]$ is the difference between $x[n]$ and $\hat{x}[n]$, minimising the sum of squares of $e[n]$ is equivalent to finding the $\{a_k\}$ that make $\hat{x}[n]$ as close as possible to $x[n]$. Therefore, $\hat{x}[n]$ is an estimate, or prediction, of $x[n]$ in terms of a linear combination of the p previous values of $x[n]$. Consequently, the error $e[n]$ is the linear prediction error and the $\{a_k\}$ are the linear prediction coefficients. Furthermore, since $x[n]*a_n = e[n]$, then $A(z)$ is called the prediction error filter.

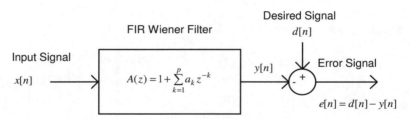

Figure 4.3. Wiener Filter Block Diagram

4.2.3.3 Digital Wiener Filter

The indirect signal model of Figure 4.2 can be modified to that shown in Figure 4.3, where the $\{b_k\}$ are replaced with some desired signal $d[n]$. This now represents a Wiener shaping filter. The problem is to find a FIR filter whose output $y[n]$ best approximates $d[n]$ in the sense that the mean square norm of the error is minimised. However, if $d[n]$ is a unit variance white Gaussian noise signal then $|X(\omega)A(\omega)|^2 \approx 1$, and therefore $A(z)$ can be used for the poles of a model of signal $x[n]$.

The optimum Wiener filter is called a whitening filter and is found by solving the normal equation for the least squares problem. In this case the error signal can be defined by Equation 4.33.

$$e[n] = d[n] - y[n] = d[n] - \sum_{k=0}^{p} a_k x[n-k] \tag{4.33}$$

The normal equations are obtained by the projection theorem which requires that the optimum error signal must be orthogonal to the basis signals $x_j[n] = x[n-j]$, $j = 0, 1,\ldots, p$. This requirement is expressed by Equation 4.34, which defines $p+1$ optimum filter equations.

$$E\{e[n]x_j[n]\} = 0, \quad j = 0,1,\ldots, p$$

$$E\left\{[\sum_{k=0}^{p} d[n] - a_k x[n-k]]x[n-j]\right\} = 0$$

$$\sum_{k=0}^{p} a_k E\{x[n-k]x[n-j]\} = E\{d[n]x[n-j]\} \tag{4.34}$$

$$\sum_{k=0}^{p} a_k r_x(j-k) = r_{dx}(j), \quad j = 0,1,\ldots, p$$

Equation 4.34 is more economically expressed in matrix form as Equation 4.35.

$$\mathbf{R}_x \begin{bmatrix} a_0 \\ a_1 \\ \vdots \\ a_p \end{bmatrix} = \begin{bmatrix} r_{dx}(0) \\ r_{dx}(1) \\ \vdots \\ r_{dx}(p) \end{bmatrix} \tag{4.35}$$

The autocorrelation matrix $\mathbf{R}_x$ is a $(p+1) \times (p+1)$ Toeplitz matrix with coefficients $r_x(j\text{-}k)$. Equation 4.35 can be solved very efficiently using the Levinson-Durbin recursion. The normal equations for the Wiener filter are similar to the equations for the autocorrelation method (discussed in the next Section) but they are more general. The autocorrelation normal equations are obtained by taking $d[n]$ to be unit variance white Gaussian noise. Therefore, $r_{dx}(0) = 1$ and $r_{dx}(j) = 0$ for $j \neq 0$ and Equation 4.35 becomes Equation 4.36.

$$\mathbf{R}_x \begin{bmatrix} a_0 \\ a_1 \\ \vdots \\ a_p \end{bmatrix} = \begin{bmatrix} 1 \\ 0 \\ \vdots \\ 0 \end{bmatrix} \tag{4.36}$$

Equation 4.36 is nearly identical to the autocorrelation equations (to within the scale factor a_0) and the solution is obtained in two steps. First Equation 4.37 is solved, and then $a_j = \alpha_j / \varepsilon_\alpha$ for $j = 1, 2, \ldots, p$.

$$\mathbf{R}_x \begin{bmatrix} 1 \\ \alpha_1 \\ \vdots \\ \alpha_p \end{bmatrix} = \begin{bmatrix} \varepsilon_\alpha \\ 0 \\ \vdots \\ 0 \end{bmatrix} \tag{4.37}$$

In theory, the solutions to all the Prony equations discussed above require infinite signal lengths. However for practical solutions a finite limit must be imposed on the signal length. This means that the statistical autocorrelation and cross-correlation functions, which are defined over an infinite time range, must be estimated from finite data samples. There are two ways of imposing this limit, each of which result in solution methods with quite different properties. For the case of all-pole modelling the two ways produce the so called autocorrelation and the covariance methods.

4.2.4 Autocorrelation and Covariance Methods

The autocorrelation and covariance methods of all-pole modelling are modifications to the all-pole Prony method for the case in which only a finite length data record is available. A fast algorithm, called the Levinson-Durbin recursion method, is available to solve the linear equations related to the autocorrelation method.

The first way to deal with finite signal lengths is to simply assume that the signals are infinite by padding them with an infinite number of zeros on their ends. In that case, for a signal $x[n]$ of length N, the least squares error Equation 4.20 becomes Equation 4.38.

$$\varepsilon = \mathbf{e}^T \mathbf{e} = \sum_{n=q+1}^{N+p-1} e^2[n] = \sum_{n=q+1}^{N+p-1} \left[x[n] + \sum_{k=1}^{p} a_k x[n-k] \right]^2 \tag{4.38}$$

The error is zero for $n \geq N+p$ and $x[n]$ is zero between the range N to $N+p-1$. However, $x[n]$ is still being approximated by a linear combination of terms $x[n-k]$, some of which are nonzero. Unless all the $\{a_k\}$ are set to zero there is no way to make the error equal to zero over this entire range. The result of this is a type of edge effect that introduces a bias in the coefficients $\{a_k\}$.

For the special case of all-pole modelling, i.e., $q = 0$, an important simplification is possible. The general finite signal length all-pole problem is now defined by Equation 4.39 or matrix Equation 4.39a, where the signal length is N.

$$\begin{bmatrix} x[0] & 0 & 0 & .. & 0 \\ x[1] & x[0] & 0 & .. & 0 \\ x[2] & x[1] & x[0] & \ddots & 0 \\ \vdots & \vdots & \vdots & \ddots & x[0] \\ x[N-1] & \vdots & \vdots & \ddots & x[1] \\ 0 & \vdots & \vdots & \ddots & x[2] \\ \vdots & \vdots & \vdots & \ddots & \vdots \\ 0 & 0 & 0 & 0 & x[N-1] \end{bmatrix} \begin{bmatrix} 1 \\ a_1 \\ a_2 \\ \vdots \\ a_p \end{bmatrix} = \begin{bmatrix} b_0 \\ 0 \\ \vdots \\ 0 \end{bmatrix}, \quad \text{for } N > p \tag{4.39}$$

$$\mathbf{Xa} = \begin{bmatrix} b_0 \\ \mathbf{0} \end{bmatrix}$$

$$\begin{bmatrix} \mathbf{X}_1 \\ \mathbf{X}_2 \end{bmatrix} \begin{bmatrix} 1 \\ \tilde{\mathbf{a}} \end{bmatrix} = \begin{bmatrix} b_0 \\ \mathbf{0} \end{bmatrix} \tag{4.39a}$$

where:

$\mathbf{X}_1$ is the first row of $\mathbf{X}$.

$\mathbf{X}_2$ is the bottom part of $\mathbf{X}$.

In this case, $r_x(i, j)$ in the correlation Equation 4.24 is only a function of the difference $(i - j)$ as can be seen by writing out the $(i+1, j+1)$st entry of Equation 4.24 as defined by Equation 4.40.

$$r_x(i+1, j+1) = \sum_{m=q+1}^{N+p-1} x[m-i-1]x^*[m-j-1]$$

$$= x[q-i]x^*[q-j] + r_x(i, j) \tag{4.40}$$

$$= r_x(i, j) = r_x(i-j), \text{ except for } i \leq 0 \text{ and } j \leq 0$$

The result is more tidy if it is also applied for $i = 0$ and $j = 0$. Fortunately this can easily be achieved by defining a constant error term $e^2[0] = x^2[0]$, in which case the sums in Equations 4.38 and 4.40 start with 0 instead of 1 and the new error norm is $\varepsilon_p = \sum_{n=0}^{N+p-1} e^2[n]$. The result is defined by Equation 4.41.

$$r_x(i+1, j+1) = r_x(i, j) \stackrel{\text{def}}{=} r_x(i-j), \text{ for } i, j = 0,1,2,..., p \qquad (4.41)$$

Therefore, the corresponding $(p+1) \times (p+1)$ autocorrelation matrix, $\mathbf{R}_x = \mathbf{X}_2^T \mathbf{X}_2$, has equal valued entries. This matrix is called Hermitian Toeplitz and is defined by Equation 4.42.

$$\mathbf{R}_x = \begin{bmatrix} r_x(0) & r_x^*(1) & .. & r_x^*(p) \\ r_x(1) & r_x(0) & .. & r_x^*(p-1) \\ : & : & : & : \\ r_x(p) & r_x(p-1) & : & r_x(0) \end{bmatrix} \qquad (4.42)$$

where:

$$r_x(m) = r_x^*(-m)$$

The so called autocorrelation method comes from the fact that the entries in Equation 4.42 are simply the first $p+1$ autocorrelation coefficients of the signal $x[n]$. Because of the Toeplitz structure the AutoCorrelation Normal Equations (ACNE), $\mathbf{R}_x \mathbf{a} = \begin{bmatrix} \varepsilon_p \\ \mathbf{0} \end{bmatrix}$, can be solved simply by using the Levinson-Durbin method. $\mathbf{R}_x$ is guaranteed to be invertible because the columns of $\mathbf{X}_2$ are always linearly independent for the all-pole case.

Strictly, the ACNE solution for the coefficients $\{a_k\}$ is incorrect because the implicit rectangular window placed over the signal $x[n]$ from $n = 0$ to $n = N-1$ distorts the true autocorrelation estimates. This error occurs whether the signal is deterministic or stochastic. Using other standard windows with tapered ends can help improve the quality of the model as is also the case for spectral estimation.

A summary of the all-pole autocorrelation modelling method equations is as follows,

1.   The all-pole autocorrelation normal equations are,

$$\sum_{m=1}^{p} a_m r_x(k-m) = -r_x(k); \quad k = 1,2,....., p$$

$$r_x(k) = \sum_{n=k}^{N} x[n]x^*[n-k]; \quad k \geq 0$$

2.   The minimum error is,

$$\{\varepsilon_p\}_{\min} = r_x(0) + \sum_{k=1}^{p} a_k r_x^*(k)$$

The covariance method, however, is able to compute a correct model from a finite segment of signal data. In the covariance method Equations 4.18 and 4.21 are restricted to a finite set based on the finite signal length for $n = 0$ to $n = N$-1. The matrix $\mathbf{X}_2$ now becomes the matrix defined by Equation 4.43.

$$\mathbf{X}_2 = \begin{bmatrix} x[q+1) & x[q] & x[q-1] & .. & x[q-p+1] \\ x[q+2] & x[q+1] & x[q] & .. & x[q-p+2] \\ \vdots & \ddots & \ddots & \ddots & \vdots \\ x[N-1] & x[N-2] & x[N-3] & .. & x[N-p-1] \end{bmatrix} \qquad (4.43)$$

Stated another way, the minimisation of the squared error norm is restricted to the range [$q+1$, N-1] and so the error $\varepsilon_p^c = \sum_{n=q+1}^{N-1} e^2[n]$ is never evaluated outside the finite range of the signal data. The main disadvantage of the covariance method is that the normal equations do not have the Toeplitz matrix structure, even in the all-pole case, which means that the guaranteed stability in the all-pole model is lost. The covariance and Prony's method equations are identical except that the computation of the autocorrelation sequence, $r_x(i, j)$ of Equation 4.24 (matrix $\mathbf{X}_2^T\mathbf{X}_2$), is modified as defined by Equation 4.44.

$$r_x(i, j) = \sum_{m=q+1}^{N-1} x[m-i]x^*[m-j] = r_x^*(j,i), \quad i, j = 0,1,2,.., p \qquad (4.44)$$

The fast Levinson-Durbin recursion cannot be used to solve the all-pole ($q = 0$) covariance method normal equations but a fast covariance algorithm is available, which is faster than Gaussian elimination.

A summary of the all-pole covariance modelling method equations is as follows,

1.   The all-pole covariance normal equations are,

$$\sum_{m=1}^{p} a_m r_x(k,m) = -r_x(k,0); \quad k = 1,2,....,p$$

$$r_x(k,m) = \sum_{n=p}^{N} x[n-m]x^*[n-k]; \quad k,m \geq 0$$

2.   The minimum error is,

$$\{\varepsilon_p^C\}_{min} = r_x(0,0) + \sum_{k=1}^{p} a_k r_x(0,k)$$

4.3 Stochastic Signal Modelling

Stochastic signals are different from deterministic signals in that they must be described statistically and the values of $x[n]$ are only known in a probabilistic sense. Since the values of $x[n]$ are unknown until they arise it is no longer possible to minimise the deterministic squared error as before. Also, the input signal to the signal model can no longer be a unit sample, rather it must be a random process. To model a wide sense stationary random process it is possible to use an ARMA model, defined by Equation 4.1, driven by white noise. It is possible to use a similar method for stochastic signal modelling as was done for the all-pole modelling of deterministic signals. Other methods are also available for the all-pole model.

4.3.1 Autoregressive Moving Average Models

A random process $x[n]$ may be modelled as an ARMA process by using a unit variance white noise source $v[n]$ as the input to the ARMA system as shown in Figure 4.4.

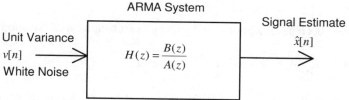

Figure 4.4. ARMA Random Signal Modelling

Generally, if the power spectrum of $v[n]$ is $P_v(z) = \sigma^2$, then the power spectrum of the order (p, q) ARMA system output $P_x(z)$ is defined by Equation 4.45.

$$P_x(z) = \sigma_v^2 \frac{B(z)B^*(\frac{1}{z^*})}{A(z)A^*(\frac{1}{z^*})},$$

$$P_x(e^{j\omega}) = \sigma_v^2 \frac{\left|B(e^{j\omega})\right|^2}{\left|A(e^{j\omega})\right|^2}, \quad \text{in terms of } \omega.$$

(4.45)

The power spectrum has $2p$ poles and $2q$ zeros with reciprocal symmetry. The variables $x[n]$ and $v[n]$ are related by the system difference equations as defined by Equation 4.46.

$$x[n] + \sum_{m=1}^{p} a_m x[n-m] = \sum_{m=0}^{q} b_m v[n-m] \tag{4.46}$$

If Equation 4.46 is multiplied by $x*[n-k]$ and expectations are taken then another equation similar to Equation 4.46 is formed, which relates the autocorrelation of $x[n]$ and the cross-correlation between $x[n]$ and $v[n]$, as defined by Equation 4.47.

$$r_x(k) + \sum_{m=1}^{p} a_m r_x(k-m) = \sum_{m=0}^{q} b_m r_{vx}(k-m) \tag{4.47}$$

The cross-correlation term $r_{vx}(k-m)$ needs to be broken into a function of the autocorrelation function $r_x(k)$ and the impulse response of the filter $h[k]$. This is done by reducing the cross-correlation term down to $r_{vx}(k-m) = E\{v[n-m]x*[n-k]\} = \sigma_v^2 h*[m-k]$ as detailed in Equation 4.48, given, $x[n] = h[n]*[n] = \sum_{j=-\infty}^{\infty} v[j]h[n-j]$.

$$
\begin{aligned}
E\{v[n-m]x^*[n-k]\} &= E\left\{ \sum_{j=-\infty}^{\infty} v[n-m]v^*[j]h^*[n-k-j] \right\} \\
&= \sum_{j=-\infty}^{\infty} E\{v[n-m]v^*[j]\}h^*[n-k-j] \\
&= \sigma_v^2 h^*[m-k]
\end{aligned}
\tag{4.48}
$$

Equation 4.47 becomes Equation 4.49 by using the substitution of Equation 4.48.

$$r_x(k) + \sum_{m=1}^{p} a_m r_x(k-m) = \sigma_v^2 \sum_{m=0}^{q} b_m h[m-k] \tag{4.49}$$

If it is assumed that $h[n]$ is causal the sum on the right side of Equation 4.49, denoted by $c(k)$, can be defined by Equation 4.50.

$$c(k) = \sum_{m=k}^{q} b_m h^*[m-k] = \sum_{m=0}^{q-k} b_{m+k} h^*[m] \tag{4.50}$$

where:
$c(k) = 0$ for $k > q$.

Equation 4.49 can now be written as Equation 4.51 for $k \geq 0$, which represents the so called *Yule-Walker* equations. These equations provide a relationship between the filter coefficients and the autocorrelation sequence.

$$r_x(k) + \sum_{m=1}^{p} a_m r_x(k-m) = \begin{cases} \sigma_v^2 c(k), 0 \le k \le q \\ 0, \qquad k > q \end{cases} \qquad (4.51)$$

In matrix form the *Yule-Walker* equations are defined by Equations 4.52 and 4.52a.

$$\mathbf{R}_x \mathbf{a} = \sigma_v^2 \begin{bmatrix} \mathbf{c} \\ \mathbf{0} \end{bmatrix} \qquad (4.52)$$

$$\begin{bmatrix} r_x(0) & r_x(-1) & .. & r_x(-p) \\ r_x(1) & r_x(0) & .. & r_x(-p+1) \\ \vdots & \vdots & \ddots & \vdots \\ r_x(q) & r_x(q-1) & .. & r_x(q-p) \\ r_x(q+1) & r_x(q) & .. & r_x(q-p+1) \\ \vdots & \vdots & \ddots & \vdots \\ r_x(q+p) & r_x(q+p-1) & .. & r_x(q) \end{bmatrix} \begin{bmatrix} 1 \\ a_1 \\ \vdots \\ a_p \end{bmatrix} = \sigma_v^2 \begin{bmatrix} c(0) \\ c(1) \\ \vdots \\ c(q) \\ 0 \\ \vdots \\ 0 \end{bmatrix} \qquad (4.52a)$$

Equation 4.51 defines a recursion for the autocorrelation sequence in terms of the filter coefficients, therefore it can be used to extrapolate the autocorrelation sequence from a finite set of values of $r_x(k)$ for $k \ge p$ where $p \ge q$. If $h[n]$ is causal the coefficients $\{a_k\}$ can be solved using the *Yule-Walker* Equation 4.52 for $k > q$, as defined by the modified *Yule-Walker* Equation 4.53.

$$\begin{bmatrix} r_x(q) & r_x(q-1) & .. & r_x(q-p+1) \\ r_x(q+1) & r_x(q) & .. & r_x(q-p+2) \\ \vdots & \vdots & \ddots & \vdots \\ r_x(q+p-1) & r_x(q+p-2) & .. & r_x(q) \end{bmatrix} \begin{bmatrix} a_1 \\ a_2 \\ \vdots \\ a_p \end{bmatrix} = - \begin{bmatrix} r_x(q+1) \\ r_x(q+2) \\ \vdots \\ r_x(q+p) \end{bmatrix} \qquad (4.53)$$

If the autocorrelations in this modified *Yule-Walker* Equation 4.53 are unknown then they may be replaced with estimated values using a sample signal realisation. The form of the *Yule-Walker* equations is the same as the Padé equations and the solutions go the same way. Therefore, once the $\{a_k\}$ coefficients are solved the $\{b_k\}$ are found by substituting into the top part of Equation 4.52, i.e., $\mathbf{R}_x \mathbf{a} = \sigma_v^2 \mathbf{c}$, where $\sigma_v^2 = 1$.

4.3.2 Autoregressive Models

A wide sense autoregressive process of order p is a special case of the ARMA process for which $q = 0$, i.e., the all-pole or AR model defined by Equation 4.54.

$$H(z) = \frac{b_0}{A(z)} = \frac{b_0}{1 + \sum_{k=1}^{p} a_k z^{-k}} \qquad (4.54)$$

The autocorrelation sequence of an AR process satisfies the *Yule-Walker* equations as defined by Equation 4.55.

$$r_x(k) + \sum_{m=1}^{p} a_m r_x(k-m) = |b_0|^2 \delta(k), \quad k \geq 0 \tag{4.55}$$

In matrix form Equation 4.55 for $k = 1,2,..., p$ is defined by Equation 4.56 if the conjugate symmetry of $r_x(k)$ is used.

$$\begin{bmatrix} r_x(0) & r_x^*(1) & .. & r_x^*(p-1) \\ r_x(1) & r_x(0) & .. & r_x^*(p-2) \\ : & : & \ddots & : \\ r_x(p-1) & r_x(p-2) & .. & r_x(0) \end{bmatrix} \begin{bmatrix} a_1 \\ a_2 \\ : \\ a_p \end{bmatrix} = - \begin{bmatrix} r_x(1) \\ r_x(2) \\ : \\ r_x(p) \end{bmatrix} \tag{4.56}$$

Equation 4.56 is the same as the normal equations for all-pole modelling of a deterministic signal using Prony's method, Equation 4.29. The only difference between them is how the autocorrelation $r_x(k)$ is defined. In Equation 4.29 $r_x(k)$ is a deterministic autocorrelation and in Equation 4.56 it is a statistical autocorrelation. Equation 4.56 is solved via the so called *Yule-Walker* method given the autocorrelations $r_x(k)$ for $k = 1, 2,..., p$. The coefficient b_0 is solved by taking $k = 0$ in Equation 4.55 as defined by Equation 4.57.

$$r_x(0) + \sum_{m=1}^{p} a_m r_x(m) = |b_0|^2 \tag{4.57}$$

In most situations the statistical autocorrelation is unknown and therefore must be estimated from a finite sample realisation of the process, i.e., $x[n]$, for $0 \leq n < N$. The required sample estimate for $r_x(k)$ is defined by Equation 4.58.

$$\hat{r}_x(k) = \frac{1}{N} \sum_{n=1}^{N-1} x[n] x^*[n-k] \tag{4.58}$$

However, once the autocorrelation must be estimated from a sample the method is actually the autocorrelation method and it can be said that the two methods, the deterministic and stochastic all-pole signal modelling methods, become equivalent.

4.3.3 Moving Average Models

A wide sense moving average process of order q is a special case of the ARMA process for which $p = 0$, i.e., the all-zero or MA model (FIR filter) defined by Equation 4.59.

$$H(z) = B(z) = \sum_{k=0}^{q} b_k z^{-k} \tag{4.59}$$

A MA model for a process $x[n]$ may be developed using Durbin's method. Durbin's method firstly finds a high pth order all-pole model $A(z)$ for the MA model of

Equation 4.59. The resulting coefficients of the all-pole model a_k become a new data set for which another qth order all-pole model is found, which is then defined as the all-zero model of Equation 4.59.

This is achieved by first letting $x[n]$ be a qth order MA process defined by Equation 4.60 where $w[n]$ is white noise.

$$x[n] = \sum_{k=0}^{q} b_k w[n-k] \tag{4.60}$$

The pth order all-pole model for $x[n]$ can be defined by Equation 4.61 if p is made large enough.

$$B(z) \approx \frac{1}{A(z)} = \frac{1}{a_0 + \sum_{k=1}^{p} a_k z^{-k}} \tag{4.61}$$

The required MA coefficients b_k can be estimated by finding the qth order all-pole model of $A(z)$. This works because the inverse of Equation 4.61 is Equation 4.62 and the coefficients for $A(z)$ can be taken as the coefficients of the MA model.

$$A(z) \approx \frac{1}{B(z)} = \frac{1}{b_0 + \sum_{k=1}^{q} b_k z^{-k}} \tag{4.62}$$

Durbin's method can be performed in two steps as follows,

1. Given $x[n]$ for $n = 0,1,..., N\text{-}1$ or $r_x(k)$ for $k = 0,1,..., N\text{-}1$ find a pth order all-pole model and normalise the coefficients according to Equation 4.61, i.e., divide by the gain term. Typically the all-pole model order p is chosen to be at least four times the order q of the MA process.

2. Using the p coefficients of the all-pole model of Step 1 as data find a qth order all-pole model. The resulting coefficients after normalisation, according to Equation 4.62, become the estimated coefficients of the MA process.

4.4 The Levinson-Durbin Recursion and Lattice Filters

In the preceding Sections there were a number signal modelling problems that involved the solution of a set of linear equations of the form defined by Equation 4.63, where $\mathbf{R}_x$ is a Toeplitz matrix.

$$\mathbf{R}_x \mathbf{a}_p = \mathbf{b} \tag{4.63}$$

In the Padé approximation method and the *Yule-Walker* equations $\mathbf{R}_x$ is a non-symmetric Toeplitz matrix. Hermitian Toeplitz equations occur in the all-pole modelling of deterministic and stochastic signals. Toeplitz equations will also be encountered in a later Chapter in relation to the design of FIR Wiener filters.

The Levinson-Durbin recursion is one important efficient solution to a special form of Equation 4.63. The form of this solution leads to a number of interesting results including the so called lattice filter structure and the Cholesky decomposition of a Toeplitz matrix, and an important procedure for the recursive computation of the Toeplitz matrix inverse. In 1947 Levinson developed a recursive algorithm for solving a set of linear symmetric Toeplitz equations generally defined by Equation 4.63. Then, in 1961 Durbin improved the Levinson recursion for the special case in which the vector **b** is a unit vector. This recursion is known as the Levinson-Durbin recursion. The number of divisions and multiplications that it requires is proportional to p^2 for a set of p equations in p unknowns. This compares very favourably with the Gaussian elimination technique, which requires divisions and multiplications proportional to p^3. The Levinson-Durbin recursion also requires less data storage than Gaussian elimination. Although the Levinson-Durbin recursion is very efficient in solving the normal equations, in modelling problems the required computation of the autocorrelation sequence will dominate the computational cost.

4.4.1 The Levinson-Durbin Recursion Development

The pth order all-pole normal equations for Prony's method or the autocorrelation method were defined by Equation 4.29 previously. The model error was also defined previously by Equation 4.30. Combining these two equations results in Equation 4.64, which represents $p+1$ linear equations in $p+1$ unknowns $a_1, a_2,...,a_p$ plus ε_p.

$$\begin{bmatrix} r_x(0) & r_x^*(1) & r_x^*(2) & .. & r_x^*(p) \\ r_x(1) & r_x(0) & r_x^*(1) & .. & r_x^*(p-1) \\ r_x(2) & r_x(1) & r_x(0) & .. & r_x^*(p-2) \\ : & : & : & \ddots & : \\ r_x(p) & r_x(p-1) & r_x(p-2) & .. & r_x(0) \end{bmatrix} \begin{bmatrix} 1 \\ a_1 \\ a_2 \\ : \\ a_p \end{bmatrix} = \varepsilon_p \begin{bmatrix} 1 \\ 0 \\ 0 \\ : \\ 0 \end{bmatrix} \quad (4.64)$$

$$\mathbf{R}_p \mathbf{a}_p = \varepsilon_p \mathbf{u}_1$$

The autocorrelation matrix $\mathbf{R}_p$ in Equation 4.64 is a $(p+1) \times (p+1)$ Hermitian Toeplitz matrix. The Levinson-Durbin recursion that will solve Equation 4.64 is a recursion in the model order. This means that the coefficients of the $(j+1)$st order all-pole model $\mathbf{a}_{j+1}$ are found from the coefficients of the jth order model $\mathbf{a}_j$. In other words, given the jth order solution, $\mathbf{R}_j \mathbf{a}_j = \varepsilon_j \mathbf{u}_1$, it is possible to derive the $(j+1)$st order solution, $\mathbf{R}_{j+1}\mathbf{a}_{j+1}=\varepsilon_{j+1}\mathbf{u}_1$. This is done by first appending a zero to the vector $\mathbf{a}_j$ and multiplying the resulting vector by $\mathbf{R}_{j+1}$ as defined by Equation 4.65.

$$
\begin{bmatrix}
r_x(0) & r_x^*(1) & r_x^*(2) & .. & r_x^*(j+1) \\
r_x(1) & r_x(0) & r_x^*(1) & .. & r_x^*(j) \\
r_x(2) & r_x(1) & r_x(0) & .. & r_x^*(j-1) \\
: & : & : & \ddots & : \\
r_x(j+1) & r_x(j) & r_x(j-1) & .. & r_x(0)
\end{bmatrix}
\begin{bmatrix}
1 \\ a_1 \\ : \\ a_j \\ 0
\end{bmatrix}
=
\begin{bmatrix}
\varepsilon_j \\ 0 \\ 0 \\ : \\ \gamma_j
\end{bmatrix}
\tag{4.65}
$$

The consequential parameter γ_j is defined by Equation 4.66.

$$
\gamma_j = r_x(j+1) + \sum_{i=1}^{j} a_i r_x(j+1-i)
\tag{4.66}
$$

In the special case when the parameter $\gamma_j = 0$, then the right side of Equation 4.65 is a scaled unit vector and the vector $\mathbf{a}_{j+1} = [1, a_1,..., a_j, 0]^T$ is the $(j+1)$st order solution. However, in general $\gamma_j \neq 0$ and $[1, a_1,...,a_j,0]^T$ is not the solution. It is possible to proceed by noting that the Toeplitz property of $\mathbf{R}_{j+1}$ allows Equation 4.65 to be rewritten equivalently as Equation 4.67.

$$
\begin{bmatrix}
r_x(0) & r_x(1) & r_x(2) & .. & r_x(j+1) \\
r_x^*(1) & r_x(0) & r_x(1) & .. & r_x(j) \\
r_x^*(2) & r_x^*(1) & r_x(0) & .. & r_x(j-1) \\
: & : & : & \ddots & : \\
r_x^*(j+1) & r_x^*(j) & r_x^*(j-1) & .. & r_x(0)
\end{bmatrix}
\begin{bmatrix}
0 \\ a_j \\ : \\ a_1 \\ 1
\end{bmatrix}
=
\begin{bmatrix}
\gamma_j \\ 0 \\ 0 \\ : \\ \varepsilon_j
\end{bmatrix}
\tag{4.67}
$$

If the complex conjugate is taken of Equation 4.67 and combined with Equation 4.65, the result is Equation 4.69 for any complex constant Γ_{j+1},

$$
\mathbf{R}_{j+1}\left\{
\begin{bmatrix} 1 \\ a_1 \\ : \\ a_j \\ 0 \end{bmatrix}
+ \Gamma_{j+1}
\begin{bmatrix} 0 \\ a_j^* \\ : \\ a_1^* \\ 1 \end{bmatrix}
\right\}
=
\begin{bmatrix} \varepsilon_j \\ 0 \\ 0 \\ : \\ \gamma_j \end{bmatrix}
+ \Gamma_{j+1}
\begin{bmatrix} \gamma_j^* \\ 0 \\ 0 \\ : \\ \varepsilon_j^* \end{bmatrix}
\tag{4.68}
$$

If $\Gamma_{j+1} = -\dfrac{\gamma_j}{\varepsilon_j^*}$ is set in Equation 4.68 then the equation becomes

$\mathbf{R}_{j+1}\mathbf{a}_{j+1} = \varepsilon_{j+1}\mathbf{u}_1$, where $\mathbf{a}_{j+1}$ is defined by Equation 4.69, and ε_{j+1} is defined by Equation 4.70.

$$\mathbf{a}_{j+1} = \begin{bmatrix} 1 \\ a_1 \\ : \\ a_j \\ 0 \end{bmatrix} + \Gamma_{j+1} \begin{bmatrix} 0 \\ a_j^* \\ : \\ a_1^* \\ 1 \end{bmatrix} \tag{4.69}$$

$$\varepsilon_{j+1} = \varepsilon_j + \Gamma_{j+1}\gamma_j^* = \varepsilon_j[1 - |\Gamma_{j+1}|^2] \tag{4.70}$$

If $a_0 = 1$ and $a_{j+1} = 0$ are set then Equation 4.69, the Levinson order-update equation, may be expressed as Equation 4.71.

$$a_{i(j+1)} = a_{i(j)} + \Gamma_{j+1}a_{j-i+1(j)}^*, \quad \text{for} \quad i = 0,1,..., j+1 \tag{4.71}$$

where:
The $(j+1)$ subscript signifies the $(j+1)$st model coefficients.

The (j) subscripts signify the jth model coefficients.

To initialise the recursion the values for the order $j = 0$ model are set to $a_0 = 1$ and $\varepsilon_0 = r_x(0)$.

A summary of the Levinson-Durbin recursion is as follows,

1. Initialise the recursion by setting,
   i. $a_0 = 1$
   ii. $\varepsilon_0 = r_x(0)$

2. For $j = 0, 1,..., p-1$,

   i. $\gamma_j = r_x(j+1) + \sum_{i=1}^{j} a_{i(j)}r_x(j-i+1)$

   ii. $\Gamma_{j+1} = -\dfrac{\gamma_j}{\varepsilon_j^*}$

   iii. For $i = 1,...,j$, $\quad a_{i(j+1)} = a_{i(j)} + \Gamma_{j+1}a_{j-i+1(j)}^*$

   iv. $a_{j+1(j+1)} = \Gamma_{j+1}$

   v. $\varepsilon_{j+1} = \varepsilon_j[1 - |\Gamma_{j+1}|^2]$

3. $b_0 = \sqrt{\varepsilon_p}$

4.4.1.1 Example of the Levinson-Durbin Recursion
Solve the autocorrelation normal equations and find a third-order all-pole model for
a signal having the following autocorrelation values,

$$r_x(0) = 1, r_x(1) = 0.25, r_x(2) = 0.25, r_x(3) = 0.19.$$

The normal equations for the third-order all-pole model are,

$$\begin{bmatrix} r_x(0) & r_x^*(1) & r_x^*(2) \\ r_x(1) & r_x(0) & r_x^*(1) \\ r_x(2) & r_x(1) & r_x(0) \end{bmatrix} \begin{bmatrix} a_1 \\ a_2 \\ a_3 \end{bmatrix} = -\begin{bmatrix} r_x(1) \\ r_x(2) \\ r_x(3) \end{bmatrix}$$

$$\begin{bmatrix} 1 & 0.25 & 0.25 \\ 0.25 & 1 & 0.25 \\ 0.25 & 0.25 & 1 \end{bmatrix} \begin{bmatrix} a_1 \\ a_2 \\ a_3 \end{bmatrix} = -\begin{bmatrix} 0.25 \\ 0.25 \\ 0.20 \end{bmatrix}$$

Using the Levinson-Durbin recursion the models from first order to third-order can
be defined as follows. First-order model,

$$\gamma_0 = r_x(1) = 0.25$$

$$\Gamma_1 = -\frac{\gamma_0}{\varepsilon_0} = -\frac{r_x(1)}{r_x(0)} = -0.25$$

$$\varepsilon_1 = r_x(0)[1 - |\Gamma_1|^2] = 0.9375$$

$$\mathbf{a}_1 = \begin{bmatrix} 1 \\ a_1 \end{bmatrix} = \begin{bmatrix} 1 \\ \Gamma_1 \end{bmatrix} = \begin{bmatrix} 1 \\ -0.25 \end{bmatrix}$$

Second-order model,

$$\gamma_1 = r_x(2) + a_{1(1)} r_x(1) = 0.1875$$

$$\Gamma_2 = -\frac{\gamma_1}{\varepsilon_1} = -0.2$$

$$\varepsilon_2 = \varepsilon_1[1 - |\Gamma_2|^2] = 0.9$$

$$\mathbf{a}_2 = \begin{bmatrix} 1 \\ -0.25 \\ 0 \end{bmatrix} + \Gamma_2 \begin{bmatrix} 0 \\ -0.25 \\ 1 \end{bmatrix} = \begin{bmatrix} 1 \\ -0.2 \\ -0.2 \end{bmatrix}$$

Third-order model,

$$\gamma_2 = r_x(3) + a_{1(2)}r_x(2) + a_{2(2)}r_x(1) = 0.09$$

$$\Gamma_3 = -\frac{\gamma_2}{\varepsilon_2} = -0.1$$

$$\varepsilon_3 = \varepsilon_2[1 - |\Gamma_3|^2] = 0.891$$

$$\mathbf{a}_3 = \begin{bmatrix} 1 \\ -0.2 \\ -0.2 \\ 0 \end{bmatrix} + \Gamma_3 \begin{bmatrix} 0 \\ -0.2 \\ -0.2 \\ 1 \end{bmatrix} = \begin{bmatrix} 1 \\ -0.18 \\ -0.18 \\ -0.1 \end{bmatrix}, \quad b_0 = \sqrt{\varepsilon_3} = 0.9439279$$

The third-order all-pole model for the signal $x[n]$ becomes,

$$H_3(z) = \frac{b_0}{1 + a_1 z^{-1} + a_2 z^{-2} + a_3 z^{-3}}$$

$$= \frac{0.9439279}{1 - 0.18z^{-1} - 0.18z^{-2} - 0.1z^{-3}}$$

Equation 4.66 can be used to compute the values for $r_x(k)$ for $k > 3$ if it is noted that $\Gamma_k = 0$, $\gamma_k = 0$, for $k > 3$. The relevant equation derivation is then defined by,

$$r_x(k) = -\sum_{i=1}^{3} a_{i(3)}r_x(k-i)$$

4.4.2 The Lattice Filter

The lattice digital filter structure is derived from the Levinson-Durbin recursion. Lattice filters have a number of significant properties including a modular structure, low sensitivity to parameter quantization effects, and a simple method to ensure minimum phase and stability. There are many possible lattice filter structures including all-pole and pole-zero lattice filters but only the all-zero lattice filter for FIR digital filters will be derived from the Levinson-Durbin recursion in this Section.

The lattice filter derivation begins with the Levinson order-update Equation 4.71. In the derivation of the Levinson-Durbin recursion a reciprocal vector, $\mathbf{a}_j^R$ to vector $\mathbf{a}_j$ was implied but not defined. These vectors are now defined by Equations 4.72.

$$\mathbf{a}_j = \begin{bmatrix} 1 \\ a_1 \\ \vdots \\ a_j \\ 0 \end{bmatrix}, \quad \mathbf{a}_j^R = \begin{bmatrix} 0 \\ a_j^* \\ \vdots \\ a_1^* \\ 1 \end{bmatrix} \tag{4.72}$$

$$a_{i(j)}^R = a_{j-i(j)}^*, \quad \text{for } i = 0,1,..., j$$

The reciprocal vector $\mathbf{a}_j^R$ is created by reversing the order of the elements in vector $\mathbf{a}_j$ and taking the complex conjugate. By substituting the reciprocal vector notation into Equation 4.71 the Levinson order-update becomes Equation 4.73.

$$a_{i(j+1)} = a_{i(j)} + \Gamma_{j+1} a_{i-1(j)}^R, \quad \text{for } i = 0,1,..., j+1 \tag{4.73}$$

where:

The $(j+1)$ subscript signifies the $(j+1)$st model coefficients.

The (j) subscripts signify the jth model coefficients.

Taking the z-Transform of Equations 4.72 and 4.73 results in Equations 4.74 and 4.75 respectively.

$$A_j^R(z) = z^{-1} A_j^*(\frac{1}{z^*}) \tag{4.74}$$

where:

$A_j(z)$ is the z $-$ Transform of $a_{i(j)}$.

$A_j^R(z)$ is the z $-$ Transform of $a_{i(j)}^R$.

$$A_{j+1}(z) = A_j(z) + \Gamma_{j+1} z^{-1} A_j^R(z) \tag{4.75}$$

Equation 4.75 is the order update for $A_j(z)$. To derive the order update for $a_{i(j+1)}^R$ and $A_{j+1}^R(z)$ take the complex conjugate of Equation 4.73 and replace i with $j - i + 1$ to produce Equation 4.76.

$$a_{j-i+1(j+1)}^* = a_{j-i+1(j)}^* + \Gamma_{j+1}^* a_{i(j)} \tag{4.76}$$

If Equation 4.74 is substituted into $a_{i(j+1)}^R = a_{i-1(j)}^R + \Gamma_{j+1}^* a_{i(j)}$ the z-Transform is,

$$A_{j+1}^R(z) = z^{-1} A_j^R(z) + \Gamma_{j+1}^* A_j(z).$$

In summary, it is possible to express the pair of coupled order update equations as difference and z-domain equations as follows,

$$a_{n(j+1)} = a_{n(j)} + \Gamma_{j+1} a^R_{n-1(j)}$$
$$a^R_{n(j+1)} = a^R_{n-1(j)} + \Gamma^*_{j+1} a_{n(j)}$$
$$A_{j+1}(z) = A_j(z) + \Gamma_{j+1} z^{-1} A^R_j(z)$$
$$A^R_{j+1}(z) = z^{-1} A^R_j(z) + \Gamma^*_{j+1} A_j(z)$$

These order update equations can be written in a matrix form as defined by Equation 4.77 and consequently can be implemented as a two-port network as shown in Figure 4.5.

$$\begin{bmatrix} A_{j+1}(z) \\ A^R_{j+1}(z) \end{bmatrix} = \begin{bmatrix} 1 & \Gamma_{j+1} z^{-1} \\ \Gamma^*_{j+1} & z^{-1} \end{bmatrix} \begin{bmatrix} A_j(z) \\ A^R_j(z) \end{bmatrix} \qquad (4.77)$$

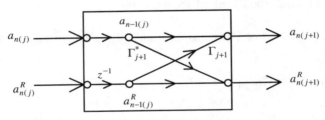

Figure 4.5. Two-port Network for the Order Update Equations

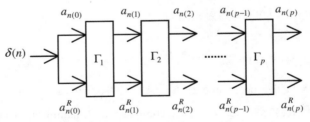

Figure 4.6. pth Order FIR Lattice Filter

The two-port network of Figure 4.5 is the basic module that is used to implement the FIR lattice filter. A pth order lattice filter can be formed by cascading p lattice filter modules with reflection coefficients Γ_1, Γ_2,..., Γ_p as shown in Figure 4.6. With an impulse input of $\delta[n]$ the output at $a_{n(p)}$ is the impulse response of the system function $A_p(z)$ and the output at $a_{n(p)}$ is the impulse response of the system function $A^R_p(z)$. Therefore, this filter is an all-zero lattice filter for a FIR digital filter represented by the z-Transform $A_p(z)$. In this signal modelling context this filter can be interpreted as the forward prediction error filter. If the filter is fed by the input signal $x[n]$ it gives the error between the signal and the signal estimate, i.e., $e_p[n] = x[n] - \hat{x}[n]$.

One important advantage of a lattice filter structure over a direct form filter is the modularity. The order of the filter can be increased or decreased in a linear prediction or all-pole signal modelling application without having to recompute the reflection coefficients. Another advantage is that the lattice filter will be minimum phase filter if and only if the reflection coefficient magnitude is bounded by one, i.e., $\left| \Gamma_j \right| < 1$. For IIR lattice filters this constraint also ensures stability, which is crucial in an adaptive filtering situation, where the reflection coefficients are changing with time.

4.4.3 The Cholesky Decomposition

The Levinson-Durbin recursion may be used to perform the Cholesky decomposition of the Hermitian Toeplitz autocorrelation matrix, $\mathbf{R}_p$. The Cholesky (LDU) decomposition is useful because it allows the derivation of a closed form expression for the inverse of the autocorrelation matrix as well a recursive algorithm for inverting the Toeplitz matrix. It also allows the establishment of the equivalence between the positive definiteness of $\mathbf{R}_p$, the positivity of the error ε_j sequence, and the unit magnitude constraint on the reflection coefficients Γ_j.

The Cholesky decomposition of a Hermitian matrix $\mathbf{C}$ is a factorisation of the form,

$$\mathbf{C} = \mathbf{L}\mathbf{D}\mathbf{L}^H$$

where, $\mathbf{L}$ is a lower triangular matrix with ones along the diagonal and $\mathbf{D}$ is a diagonal matrix. If the terms of $\mathbf{D}$ are nonnegative, i.e., $\mathbf{C}$ is positive definite, then $\mathbf{D}$ may be split into a product of two matrices by simply taking square roots as follows,

$$\mathbf{D} = \mathbf{D}^{1/2}\mathbf{D}^{1/2}$$

To derive the Cholesky decomposition start with a $(p+1) \times (p+1)$ upper triangle matrix $\mathbf{A}_p$ formed from the vectors, $\mathbf{a}_0, \mathbf{a}_1, ..., \mathbf{a}_p$. These vectors are produced when the Levinson-Durbin recursion is applied to the autocorrelation sequence $r_x(0), r_x(1), ..., r_x(p)$, as defined by Equation 4.78.

$$\mathbf{A}_p = \begin{bmatrix} 1 & a^*_{1(1)} & a^*_{2(2)} & .. & a^*_{p(p)} \\ 0 & 1 & a^*_{1(2)} & .. & a^*_{p-1(p)} \\ \vdots & \ddots & \ddots & \ddots & \vdots \\ 0 & 0 & 0 & .. & 1 \end{bmatrix}$$ (4.78)

The jth column of $\mathbf{A}_p$ has the filter coefficients $\mathbf{a}^R_{j-1}$ padded with zeros. Since, from the Hermitian Toeplitz property of $\mathbf{R}_p$, $\mathbf{R}_j \mathbf{a}^R_j = \varepsilon_j \mathbf{u}_j$, where $\mathbf{u}_j = [0,0,...,1]^T$ is a unit vector of length $j+1$, then $\mathbf{R}_p \mathbf{A}_p$ is defined by Equation 4.79.

$$\mathbf{R}_p \mathbf{A}_p = \begin{bmatrix} \varepsilon_0 & 0 & 0 & .. & 0 \\ ? & \varepsilon_1 & 0 & .. & 0 \\ \vdots & \ddots & \ddots & \ddots & \vdots \\ ? & ? & ? & .. & \varepsilon_p \end{bmatrix}$$ (4.79)

Equation 4.79 is a lower triangle matrix with the prediction errors along the diagonal. If $\mathbf{J}\mathbf{R}^*_p\mathbf{J} = \mathbf{R}_p$ then Equation 4.80 can be defined.

$$\mathbf{R}_j \mathbf{a}_j = (\mathbf{J}\mathbf{R}^*_j\mathbf{J})\mathbf{a}_j = \varepsilon_j \mathbf{u}_j$$ (4.80)

where:

$$\mathbf{J}^2 = \mathbf{I} \text{ and } \mathbf{J}\mathbf{a}_j = (\mathbf{a}^R_j)^*$$

If $\mathbf{R}_p \mathbf{A}_p$ is multiplied on the left by the lower triangle matrix $\mathbf{A}^H_p$ the result is another lower triangle matrix $\mathbf{A}^H_p \mathbf{R}_p \mathbf{A}_p$. Because the terms on the diagonal of $\mathbf{A}_p$ are equal to one, the diagonal of $\mathbf{A}^H_p \mathbf{R}_p \mathbf{A}_p$ will be the same as that of $\mathbf{R}_p \mathbf{A}_p$ and the resulting Hermitian matrix $\mathbf{A}^H_p \mathbf{R}_p \mathbf{A}_p$ has the form defined by Equation 4.81.

$$\mathbf{A}^H_p \mathbf{R}_p \mathbf{A}_p = \begin{bmatrix} \varepsilon_0 & 0 & 0 & .. & 0 \\ 0 & \varepsilon_1 & 0 & .. & 0 \\ \vdots & \ddots & \ddots & \ddots & \vdots \\ 0 & 0 & 0 & .. & \varepsilon_p \end{bmatrix} = \mathbf{D}_p$$ (4.81)

where:

$$\mathbf{D} = \text{diag}\{\varepsilon_0, \varepsilon_1, ..., \varepsilon_p\}$$

The desired Cholesky factorisation, Equation 4.82, is achieved by multiplying Equation 4.81 on the left by $\mathbf{L}_p = (\mathbf{A}_p^H)^{-1}$ and on the right by $\mathbf{L}_p^H$, where $\mathbf{A}_p^H$ and $(\mathbf{A}_p^H)^{-1}$ are both lower triangle matrices and $\det(\mathbf{A}_p^H) = 1$, since $\mathbf{A}_p^H$ is nonsingular with ones along its diagonal.

$$\mathbf{R}_p = \mathbf{L}_p \mathbf{D}_p \mathbf{L}_p^H \tag{4.82}$$

The determinant of $\mathbf{R}_p = \prod\limits_{k=0}^{p} \varepsilon_k$, the product of the modelling errors.

The inverse of a Toeplitz matrix $\mathbf{R}_p$ is easily found by using the decomposition of Equation 4.81 and taking the inverse as defined by Equation 4.83.

$$(\mathbf{A}_p^H \mathbf{R}_p \mathbf{A}_p)^{-1} = \mathbf{A}_p^{-1} \mathbf{R}_p^{-1} \mathbf{A}_p^{-H} = \mathbf{D}_p^{-1} \tag{4.83}$$

If Equation 4.83 is multiplied by $\mathbf{A}_p$ on the left and by $\mathbf{A}_p^H$ on the right this gives the inverse as defined by Equation 4.84.

$$\mathbf{R}_p^{-1} = \mathbf{A}_p \mathbf{D}_p^{-1} \mathbf{A}_p^H \tag{4.84}$$

The matrix $\mathbf{A}_p$ is formed by applying the Levinson-Durbin recursion on the autocorrelation sequence $r_x(0), r_x(1), \ldots, r_x(p)$, and the inverse of $\mathbf{D}_p$ is easy to find since $\mathbf{D}_p$ is a diagonal matrix.

4.4.4 The Levinson Recursion

The Levinson-Durbin recursion allows an efficient solution to the all-pole normal equations in the form of Equation 4.64 and Equation 4.85. However, it is often necessary to solve a more general form of Toeplitz equations as defined by Equation 4.86 , where vector $\mathbf{b}$ is arbitrary.

$$\begin{bmatrix} r_x(0) & r_x^*(1) & r_x^*(2) & .. & r_x^*(j) \\ r_x(1) & r_x(0) & r_x^*(1) & .. & r_x^*(j-1) \\ r_x(2) & r_x(1) & r_x(0) & .. & r_x^*(j-2) \\ \vdots & \vdots & \vdots & \ddots & \vdots \\ r_x(j) & r_x(j-1) & r_x(j-2) & .. & r_x(0) \end{bmatrix} \begin{bmatrix} 1 \\ a_{1(j)} \\ a_{2(j)} \\ \vdots \\ a_{j(j)} \end{bmatrix} = \varepsilon_j \begin{bmatrix} 1 \\ 0 \\ 0 \\ \vdots \\ 0 \end{bmatrix} \tag{4.85}$$

$$\mathbf{R}_j \mathbf{a}_j = \varepsilon_j \mathbf{u}_1$$

$$\begin{bmatrix} r_x(0) & r_x^*(1) & r_x^*(2) & .. & r_x^*(j) \\ r_x(1) & r_x(0) & r_x^*(1) & .. & r_x^*(j-1) \\ r_x(2) & r_x(1) & r_x(0) & .. & r_x^*(j-2) \\ : & : & : & \ddots & : \\ r_x(j) & r_x(j-1) & r_x(j-2) & .. & r_x(0) \end{bmatrix} \begin{bmatrix} x_j[0] \\ x_j[1] \\ x_j[2] \\ : \\ x_j[j] \end{bmatrix} = \begin{bmatrix} b_0 \\ b_1 \\ b_2 \\ : \\ b_j \end{bmatrix} \tag{4.86}$$

$$\mathbf{R}_j \mathbf{x}_j = \mathbf{b}$$

Levinson's original recursion has the Levinson-Durbin recursion embedded in it and it can simultaneously solve Equations 4.85 and 4.86 for $j = 0,1,...,p$. In fact Levinson's recursion is very similar to the Levinson-Durbin recursion.

Levinson's recursion starts by finding the solution to Equations 4.85 and 4.86 for $j = 0$, which is simply $\varepsilon_0 = r_x(0)$ and $x_0[0] = b_0/r_x(0)$. Given that the solutions to the jth order equations are known it is possible to derive the $(j+1)$st order equations as follows. Having the solution $\mathbf{a}_j$ to Equation 4.85 the Levinson-Durbin update Equation 4.69 gives the solution to $\mathbf{a}_{j+1}$ and to the $(j+1)$st equations as defined by Equation 4.87.

$$\begin{bmatrix} r_x(0) & r_x^*(1) & r_x^*(2) & .. & r_x^*(j+1) \\ r_x(1) & r_x(0) & r_x^*(1) & .. & r_x^*(j) \\ r_x(2) & r_x(1) & r_x(0) & .. & r_x^*(j-1) \\ : & : & : & \ddots & : \\ r_x(j+1) & r_x(j) & r_x(j-1) & .. & r_x(0) \end{bmatrix} \begin{bmatrix} 1 \\ a_{1(j+1)} \\ a_{2(j+1)} \\ : \\ a_{j+1(j+1)} \end{bmatrix} = \begin{bmatrix} \varepsilon_{j+1} \\ 0 \\ 0 \\ : \\ 0 \end{bmatrix} \tag{4.87}$$

Equation 4.87 can be rewritten as Equation 4.88 by taking the complex conjugate and reversing of the order of the rows and columns of the Hermitian Toeplitz matrix $\mathbf{R}_{j+1}$.

$$\begin{bmatrix} r_x(0) & r_x^*(1) & r_x^*(2) & .. & r_x^*(j+1) \\ r_x(1) & r_x(0) & r_x^*(1) & .. & r_x^*(j) \\ r_x(2) & r_x(1) & r_x(0) & .. & r_x^*(j-1) \\ : & : & : & \ddots & : \\ r_x(j+1) & r_x(j) & r_x(j-1) & .. & r_x(0) \end{bmatrix} \begin{bmatrix} a_{j+1(j+1)}^* \\ a_{j(j+1)}^* \\ a_{j-1(j+1)}^* \\ : \\ 1 \end{bmatrix} = \begin{bmatrix} 0 \\ 0 \\ 0 \\ : \\ \varepsilon_{j+1}^* \end{bmatrix} \tag{4.88}$$

As the solution to Equation 4.86 is $\mathbf{x}_j$ it is possible to append a zero to vector $\mathbf{x}_j$ and multiply it by $\mathbf{R}_{j+1}$, as done in the Levinson-Durbin recursion, to arrive at

Equation 4.89, where $\delta_j = \sum_{i=0}^{j} r_x(j+1)x_j[i]$.

$$\begin{bmatrix} r_x(0) & r_x^*(1) & r_x^*(2) & .. & r_x^*(j+1) \\ r_x(1) & r_x(0) & r_x^*(1) & .. & r_x^*(j) \\ \vdots & \vdots & \vdots & \ddots & \vdots \\ r_x(j) & r_x(j-1) & r_x(j-2) & .. & r_x^*(1) \\ r_x(j+1) & r_x(j) & r_x(j-1) & .. & r_x(0) \end{bmatrix} \begin{bmatrix} x_j[0] \\ x_j[1] \\ \vdots \\ x_j[j] \\ 0 \end{bmatrix} = \begin{bmatrix} b_0 \\ b_1 \\ \vdots \\ b_j \\ \delta_j \end{bmatrix} \quad (4.89)$$

The extended vector $[\mathbf{x}_j, 0]^T$ is generally not the solution to the $(j+1)$st order equations because generally $\delta_j \neq b_{j+1}$. However, the sum defined by Equation 4.90 can be formed, where $q_{j+1} = (b_{j+1} - \delta_j) / \varepsilon_{j+1}^*$ is an arbitrary complex constant.

$$\mathbf{R}_{j+1} \left\{ \begin{bmatrix} x_j[0] \\ x_j[1] \\ \vdots \\ x_j[j] \\ 0 \end{bmatrix} + q_{j+1} \begin{bmatrix} a_{j+1(j+1)}^* \\ a_{j(j+1)}^* \\ a_{j-1(j+1)}^* \\ \vdots \\ 1 \end{bmatrix} \right\} = \begin{bmatrix} b_0 \\ b_1 \\ \vdots \\ b_j \\ \delta_j + q_{j+1}\varepsilon_{j+1}^* \end{bmatrix} \quad (4.90)$$

Then, it follows that the solution is,

$$b_{j+1} = \delta_j + q_{j+1}\varepsilon_{j+1}^*$$

$$\mathbf{x}_{j+1} = \left\{ \begin{bmatrix} x_j[0] \\ x_j[1] \\ \vdots \\ x_j[j] \\ 0 \end{bmatrix} + q_{j+1} \begin{bmatrix} a_{j+1(j+1)}^* \\ a_{j(j+1)}^* \\ a_{j-1(j+1)}^* \\ \vdots \\ 1 \end{bmatrix} \right\}$$

A summary of the Levinson recursion is as follows,

1. Initialise the recursion by setting,
    i. $a_0 = 1$
    ii. $x_0[0] = b_0 / r_x(0)$
    iii. $\varepsilon_0 = r_x(0)$

2. For $j = 0, 1, ..., p - 1$ set,

    i. $\gamma_j = r_x(j+1) + \sum_{i=1}^{j} a_{i(j)} r_x(j-i-1)$

    ii. $\Gamma_{j+1} = -\dfrac{\gamma_j}{\varepsilon_j}$

iii. For $i = 1,...,j$, $a_{i(j+1)} = a_{i(j)} + \Gamma_{j+1}a^*_{j-i+1(j)}$

iv. $a_{j+1(j+1)} = \Gamma_{j+1}$

v. $\varepsilon_{j+1} = \varepsilon_j[1 - |\Gamma_{j+1}|^2]$

vi. $\delta_j = \sum_{i=0}^{j} r_x(j-i+1)x_j[i]$

vii. $q_{j+1} = (b_{j+1} - \delta_j)\big/\varepsilon^*_{j+1}$

viii. For $i = 0, 1,...,j$, $x_{j+1}[i] = x_j[i] + q_{j+1}a^*_{j-i+1(j+1)}$

ix. $x_{j+1}[j+1] = q_{j+1}$

4.5 Exercises

The following Exercises identify some of the basic ideas presented in this Chapter.

4.5.1 Problems

4.1.   Under what conditions are Equations 4.1 and 4.2 referred to as a Moving-Average (MA), Autoregressive (AR) and Autoregressive Moving-Average (ARMA) model?

4.2.   Which models, MA, AR or ARMA, are the easiest to solve for their unknown parameters?

4.3.   Show that Partial differentiation of Equation 4.5 with respect to variables a^*_k results in Equation 4.6.

4.4.   Explain why the Padé "approximation" method is badly named.

4.5.   Explain the main idea behind the Padé approximation method and why it is likely to be problematic in practice.

4.6.   Given the signal, $\mathbf{x} = \begin{bmatrix} 1 & 1.5 & 0.75 \end{bmatrix}^T$, find the Padé approximation model for,

      a. $p = 2$ and $q = 0$
      b. $p = 0$ and $q = 2$
      c. $p = q = 1$

4.7.　Given the signal, $x[n] = \begin{cases} 1, & n = 0,1,..,20 \\ 0, & \text{otherwise} \end{cases}$, use Prony's method to model $x[n]$ with an ARMA, pole-zero, model with $p = q = 1$.

4.8.　Find,
　　　a.　the first-order, and,
　　　b.　the second-order all-pole models for the signal,
　　　　　$x[n] = \delta[n] - \delta[n-1]$.

4.9.　Find the second-order all-pole model of a signal $x[n]$ whose first $N = 20$ values are $\mathbf{x} = [1,-1,1,-1,...,1,-1]^T$ by using,
　　　a.　the autocorrelation method, and,
　　　b.　the covariance method.

4.10.　Use the modified *Yule-Walker* equations to find the first order ARMA model ($p = q = 1$) of a real valued stochastic process having the following autocorrelation values,

$$r_x(0) = 26, \quad r_x(1) = 7, \quad r_x(2) = \frac{7}{2}$$

4.11.　Solve the autocorrelation normal equations using the Levinson-Durbin recursion to find a third-order all-pole model for a signal having the following autocorrelation values,

$$r_x(0) = 1, \ r_x(1) = 0.5, \ r_x(2) = 0.5, \ r_x(3) = 0.25$$

4.12.　For Problem 4.11 compute the next two values in the autocorrelation sequence, i.e., $r_x(4)$ and $r_x(5)$.

PART III. CLASSICAL FILTERS and SPECTRAL ANALYSIS

Classical linear filters are used to estimate a specific signal from another noisy or linearly corrupted signal. Often a classical lowpass, bandpass, highpass or bandreject filter can be employed to achieve an acceptable result in basic problems. In more complex problems, however, these types of filters fall short of achieving an optimum or best estimate of the desired signal. To achieve an optimum linear filtering solution a Wiener or Kalman filter may be used. The optimality criteria used in both the Wiener and Kalman filters is the minimisation of the mean square error with respect to the desired signal.

In relation to discrete Wiener filters, the four following problems of filtering, smoothing, prediction and deconvolution are the most important ones (Hayes 1996).

1. **Filtering** - given $x[n] = s[n] + v[n]$ the goal is to estimate $s[n]$ using a causal Wiener filter (a filter using only current and past values of $x[n]$).

2. **Smoothing** - the same as filtering except the filter may be noncausal (using all the available data).

3. **Prediction** - the causal Wiener filter produces a prediction of $x[n+1]$ in terms of a linear combination of previous values of $x[n]$.

4. **Deconvolution** - when $x[n] = s[n] * g[n] + v[n]$, where $g[n]$ is the unit sample response of a linear shift-invariant filter, the Wiener filter becomes a deconvolution filter.

The discrete Wiener filter is linear shift-invariant and meant for estimating stationary processes. The discrete form of the Wiener-Hopf equations specify the filter coefficients of the optimum FIR filter. It is also possible to design causal IIR Wiener filters but it requires the solution to a nonlinear problem that requires a spectral factorisation of the power spectrum of the input process.

Recursive approaches to signal estimation lead to the derivation of the discrete Kalman filter. The Kalman filter is a shift-varying filter and therefore applicable to nonstationary as well as stationary processes. It can be described as a deterministic system with stochastic excitation and observation errors. The general framework for all of the known algorithms that constitute the recursive least-squares family of adaptive filters is provided by the Kalman filter (Sayed and Kailath 1994).

The problem of estimating the power spectrum of wide-sense stationary processes is very important to the development of filter systems. The power spectrum is the Fourier transform of the autocorrelation function, therefore estimating the power spectrum is equivalent to estimating the autocorrelation. However, in practice only finite and often noisy data records are available to estimate the power spectrums of what are essentially infinite data processes. This estimation problem can be solved using various parametric and nonparametric techniques, each of which have their advantages and disadvantages that determine their appropriateness for specific applications.

5. Optimum Wiener Filter

The optimum Wiener filter, in theory, provides the best linear method to remove stationary Gaussian noise added to a linear process and it is a form of the stochastic estimation model. Wiener developed his continuous-time filtering ideas based on wide-sense stationary statistics in the early 1940s for application to anti-aircraft fire control systems. He considered the problem of designing a linear filter that would produce the minimum mean square error estimate with respect to the desired signal. Kolmogorov later developed similar filtering approaches utilising mean square theory for discrete stochastic processes.

Given a process $x(t) = s(t) + v(t)$ where $s(t)$ is the result of the convolution of an unknown original signal $d(t)$ with a known linear system having an impulse response $g(t)$ then it is possible to estimate $d(t)$ using a Wiener filter. Refer to Figure 5.1. If $s(t)$ and $v(t)$ are zero mean with stationary statistics and they are uncorrelated the optimum Wiener filter $\Phi(f)$ is defined by Equation 5.1 in the frequency domain (Press *et al* 1986).

$$\Phi(f) = \frac{|S(f)|^2}{|S(f)|^2 + |V(f)|^2} \tag{5.1}$$

Since $s(t)$ and $v(t)$ are uncorrelated $|S(f)|^2 + |V(f)|^2$ can be estimated by $|X(f)|^2$ from a long sample of $x(t)$ using a spectral estimation technique. If the statistics of $v(t)$ are known then $|V(f)|^2$ can be subtracted from $|X(f)|^2$, else it is often possible to deduce both $|S(f)|^2$ and $|V(f)|^2$ from an inspection of $|X(f)|^2$. The filter is useful even if it is not completely accurate because it is derived by least squares minimisation, consequently the output differs from the true optimum by only a second-order amount.

The signal spectrum of the unknown signal $D(f)$ is estimated by Equation 5.2.

$$D(f) \approx \frac{X(f)\Phi(f)}{G(f)} \tag{5.2}$$

The Wiener filter requirement that the signal statistics be stationary limits its application and this naturally led to the development of the Kalman filter for dealing with the nonstationary case. When the measurement noise is white, the signal model is finite dimensional, and all processes are stationary then Wiener and Kalman filtering theories lead to the same result. Otherwise the Kalman filter is expected to produce superior results.

5.1 Derivation of the Ideal Continuous-time Wiener Filter

The general ideal continuous-time Wiener filter defined by Equation 5.1 can be derived starting with the system configuration shown in Figure 5.1.

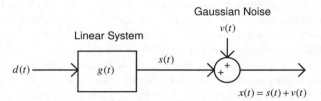

Figure 5.1. Wiener Filter Derivation

The signal $s(t)$ is the output of the linear filter $G(f)$, which is the convolution of the unknown signal $d(t)$ and the filter impulse response $g(t)$. The signal $x(t)$ is the signal $s(t)$ plus some random noise $v(t)$ as follows,

$$s(t) = \int_{-\infty}^{+\infty} g(t)d(t-\tau)d\tau, \text{ i.e., } S(f) = G(f)D(f),$$

and,

$$X(f) = S(f) + V(f)$$

The problem is to find a filter $\Phi(f)$ to estimate $D(f)$ from $X(f)$, i.e.,

$$\hat{D}(f) = \frac{X(f)}{G(f)}\Phi(f)$$

If $\hat{D}(f)$ is to be close to $D(f)$ in the least squares sense then it is necessary to minimise the squared difference function ξ between the signal $d(t)$ and its estimate $\hat{d}(t)$ along the following lines,

$$\xi = \int_{-\infty}^{+\infty}\left|\hat{d}(t) - d(t)\right|^2 dt = \int_{-\infty}^{+\infty}\left|\hat{D}(f) - D(f)\right|^2 df$$

$$\xi = \int_{-\infty}^{+\infty} \left| \frac{(S(f)+V(f))\Phi(f)}{G(f)} - \frac{S(f)}{G(f)} \right|^2 df =$$

$$\int_{-\infty}^{+\infty} |G(f)|^2 \left\{ |S(f)|^{-2} |1-\Phi(f)|^2 + |V(f)|^2 |\Phi(f)|^2 \right\} df$$

Since the signal $S(f)$ and noise $V(f)$ are uncorrelated their cross-product is zero when integrated over frequency. (Noise by definition is that part that does not correlate with the signal). The error function ξ can be minimised with respect to $\Phi(f)$ at every value of f by differentiating it with respect to $\Phi(f)$ and equating to zero. This results in the following optimum Wiener filter solution,

$$\Phi(f) = \frac{|S(f)|^2}{|S(f)|^2 + |V(f)|^2}$$

Statistically speaking $\Phi(f)$ is the optimum Wiener filter, which means it is best in the least squares sense for Gaussian noise. The reason that $\Phi(f)$ is the optimum filter is that Gaussian noise is fully described by its second-order statistics (its variance if the mean is zero) and least squares is a second-order minimisation technique. Note that the optimum filter can be found independently of the deconvolution function which relates $S(f)$ and $D(f)$.

In practice it is necessary to have a way to estimate $|S(f)|^2$ and $|V(f)|^2$ if they are not known *a priori*. Often, in simple problems this can be done from $X(f)$ alone without other information or assumptions. If a long stretch of $x(t)$ is sampled and its power spectral density $P_x(f)$ is estimated from it then,

$$P_x(f) = |X(f)|^2 \approx |S(f)|^2 + |V(f)|^2, \text{ since } S(f) \text{ and } V(f) \text{ are uncorrelated.}$$

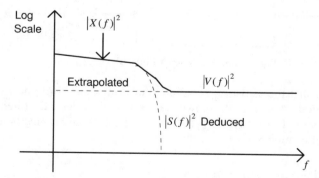

Figure 5.2. Wiener Filter Spectrum Example

For example, a typical filtering problem may be described by the spectrums shown in Figure 5.2. Usually the power spectral plot will show the spectral signature of a signal sticking up above a continuous noise spectrum. The noise spectrum may be flat, or tilted, or smoothly varying; it doesn't matter, provided a guess can be made at a reasonable hypothesis of it. One way to do this is to draw a smooth curve through the noise spectrum and extrapolate it into the signal region. The difference between $|X(f)|^2$ and $|V(f)|^2$ gives $|S(f)|^2$. Extend the $\Phi(f)$ to negative values of frequency by making $\Phi(-f) = \Phi(f)$. The resulting Wiener filter $\Phi(f)$ will be close to unity where the noise is negligible and close to zero where the noise is dominant. The model is still useful even if $|S(f)|^2$ is not completely accurate because the optimum filter results from a minimisation where the quality of the result differs from the true optimum by only a second-order amount in precision. Even a 10% error in $|S(f)|$ can still give excellent results. However, it is not possible to improve on the estimate of $|D(f)|$ any further by using the resulting $\Phi(f)$ to make a better guess at $\hat{D}(f)$ and then reapplying the same filtering technique to it, and so on. Doing this will only result in a convergence to $S(f) = 0$.

5.2 The Ideal Discrete-time FIR Wiener Filter

The general Wiener filtering problem can be stated as follows. Given two wide-sense stationary processes $x[n]$ and $s[n]$ that are statistically related to each other, the Wiener filter $W(z)$ produces the minimum mean square estimate $y[n]$ of $s[n]$. A FIR filter whose output $y[n]$ best approximates the desired signal $s[n]$ in the sense that the mean square norm of the error is minimised is called the optimum FIR Wiener filter. Figure 5.3 illustrates the discrete-time FIR Wiener filtering problem.

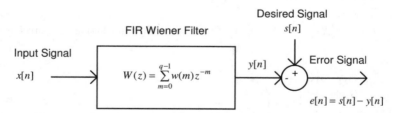

Figure 5.3. FIR Wiener Filter Block Diagram

It is assumed that the statistics including autocorrelations $r_x(k)$ and $r_s(k)$ and the cross-correlation $r_{sx}(k)$ are all known. For a $(q-1)$th order FIR filter having an impulse response of $\mathbf{w} = [w[0], w[1],...., w[q-1]]^T$, the system convolution function is defined by Equation 5.3.

$$y[n] = \sum_{m=0}^{q-1} w[m]x[n-m] \tag{5.3}$$

It is necessary to find the filter coefficients $w[k]$ that minimise the mean square error ξ as defined by Equation 5.4.

$$\xi = E\{|e[n]|^2\} = E\{|s[n] - y[n]|^2\}$$
(5.4)

To minimise ξ Equation 5.4 is differentiated with respect to $w^*[k]$ and set to zero and solved for $k = 0,1,....,q-1$ as defined by Equations 5.5 and 5.6 (Hayes 1996).

$$\frac{\partial \xi}{\partial w^*[k]} = \frac{\partial}{\partial w^*[k]} E\{e[n]e^*[n]\} = E\left\{e[n]\frac{\partial}{\partial w^*[k]}e^*[n]\right\} = 0$$
(5.5)

$$\frac{\partial}{\partial w^*[k]}e^*(n) = \frac{\partial}{\partial w^*[k]}\left[s^*[n] - \sum_{m=0}^{q-1}w^*[m]x^*[n-m]\right] = -x^*[n-k]$$
(5.6)

Since the error is defined by Equation 5.7, Equation 5.5 becomes Equation 5.8 and consequently 5.9 according to the orthogonality principle.

$$e[n] = s[n] - \sum_{m=0}^{q-1}w[m]x[n-m]$$
(5.7)

$$E\{e[n]x^*[n-k]\} = 0, \quad \text{for } k = 0,1,...,q-1$$
(5.8)

i.e.,

$$E\{s[n]x^*[n-k]\} - \sum_{m=0}^{q-1}w[m]E\{x[n-m]x^*[n-k]\} = 0$$
(5.9)

As, $E\{s[n]x^*[n-k]\} = r_{sx}(k)$ and $E\{x[n-m]...x^*[n-k]\} = r_x(k-m)$ Equation 5.9 becomes a set of q linear equations in the q unknowns $w[k]$ for $k = 0,1,....,q-1$, and then the required solution can be defined by Equation 5.10.

$$\sum_{m=0}^{q-1}w[k]r_x(k-m) = r_{sx}(k), \quad k = 0,1,..,q-1$$
(5.10)

Equation 5.10 is the so called Wiener-Hopf solution. It can be expressed in matrix form as defined by Equation 5.11, $\mathbf{R}_x\mathbf{w} = \mathbf{r}_{sx}$, by using the fact that the autocorrelation sequence is conjugate symmetric, i.e., $r_x(k) = r_x^*(-k)$.

$$\begin{bmatrix} r_x(0) & r_x^*(1) & .. & r_x^*(q-1) \\ r_x(1) & r_x(0) & .. & r_x^*(q-2) \\ \vdots & \vdots & \ddots & \vdots \\ r_x(q-1) & r_x(q-2) & .. & r_x(0) \end{bmatrix}\begin{bmatrix} w[0] \\ w[1] \\ \vdots \\ w[q-1] \end{bmatrix} = \begin{bmatrix} r_{sx}(0) \\ r_{sx}(1) \\ \vdots \\ r_{sx}(q-1) \end{bmatrix}$$
(5.11)

The minimum mean square error of the estimate of $s[n]$ may be determined by Equation 5.4 as defined by Equation 5.12.

$$\xi_{min} = E\{e[n]e^*[n]\} = E\left\{e[n]\left[s[n] - \sum_{m=0}^{q-1} w[m]x[n-m]\right]^*\right\}$$

$$= E\{e[n]s^*[n]\} - \sum_{m=0}^{q-1} w^*[m]E\{e[n]x^*[n-m]\}$$ (5.12)

$$= E\{e[n]s^*[n]\}, \text{ since by equation (5.8), } E\{e[n]x^*[n-m]\} = 0$$

$$= E\left\{\left[s[n] - \sum_{m=0}^{q-1} w[m]x[n-m]\right]s^*[n]\right\} = r_s(0) - \sum_{m=0}^{q-1} w[m]r_{sx}^*(m)$$

Using vector notation the minimum mean square error can then be expressed as defined Equation 5.13.

$$\xi_{min} = r_s(0) - \mathbf{r}_{sx}^H \mathbf{w}$$

$$= r_s(0) - \mathbf{r}_{sx}^H \mathbf{R}_x^{-1} \mathbf{r}_{sx}$$ (5.13)

A summary of the Wiener-Hopf equations for the FIR Wiener filter is as follows,

1.   Wiener-Hopf equations: $\sum_{m=0}^{q-1} w[m]r_x(k-m) = r_{sx}(k), \quad k = 0,1,...,q-1$

2.   Correlations: $r_x(k-m) = E\{x[n-m]x^*[n-k]\}$
       $r_{sx}(k) = E\{s[n-m]x^*[n-k]\}$

3.   Minimum error: $\xi_{min} = r_s(0) - \sum_{m=0}^{q-1} w[m]r_{sx}^*(m)$

5.2.1 General Noise FIR Wiener Filtering

In general noise filtering problems there are the following data sets,

{$s[n]$}, a desired signal process,
{$v[n]$}, a noise process, not necessarily white,
{$x[n] = s[n+c]+v[n]$}, a measurement process,

where, {$s[n]$} and {$v[n]$} are independent, zero mean and stationary and their initial time is in the infinitely remote past. Here, the desired signal $s[n+c]$ must be estimated from the noisy measurement $x[n]$. This problem may be best represented by the model shown in Figure 5.4.

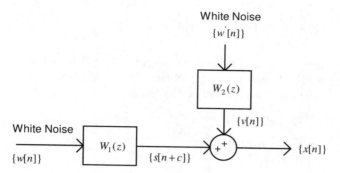

Figure 5.4. Wiener Noise Filtering Model

Systems $W_1(z)$ and $W_2(z)$ are real rational transfer function matrices with all their poles inside the unit circle, i.e., $|z| < 1$ and $\{w[n]\}$ and $\{w'[n]\}$ are both independent zero mean unit variance white noise processes. From this one model it is possible to pose three important problem definitions. Forming the expectation $E\{s[n+c]\,|\,x[n]\}$ for zero, negative and positive integers of c defines the filtering, smoothing and prediction problems respectively.

Using the results from the previous Section it is possible to derive the optimum FIR Wiener filter, i.e., the special case for $c = 0$. Since $s[n]$ and $v[n]$ are uncorrelated the cross-correlation between $s[n]$ and $x[n]$ is simply the autocorrelation of $s[n]$ as defined by Equation 5.14.

$$r_{sx}(k) = E\{s[n]x^*[n-k]\}$$
$$= E\{s[n]s^*[n-k]\} + E\{s[n]v^*[n-k]\} \qquad (5.14)$$
$$= r_s(k)$$

It also follows that $r_x(k)$ is defined by Equation 5.15.

$$r_x(k) = E\{x[n+k]x^*[n]\}$$
$$= E\{[s[n+k]+v[n+k]][s[n]+v[n]]^*\} \qquad (5.15)$$
$$= r_s(k) + r_v(k)$$

Therefore, the Wiener-Hopf equations for this filter are defined by Equation 5.16.

$$[\mathbf{R}_s + \mathbf{R}_v]\mathbf{w} = \mathbf{r}_s \qquad (5.16)$$

Equation 5.16 may be simplified further if more specific statistical information about the signal and noise is available.

5.2.2 FIR Wiener Linear Prediction

In the special case of noise-free measurements of a signal $x[n]$, linear prediction is concerned with the prediction or estimation of $x[n+c]$ in terms of a linear combination of the current and previous values of $x[n]$ as defined by Equation 5.17.

$$\hat{x}[n+c] = \sum_{m=0}^{q-1} w[k]x[n-m], \quad \text{typically } c = 1. \tag{5.17}$$

If the desired signal is defined to be $s[n] = x[n+c]$, for $c = 1$, the linear predictor can be treated as a FIR Wiener filtering problem. The cross-correlation between $s[n]$ and $x[n]$ can then be evaluated as defined by Equation 5.18.

$$
\begin{aligned}
r_{sx}(k) &= E\{s[n]x^*[n-k]\} \\
&= E\{x[n+1]x^*[n-k]\} \\
&= r_x(k+1)
\end{aligned}
\tag{5.18}
$$

From this, the Wiener-Hopf equations for the optimum predictor are defined by Equation 5.19 (Hayes 1996).

$$\mathbf{R}_x\mathbf{w} = \mathbf{r}_{sx(1)} = \mathbf{r}_{x(1)}$$

$$
\begin{bmatrix}
r_x(0) & r_x^*(1) & .. & r_x^*(q-1) \\
r_x(1) & r_x(0) & .. & r_x^*(q-2) \\
\vdots & \vdots & \ddots & \vdots \\
r_x(q-1) & r_x(q-2) & .. & r_x(0)
\end{bmatrix}
\begin{bmatrix}
w[0] \\
w[1] \\
\vdots \\
w[q-1]
\end{bmatrix}
=
\begin{bmatrix}
r_x(1) \\
r_x(2) \\
\vdots \\
r_x(q)
\end{bmatrix}
\tag{5.19}
$$

The corresponding minimum mean square error for this predictor is defined by Equation 5.20.

$$\xi_{\min} = r_x(0) - \sum_{m=0}^{q-1} w[m]r_x^*(m+1) = r_x(0) - \mathbf{r}_{x(1)}^H\mathbf{w} \tag{5.20}$$

The set of Equations 5.19 are essentially the same as the Prony all-pole normal Equations 4.29 except for the fact that in the Prony equations $r_x(k)$ is a deterministic autocorrelation rather than a stochastic one, as is the case here.

Linear prediction for $c > 1$, where the desired signal is to be $s[n] = x[n+c]$, is referred to as multistep prediction as opposed to $c = 1$, which is single step prediction. The development of the equations for multistep linear prediction is similar to single step linear prediction. The only difference is in the crosscorrelation vector $\mathbf{r}_{sx}$. Since $s[n] = x[n+c]$ then $r_{sx}(k)$ is defined by Equation 5.21.

$$
\begin{aligned}
r_{sx}(k) &= E\{s[n]x^*[n-k]\} \\
&= E\{x[n+c]x^*[n-k]\} \\
&= r_x(k+c)
\end{aligned}
\tag{5.21}
$$

The Wiener-Hopf equations for the multistep predictor are consequently defined by Equation 5.22.

$$\mathbf{R}_x \mathbf{w} = \mathbf{r}_{sx}(c) = \mathbf{r}_x(c)$$

$$\begin{bmatrix} r_x(0) & r_x^*(1) & .. & r_x^*(q-1) \\ r_x(1) & r_x(0) & .. & r_x^*(q-2) \\ : & : & \ddots & : \\ r_x(q-1) & r_x(q-2) & .. & r_x(0) \end{bmatrix} \begin{bmatrix} w[0] \\ w[1] \\ : \\ w[q-1] \end{bmatrix} = \begin{bmatrix} r_x(c) \\ r_x(c+1) \\ : \\ r_x(c+q-1) \end{bmatrix} \tag{5.22}$$

The corresponding minimum mean square error for the multistep predictor is defined by Equation 5.23.

$$\xi_{\min} = r_x(0) - \sum_{m=0}^{q-1} w[m]r_x^*(m+c) = r_x(0) - \mathbf{r}_{x(c)}^H \mathbf{w} \tag{5.23}$$

5.3 Discrete-time Causal IIR Wiener Filter

The problem formulation for a discrete-time causal filter design is the same for both the FIR and IIR Wiener filters except that the FIR has a finite number of filter coefficients, whereas the IIR has an infinite number of unknowns. If no constraints are placed on the solution the optimum IIR filter has a simple closed form expression for the frequency response, but it will usually be noncausal. It is possible to design a realisable causal IIR filter by forcing the filter's unit sample response $h[n]$ to be zero for $n \leq 0$. However, for the causal filter, it is only possible to specify the system function in terms of a spectral factorisation.

The estimate of the desired signal $s[n]$ for a causal filter with a unit sample response $h[n]$ takes the form as defined by Equation 5.24.

$$\hat{s}[n] = y[n] = x[n] * h[n] = \sum_{k=0}^{\infty} h[k]x[n-k] \tag{5.24}$$

To find the filter coefficients $h[k]$ the mean square error ξ, as defined by Equation 5.4, is minimised by differentiating with respect to $h^*[n]$, setting it to zero and solving it for $k \geq 0$. By a similar process to the FIR Wiener solution this results in the Wiener-Hopf equations for the causal IIR Wiener filter as defined by Equation 5.25.

$$\sum_{m=0}^{\infty} h[m]r_x(k-m) = r_{sx}(k), \quad 0 \leq k < \infty \tag{5.25}$$

The equations for the noncausal IIR Wiener filter are the same as Equation 5.25 except that the lower limit for k is $-\infty$ instead of 0.

To solve Equation 5.25 it is necessary to study the special case where the input to a Wiener filter $g[n]$ is unit variance white noise $\varepsilon[n]$, resulting in the Wiener-Hopf Equations 5.26 (Hayes 1996).

$$\sum_{m=0}^{\infty} g[m]r_{\varepsilon}(k-m) = r_{s\varepsilon}(k), \quad 0 \le k < \infty$$

$$\text{since,} \quad r_{\varepsilon}(k) = \delta(k) \text{ and } g[n] = 0, \quad \text{for } n < 0 \tag{5.26}$$

$$g[n] = r_{s\varepsilon}(n)u[n]$$

The solution to Equations 5.26 in the z-domain is $G(z) = [P_{s\varepsilon}(z)]_+$, where the "+" denotes the positive-time part of the sequence whose z-Transform is contained inside the brackets.

In a real filtering problem the input is unlikely to be white noise. For a random process input $x[n]$ having a rational power spectrum with no poles or zeros on the unit circle it is possible to perform a spectral factorisation of $P_x(z)$ as defined by Equation 5.27.

$$P_x(z) = \sigma_0^2 Q(z)Q^*(\frac{1}{z^*}) \tag{5.27}$$

$Q(z)$ is a minimum phase rational function composed of a minimum phase numerator $N(z)$ and denominator $D(z)$, which is a monic (the coefficient of the highest power is unity) polynomial as defined by Equation 5.28.

$$Q(z) = \frac{N(z)}{D(z)} = 1 + q[1]z^{-1} + q[2]z^{-2} + ... \tag{5.28}$$

If $x[n]$ is filtered by a whitening filter $F(z) = 1/\sigma_0 Q(z)$ the power spectrum of the resulting white noise output process $\varepsilon[n]$ is $P_\varepsilon(z) = P_x(z)F(z)F^*(1/z^*) = 1$. As $Q(z)$ is minimum phase $F(z)$ is casual and stable having a causal inverse. Therefore, $x[n]$ may be fully recovered from $\varepsilon[n]$ by filtering with the inverse filter.

When $x[n]$ has a rational power spectrum the optimum causal Wiener filter $H(z)$ can be derived by the following argument. Suppose that $x[n]$ is filtered with a cascade of the three filters, $F(z)$, $F^{-1}(z)$, and $H(z)$ as shown in Figure 5.5.

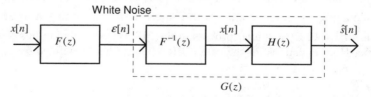

Figure 5.5. A Causal Wiener Filtering Model

The cascade causal filter $G(z) = F^{-1}(z)H(z) = [P_{s\varepsilon}(z)]_+$ is the causal Wiener filter that produces the minimum mean square estimate of the desired signal $s[n]$ from the white noise process $\varepsilon[n]$. The cross-correlation between $s[n]$ and $\varepsilon[n]$ is defined by Equation 5.29.

$$r_{sx}(k) = E\{s[n]\varepsilon^*[n-k]\}$$

$$= E\left\{s[n]\left[\sum_{m=-\infty}^{\infty} f[m]x[n-k-m]\right]^*\right\} = \sum_{m=-\infty}^{\infty} f^*[m]r_{sx}(k+m) \qquad (5.29)$$

The cross-power spectral density $P_{s\varepsilon}(z)$ is defined by Equation 5.30.

$$P_{s\varepsilon}(z) = P_{sx}(z)F^*(\frac{1}{z^*}) = \frac{P_{sx}(z)}{\sigma_0^2 Q(z)Q^*(\frac{1}{z^*})} \qquad (5.30)$$

Therefore, the required Wiener filter solution is defined by Equation 5.31.

$$H(z) = F(z)G(z) = \frac{1}{\sigma_0^2 Q(z)}\left[\frac{P_{sx}(z)}{Q^*(\frac{1}{z^*})}\right]_+ \text{, and,} \qquad (5.31)$$

$$H(z) = \frac{1}{\sigma_0^2 Q(z)}\left[\frac{P_{sx}(z)}{Q(z^{-1})}\right]_+ \text{, for real processess.}$$

The minimum mean square error for the causal IIR Wiener filter is defined by Equation 5.32.

$$\xi_{\min} = r_s(0) - \sum_{m=0}^{\infty} h[m]r_{sx}^*(m) \qquad (5.32)$$

5.3.1 Causal IIR Wiener Filtering

The system function for causal IIR Wiener filtering of $s[n]$ given $x[n] = s[n] + v[n]$ (where $v[n]$ is noise that is uncorrelated with $s[n]$) is defined by Equation 5.28. Given that the noise $v[n]$ is uncorrelated with the desired signal $s[n]$ then $P_{sx}(z) = P_s(z)$ and the causal filter can be defined by Equation 5.33.

$$H(z) = \frac{1}{\sigma_0^2 Q(z)}\left[\frac{P_s(z)}{Q^*(\frac{1}{z^*})}\right]_+ \qquad (5.33)$$

where:

$$P_x(z) = P_s(z) + P_v(z) = \sigma_0^2 Q(z)Q^*(\frac{1}{z^*})$$

Equation 5.33 may be simplified further based on having more specific statistical information about the signal and noise.

5.3.2 Wiener Deconvolution

Deconvolution is concerned with the recovery of a signal $s[n]$ that has been convolved with a filter $g[n]$, i.e., $x[n] = s[n] * g[n]$ and $s[n] = x[n] * g^{-1}[n]$, $S(e^{j\theta}) = X(e^{j\theta})/G(e^{j\theta})$. This is a rather difficult problem that arises in many applications, especially those associated with signal recording. The difficulty is due to two possible reasons. Firstly, the filter may not be precisely known in practice. Secondly, even if it is known the filter's frequency response $G(e^{j\theta})$ is often zero at one or more frequencies or, at least, it is very small over parts of the frequency band. The second problem is quite serious because $G(e^{j\theta})$ is consequently nonivertible or at least ill-conditioned. A further difficulty is introduced when the measurement process $x[n]$ includes additive white noise $w[n]$ uncorrelated with $s[n]$, i.e., $x[n] = s[n] * g[n] + w[n]$. In this case, when the inverse filter exists and is well-behaved, it turns out that the signal estimate is the sum of original signal and a residual filter noise term as defined by Equation 5.34.

$$\hat{S}(e^{j\theta}) = S(e^{j\theta}) + \frac{W(e^{j\theta})}{G(e^{j\theta})} = S(e^{j\theta}) + V(e^{j\theta}) \tag{5.34}$$

The difficulty with this is that if $G(e^{j\theta}) \approx 0$ over any frequencies the noise component $V(e^{j\theta})$ becomes large and may dominate the signal estimate.

A way to avoid these types of difficulties is to frame the estimation in the form of a Wiener filtering problem. That is, design a linear shift-invariant IIR filter $h[n]$ that produces the minimum mean square estimate of $s[n]$ from $x[n]$ as defined by Equation 5.35.

$$\hat{s}[n] = y[n] = x[n] * h[n] = \sum_{k=-\infty}^{\infty} h[k][n-k] \tag{5.35}$$

The noncausal filter coefficients that minimise the mean square error defined by Equation 5.4 are the solutions to the Wiener-Hopf equations, which can be represented in the frequency domain by Equation 5.36.

$$H(e^{j\theta}) = \frac{P_{sx}(e^{j\theta})}{P_x(e^{j\theta})} \tag{5.36}$$

Since $s[n]$ is assumed to be uncorrelated with $w[n]$ it is also uncorrelated with $s[n] * g[n]$, therefore the power spectrum of $x[n]$ is equal to the power spectrum of $s[n] * g[n]$ plus the power spectrum of $w[n]$, i.e.,

$P_x(e^{j\theta}) = P_s(e^{j\theta})\left|G(e^{j\theta})\right|^2 + P_w(e^{j\theta})$, and also the cross-power spectrum of $s[n]$

and $x[n]$, $P_{sx}(e^{j\theta}) = P_s(e^{j\theta})G^*(e^{j\theta})$. Consequently, it follows that the Wiener filter for deconvolution is defined by Equation 5.37.

$$H(e^{j\theta}) = \frac{P_s(e^{j\theta})G^*(e^{j\theta})}{P_s(e^{j\theta})\left|G(e^{j\theta})\right|^2 + P_w(e^{j\theta})} \tag{5.37}$$

5.4 Exercises

The following Exercises identify some of the basic ideas presented in this Chapter.

5.4.1 Problems

5.1. Find the optimum first-order FIR Wiener filter for estimating a signal $s[n]$ from a noisy real signal $x[n] = s[n] + v[n]$, where $v[n]$ is a white noise process with a variance σ_v^2 that is uncorrelated with $s[n]$. Assume that $s[n]$ is a first-order AR process having an autocorrelation sequence defined by $r_s(k) = \alpha^{|k|}$.

5.2. In Problem 5.1 the optimum first-order FIR Wiener filter for $\sigma_v^2 = 1$ and $\alpha = 0.8$ is, $W(z) = 0.4048 + 0.2381z^{-1}$. What is the signal to noise ratio (S/N) improvement, computed in dB, achieved by the Wiener filtering?

5.3. Find the optimum first-order linear predictor having the form $\hat{x}[n+1] = w[0]x[n] + w[1]x[n-1]$, for a first-order AR process $x[n]$ that has an autocorrelation sequence defined by $r_s(k) = \alpha^{|k|}$.

5.4. Reconsider Problem 5.3 when the measurement of $x[n]$ is contaminated with zero-mean white noise having a variance of σ_v^2, i.e., $y[n] = x[n] + v[n]$. Find the optimum first-order linear predictor in this case.

5.5. Show that the solution to Problem 5.4 approaches the solution to Problem 5.3 as $\sigma_v^2 \to 0$.

5.6. Consider a random process whose autocorrelation sequence is defined by,

$r_x(k) = \delta(k) + (0.9)^{|k|} \cos(\pi k / 4)$. The first six autocorrelation values are,

$\mathbf{r}_x = [2.0 \quad 0.6364 \quad 0 \quad -0.5155 \quad -0.6561 \quad -0.4175]^T$.

Solve the equations for a first-order three step linear predictor and compute the mean square error.

5.7.   Compute the minimum mean square error for Problem 5.6 for 1 to 5 step predictors. What do you find odd about the sequence of errors from step 1 through to step 5?

5.8.   Find the optimum causal and noncausal IIR Wiener filters for estimating a zero-mean signal $s[n]$ from a noisy real zero-mean signal $x[n] = s[n] + v[n]$, where $v[n]$ is a unit variance white noise zero-mean process uncorrelated with $s[n]$ and $s[n]$ is a first-order AR process defined by,

$s[n] = 0.8s[n-1] + w[n]$,

$w[n]$ is white noise with variance $\sigma_w^2 = 0.36$,

$r_s(k) = (0.8)^k$.

6. Optimum Kalman Filter

The discrete-time Kalman filter can be described as a linear model based finite dimensional processor that aims to provide an estimate of a system's state as it develops dynamically in time. It is an optimum linear estimator if the stochastic distributions are Gaussian and the best linear estimator if they are only symmetrical.

6.1 Background to The Kalman Filter

The Kalman filter is used for estimating or predicting the next stage of a system based on a moving average of measurements driven by white noise, which is completely unpredictable. It needs a model of the relationship between inputs and outputs to provide feedback signals but it can follow changing noise statistics quite well. The development and serious application of Kalman filters began in the 1960s following Kalman's publication (Kalman 1960) on discrete-time, recursive mean square filtering. Kalman filters were first used for control applications in aerospace and they are still used primarily for control and tracking applications related to vessels, spacecraft, radar, and target trajectories. Other applications include adaptive equalisation of telephone lines, fading dispersive communications channels, and adaptive antenna arrays. The Kalman filter represents the most widely applied and useful result to emerge from the state variable approach of "modern control theory."

The Kalman filter is an optimum estimator that estimates the state a of linear system developing dynamically through time. An optimum estimator can be defined as an algorithm that processes all the available data to yield an estimate of the "state" of a system whilst at the same time estimating some predefined optimality criterion. The Kalman filter uses a recursive algorithm whereby the parameters required at each time step are computed from the corresponding parameters at the previous time step. This means that it is not necessary to retain and operate on all the past data values. Since this process is inherently discrete it can be readily applied to vector valued processes.

6.2 The Kalman Filter

In the design of a causal Wiener filter to estimate a signal process $s[n]$ from a set of noisy observations $x[n] = s[n] + v[n]$ it is necessary that $s[n]$ and $x[n]$ be jointly wide-sense stationary processes. This is a problem that limits the usefulness of Wiener filters in practical applications since most real processes tend to be nonstationary. The Kalman filter can dispense with the stationarity requirement, thereby making it much more useful for filtering nonstationary processes.

The Kalman filter, according to (Hayes 1996), can be anticipated from a special case of the causal IIR Wiener filter for estimating a process $x[n]$ from noisy measurements, i.e., $y[n] = x[n] + v[n]$, where $v[n]$ is a zero mean white noise process. The special case is that of estimating an AR(1) process of the form defined by Equation 6.1, where $w[n]$ is a zero mean white noise process uncorrelated with $v[n]$.

$$x[n] = a_1 x[n-1] + w[n] \tag{6.1}$$

The optimum Wiener solution to this problem that minimises the mean square error between $x[n]$ and $\hat{x}[n]$ can be shown to have the recursive form defined by Equation 6.2. Refer to Example 1 in Section 6.2.1.

$$\hat{x}[n] = a_1 \hat{x}[n-1] + K[y[n] - a_1 \hat{x}[n-1]] \tag{6.2}$$

where:
  K is a constant.

The constant K is known as the Kalman gain (that minimises the mean square error, $E\{|x[n] - \hat{x}[n]|^2\}$). The part of Equation 6.2 that is multiplied by K is referred to as the innovations process, which is an estimate correction factor. This solution shows that all the observations of $y[k]$ for $k \le n$ are used to compute the estimate of $x[n]$. There are two problems with this optimum solution. Firstly, Equation 6.2 is the optimum solution only if $x[n]$ and $y[n]$ are jointly wide-sense stationary processes. However, it will later be shown that modified Equation 6.3 can be made to represent the optimum estimate if $K(n)$ is a suitably chosen time-varying gain.

$$\hat{x}[n] = a_{1(n-1)} \hat{x}[n-1] + K(n)[y[n] - a_{1(n-1)} \hat{x}[n-1]] \tag{6.3}$$

where:
  The a subscript "$1(n-1)$," represents coefficient a_1 from the previous iteration $[n-1]$.

The second problem with the Wiener solution Equation 6.2 is that it does not allow the filter to begin at time $n = 0$ because it assumes that all observations of $y[k]$ for $k \le n$ are available. On the other hand, Equation 6.3 can be shown to deal with this problem well.

It is possible to extend the problem of the estimation of an AR(1) process to the more general problem of estimation of an AR(p) process by using state variables. Assume that there is an AR(p) process generating a signal $x[n]$ according to the difference Equation 6.4, and that $x[n]$ is measured in the presence of additive noise in accordance with Equation 6.5.

$$x[n] = \sum_{k=1}^{p} a_k x[n-k] + w[n] \tag{6.4}$$

$$y[n] = x[n] + v[n] \tag{6.5}$$

In this case the vector $\mathbf{x}[n] = [x[n], x[n-1], ..., x[n-p+1]]^T$, can be defined to be a p-dimensional state vector and consequently Equations 6.4 and 6.5 may be reformulated in appropriate matrix form as defined by Equations 6.6 and 6.7 respectively

$$\mathbf{x}[n] = \begin{bmatrix} a_1 & a_2 & .. & a_{p-1} & a_p \\ 1 & 0 & .. & 0 & 0 \\ 0 & 1 & .. & 0 & 0 \\ \vdots & \vdots & \ddots & \vdots & \vdots \\ 0 & 0 & .. & 1 & 0 \end{bmatrix} \mathbf{x}[n-1] + \begin{bmatrix} 1 \\ 0 \\ 0 \\ \vdots \\ 0 \end{bmatrix} \mathbf{w}[n] \tag{6.6}$$

$$= \mathbf{A}\mathbf{x}[n-1] + \mathbf{w}[n]$$

where:
    $\mathbf{A}$ is a $p \times p$ state transition matrix.
    $\mathbf{w}[n] = [w[n], 0, ..., 0]^T$ is a vector zero-mean white noise process.

$$y[n] = [1, 0, .., 0]\mathbf{x}[n] + v[n]$$
$$= \mathbf{c}^T \mathbf{x}[n] + v[n] \tag{6.7}$$

where:
    $\mathbf{c}$ is a unit vector of length p.

A new equation for the optimum estimate of the state vector $\mathbf{x}[n]$, similar to Equation 6.3 for an AR(1), process can now be formulated for the AR(p) process as defined by Equation 6.8.

$$\hat{\mathbf{x}}[n] = \mathbf{A}\hat{\mathbf{x}}[n-1] + \mathbf{K}[y[n] - \mathbf{c}^T \mathbf{A}\hat{\mathbf{x}}[n-1]] \tag{6.8}$$

where:
    $\mathbf{K}$ is the Kalman gain vector.
    $\mathbf{c}$ is a unit vector of length p.

Equation 6.8 is still only applicable to stationary processes but it may easily be generalised to nonstationary processes by making the following modifications. Make the state variable $\mathbf{x}[n]$ evolve according to the difference Equation 6.9.

$$\mathbf{x}[n] = \mathbf{A}(n-1)\mathbf{x}[n-1] + \mathbf{w}[n] \tag{6.9}$$

where:

$\mathbf{A}(n\text{-}1)$ is a time-varying $p \times p$ state transition matrix.

The zero mean white noise process represented by vector $\mathbf{w}[n]$ has an expectation $E\{\mathbf{w}[n]\mathbf{w}^H[k]\}$ as defined by Equation 6.10.

$$E\{\mathbf{w}[n]\mathbf{w}^H[k]\} = \begin{cases} \mathbf{Q}_\mathbf{w}(n) & : k = n \\ \mathbf{0} & : k \neq n \end{cases} \tag{6.10}$$

Let $\mathbf{y}[n]$ be a q length vector of observations that are defined by Equation 6.11, where $\mathbf{C}(n)$ is a time-varying $q \times p$ matrix, and $\mathbf{v}[n]$ represents a zero mean white noise processes independent of $\mathbf{w}[n]$ and having an expectation $E\{\mathbf{v}[n]\mathbf{v}^H[k]\}$ as defined by Equation 6.12.

$$\mathbf{y}[n] = \mathbf{C}(n)\mathbf{x}[n] + \mathbf{v}[n] \tag{6.11}$$

$$E\{\mathbf{v}[n]\mathbf{v}^H[k]\} = \begin{cases} \mathbf{Q}_\mathbf{v}(n) & : k = n \\ \mathbf{0} & : k \neq n \end{cases} \tag{6.12}$$

If the result defined by Equation 6.8 is generalised the optimum linear estimator, or the discrete Kalman filter, may now be defined by Equation 6.13.

$$\hat{\mathbf{x}}[n] = \mathbf{A}(n-1)\hat{\mathbf{x}}[n-1] + \mathbf{K}(n)[\mathbf{y}[n] - \mathbf{C}(n)\mathbf{A}(n-1)\hat{\mathbf{x}}[n-1]] \tag{6.13}$$

where:

$\mathbf{K}(n)$ is the appropriate Kalman gain matrix. It remains to be shown that the optimum linear recursive estimate of $\mathbf{x}[n]$ has this form and the optimum Kalman gain $\mathbf{K}(n)$ that minimises the mean square estimation needs yet to be derived. To do this it must be assumed that the matrices $\mathbf{A}(n)$, $\mathbf{C}(n)$ $\mathbf{Q}_w(n)$ and $\mathbf{Q}_v(n)$ are all known.

If $\hat{\mathbf{x}}[n\,|\,n]$ denotes the best linear estimate of $\mathbf{x}[n]$ at time n given all the observations $\mathbf{y}[i]$ for $i = 1,2,....,n$ and $\hat{\mathbf{x}}[n\,|\,n-1]$ denotes the best estimate given all the observations up to and including time n-1 then the corresponding state estimation errors are,

$$\mathbf{e}[n\,|\,n] = \mathbf{x}[n] - \hat{\mathbf{x}}[n\,|\,n]$$
$$\mathbf{e}[n\,|\,n-1] = \mathbf{x}[n] - \hat{\mathbf{x}}[n\,|\,n-1]$$

The corresponding covariance errors are consequently defined by Equations 6.14a and 16.4b.

$$\mathbf{P}(n \mid n) = E\{\mathbf{e}[n \mid n]\mathbf{e}^{H}[n \mid n]\} \tag{6.14a}$$

$$\mathbf{P}(n \mid n-1) = E\{\mathbf{e}[n \mid n-1]\mathbf{e}^{H}[n \mid n-1]\} \tag{6.14b}$$

Given an estimate $\hat{\mathbf{x}}[0 \mid 0]$ of the state $\mathbf{x}[0]$, and if the error covariance matrix for this estimate $\mathbf{P}(0|0)$ is known, when the measurement $\mathbf{y}[1]$ becomes available the goal is then to update $\hat{\mathbf{x}}[0 \mid 0]$ and find the estimate $\hat{\mathbf{x}}[1 \mid 1]$ that minimises the mean square error defined by Equation 6.15.

$$\xi(1) = E\{\|e[1 \mid 1]\|^{2}\} = tr\{\mathbf{P}(1 \mid 1)\} = \sum_{i=0}^{p-1} E\{|e_{i}[1 \mid 1]|^{2}\} \tag{6.15}$$

The estimation is repeated for the next observation $\mathbf{y}[2]$ and so on. For each $n > 0$, given $\hat{\mathbf{x}}[n-1 \mid n-1]$ and $\mathbf{P}(n\text{-}1|n\text{-}1)$, when a new observation $\mathbf{y}[n]$ becomes available, the problem is to find the minimum mean square estimate $\hat{\mathbf{x}}[n \mid n]$ of the state vector $\mathbf{x}[n]$. This problem is solved in the following two steps. Firstly, given $\hat{\mathbf{x}}[n-1 \mid n-1]$ find $\hat{\mathbf{x}}[n \mid n-1]$, which is the best estimate of $\mathbf{x}[n]$ without the observation $\mathbf{y}[n]$. Secondly, given $\mathbf{y}[n]$ and $\hat{\mathbf{x}}[n \mid n-1]$ estimate $\mathbf{x}[n]$.

Step 1

In the first step all that is known is the evolution of $\mathbf{x}[n]$ according to the state Equation 6.9. Since $\mathbf{w}[n]$ is an unknown zero mean white noise process then $\mathbf{x}[n]$ is predicted according to Equation 6.16.

$$\hat{\mathbf{x}}[n \mid n-1] = \mathbf{A}(n-1)\hat{\mathbf{x}}[n-1 \mid n-1] \tag{6.16}$$

Equation 6.16 has an unbiased ($E\{\mathbf{e}[n \mid n-1]\} = 0$) estimation error $\mathbf{e}[n \mid n-1]$ defined by Equation 6.17.

$$\begin{aligned}
\mathbf{e}[n \mid n-1] &= \mathbf{x}[n] - \hat{\mathbf{x}}[n \mid n-1] \\
&= \mathbf{A}(n-1)\mathbf{x}[n-1] + \mathbf{w}[n] - \mathbf{A}(n-1)\hat{\mathbf{x}}[n-1 \mid n-1] \\
&= \mathbf{A}(n-1)\mathbf{e}[n-1 \mid n-1] + \mathbf{w}[n]
\end{aligned} \tag{6.17}$$

The estimation errors $\mathbf{e}[n-1 \mid n-1]$ and $\mathbf{e}[n \mid n-1]$ are uncorrelated with $\mathbf{w}[n]$ and therefore $\mathbf{P}(n \mid n-1)$ is defined by Equation 6.18.

$$\mathbf{P}(n \mid n-1) = \mathbf{A}(n-1)\mathbf{P}(n-1 \mid n-1)\mathbf{A}^{H}(n-1) + \mathbf{Q}_{w}(n) \tag{6.18}$$

Step 2

In the second step the new measurement $\mathbf{y}[n]$ is incorporated into the estimate $\hat{\mathbf{x}}[n \mid n-1]$. A linear estimate of $\mathbf{x}[n]$, which is based on $\hat{\mathbf{x}}[n \mid n-1]$ and $\mathbf{y}[n]$, can be formulated as defined by Equation 6.19.

$$\hat{\mathbf{x}}(n \mid n) = \mathbf{K}^{'}(n)\hat{\mathbf{x}}[n \mid n-1] + \mathbf{K}(n)\mathbf{y}[n] \tag{6.19}$$

where:

$\mathbf{K}^{'}(n)$ and $\mathbf{K}(n)$ are matrices yet to be specified.

$\hat{\mathbf{x}}[n \mid n]$ must be unbiased ($E\{\mathbf{e}[n \mid n]\} = 0$) and it must minimise the mean square error ($E\{\|e[n \mid n]\|^2\} = \xi_{\min}(n)$). Using Equation 6.19 the error $\mathbf{e}[n \mid n]$ can be expressed in terms of error $\mathbf{e}[n \mid n-1]$ as defined by Equation 6.20.

$$\begin{aligned}
\mathbf{e}[n \mid n] &= \mathbf{x}[n] - \mathbf{K}^{'}(n)\hat{\mathbf{x}}[n \mid n-1] - \mathbf{K}(n)\mathbf{y}[n] \\
&= \mathbf{x}[n] - \mathbf{K}^{'}(n)\big[\mathbf{x}[n] - \mathbf{e}[n \mid n-1]\big] - \mathbf{K}(n)\big[\mathbf{C}(n)\mathbf{x}[n] + \mathbf{v}[n]\big] \\
&= \big[\mathbf{I} - \mathbf{K}^{'}(n) - \mathbf{K}(n)\mathbf{C}(n)\big]\mathbf{x}[n] + \mathbf{K}^{'}(n)\mathbf{e}[n \mid n-1] - \mathbf{K}(n)\mathbf{v}[n]
\end{aligned} \tag{6.20}$$

Since $E\{\mathbf{v}[n]\} = 0$ and $E\{\mathbf{e}[n \mid n-1]\} = 0$, it can be seen from Equation 6.20 that the estimate $\hat{\mathbf{x}}[n \mid n]$ will be unbiased for any $\mathbf{x}[n]$ only if $\big[\mathbf{I} - \mathbf{K}^{'}(n) - \mathbf{K}(n)\mathbf{C}(n)\big] = 0$ or $\mathbf{K}^{'}(n) = \mathbf{I} - \mathbf{K}(n)\mathbf{C}(n)$. Equation 6.19 can now be expressed as Equation 6.21 or alternatively 6.22.

$$\hat{\mathbf{x}}[n \mid n] = \big[\mathbf{I} - \mathbf{K}(n)\mathbf{C}(n)\big]\hat{\mathbf{x}}[n \mid n-1] + \mathbf{K}(n)\mathbf{y}[n] \tag{6.21}$$

$$\hat{\mathbf{x}}[n \mid n] = \hat{\mathbf{x}}[n \mid n-1] + \mathbf{K}(n)[\mathbf{y}[n] - \mathbf{C}(n)\hat{\mathbf{x}}[n \mid n-1]] \tag{6.22}$$

The error for the estimate $\hat{\mathbf{x}}[n \mid n]$ is defined by Equation 6.23.

$$\begin{aligned}
\mathbf{e}[n \mid n] &= \mathbf{K}^{'}(n)\mathbf{e}[n \mid n-1] - \mathbf{K}(n)\mathbf{v}[n] \\
&= \big[\mathbf{I} - \mathbf{K}(n)\mathbf{C}(n)\big]\mathbf{e}[n \mid n-1] - \mathbf{K}(n)\mathbf{v}[n]
\end{aligned} \tag{6.23}$$

Since $\mathbf{v}[n]$ is uncorrelated with $\mathbf{w}[n]$ then $\mathbf{v}[n]$ is uncorrelated with $\mathbf{x}[n]$ and $\hat{\mathbf{x}}[n \mid n-1]$. Also, since $\mathbf{e}[n \mid n-1] = \mathbf{x}[n] - \hat{\mathbf{x}}[n \mid n-1]$ then $\mathbf{v}[n]$ is uncorrelated with $\mathbf{e}[n \mid n-1]$, i.e., $E\{\mathbf{e}[n \mid n-1]\mathbf{v}[n]\} = 0$. The error covariance matrix for $\mathbf{e}[n \mid n]$ is now as defined by Equation 6.24.

$$\begin{aligned}
\mathbf{P}(n \mid n) &= E\{\mathbf{e}[n \mid n]\mathbf{e}^{H}[n \mid n]\} \\
&= \big[\mathbf{I} - \mathbf{K}(n)\mathbf{C}(n)\big]\mathbf{P}(n \mid n-1)\big[\mathbf{I} - \mathbf{K}(n)\mathbf{C}(n)\big]^{H} + \mathbf{K}(n)\mathbf{Q}_{\mathbf{v}}(n)\mathbf{K}^{H}(n)
\end{aligned} \tag{6.24}$$

The Kalman gain $\mathbf{K}(n)$, which minimises the mean square error $\xi(n) = \text{tr}\big[\mathbf{P}(n \mid n)\big]$, can be found by different means but the most expedient is to differentiate with respect to $\mathbf{K}(n)$ and set the differentials to zero to solve for $\mathbf{K}(n)$. This is done according to Equation 6.25.

$$\frac{d}{d\mathbf{K}}\text{tr}[\mathbf{P}(n\mid n)] = -2[\mathbf{I}-\mathbf{K}(n)\mathbf{C}(n)]\mathbf{P}(n\mid n-1)\mathbf{C}^H(n) + 2\mathbf{K}(n)\mathbf{Q}_v(n)$$

$= 0,$ using the matrix differential formulas,

$$\frac{d}{d\mathbf{K}}\text{tr}[\mathbf{KA}] = \mathbf{A}^H, \text{ and } \frac{d}{d\mathbf{K}}\text{tr}[\mathbf{KAK}^H] = 2\mathbf{KA}$$

(6.25)

Solving for $\mathbf{K}(n)$ the result is as defined by Equation 6.26.

$$\mathbf{K}(n) = \mathbf{P}(n\mid n-1)\mathbf{C}^H(n)[\mathbf{C}(n)\mathbf{P}(n\mid n-1)\mathbf{C}^H(n) + \mathbf{Q}_v(n)]^{-1} \qquad (6.26)$$

The expression given in Equation 6.24 for the error covariance can now be simplified as defined by Equation 6.27.

$$\mathbf{P}(n\mid n) = [\mathbf{I}-\mathbf{K}(n)\mathbf{C}(n)]\mathbf{P}(n\mid n-1)$$
$$-\{[\mathbf{I}-\mathbf{K}(n)\mathbf{C}(n)]\mathbf{P}(n\mid n-1)\mathbf{C}^H(n) + \mathbf{K}(n)\mathbf{Q}_v(n)\}\mathbf{K}^H(n),$$

and from equation (6.25) it follows that, (6.27)

$$\mathbf{P}(n\mid n) = [\mathbf{I}-\mathbf{K}(n)\mathbf{C}(n)]\mathbf{P}(n\mid n-1)$$

This completes the derivation of the recursion equations for the Kalman filter. The recursion must be initialised at time $n = 0$ but the value of the initial state is unknown. Therefore the initial estimate is chosen to be $\hat{\mathbf{x}}[0\mid0] = E\{\mathbf{x}[0]\}$ and the initial value for the covariance matrix is $\mathbf{P}(0\mid0) = E\{\mathbf{x}[0]\mathbf{x}^H[0]\}$. This choice of initial values ensures that $\hat{\mathbf{x}}[0\mid0]$ is an unbiased estimate of $\mathbf{x}[0]$ and it ensures that $\hat{\mathbf{x}}[n\mid n]$ will be unbiased for all n. The Kalman gain $\mathbf{K}(n)$ and error covariance matrix $\mathbf{P}(n\mid n)$ can be computed off-line prior to filtering as they are not dependent on $\mathbf{x}[n]$.

A summary of the Kalman filter is as follows,

Given
State vector: $\mathbf{x}[n]$
Observation vector: $\mathbf{y}[n]$
Covariance matrix of process noise: $\mathbf{Q}_w(n) = \mathbf{Q}(n)$
Covariance matrix of measurement noise: $\mathbf{Q}_v(n) = \mathbf{R}(n)$
State transition matrix from n-1 to n: $\mathbf{A}(n-1,n)$
Measurement matrix: $\mathbf{C}(n)$

State Equation:

$$\mathbf{x}[n] = \mathbf{A}(n-1,n)\mathbf{x}[n-1] + \mathbf{w}[n] \qquad (6.9)$$

Observation Equation:

$$\mathbf{y}[n] = \mathbf{C}(n)\mathbf{x}[n] + \mathbf{v}[n] \tag{6.11}$$

Initialisation

Initial estimate: $\qquad\qquad\qquad\qquad$ $\hat{\mathbf{x}}[0\,|\,0] = E\{\mathbf{x}[0]\}$

Error covariance matrix for estimate $\hat{\mathbf{x}}[0\,|\,0]$: $\quad$ $\mathbf{P}(0\,|\,0) = E\{\mathbf{x}[0]\mathbf{x}^H[0]\}$

Computation: For $n = 1, 2,....$, compute all the following equations in sequence,

$$\hat{\mathbf{x}}[n\,|\,n-1] = \mathbf{A}(n-1,n)\hat{\mathbf{x}}[n-1\,|\,n-1] \tag{6.16}$$

$$\mathbf{P}(n\,|\,n-1) = \mathbf{A}(n-1,n)\mathbf{P}(n-1\,|\,n-1)\mathbf{A}^H(n-1,n) + \mathbf{Q}_\mathbf{w}(n) \tag{6.18}$$

Filter gain:

$$\mathbf{K}(n) = \mathbf{P}(n\,|\,n-1)\mathbf{C}^H(n)\left[\mathbf{C}(n)\mathbf{P}(n\,|\,n-1)\mathbf{C}^H(n) + \mathbf{Q}_\mathbf{v}(n)\right]^{-1} \tag{6.26}$$

Estimator: (best estimate at n given all observations up to $n-1$)

$$\hat{\mathbf{x}}[n\,|\,n] = \hat{\mathbf{x}}[n\,|\,n-1] + \mathbf{K}(n)\left[\mathbf{y}[n] - \mathbf{C}(n)\hat{\mathbf{x}}[n\,|\,n-1]\right] \tag{6.22}$$

Error covariance matrix:

$$\mathbf{P}(n|n) = \left[\mathbf{I} - \mathbf{K}(n)\mathbf{C}(n)\right]\mathbf{P}(n|n-1) \tag{6.27}$$

Following similar arguments the Kalman predictor can also be derived.

A summary of the Kalman predictor is as follows,

State Equation:

$$\mathbf{x}[n] = \mathbf{A}(n-1,n)\mathbf{x}[n-1] + \mathbf{w}[n] \tag{6.9}$$

Observation Equation:

$$\mathbf{y}[n] = \mathbf{C}(n)\mathbf{x}[n] + \mathbf{v}[n] \tag{6.11}$$

Initialisation

Initial estimate: $\qquad\qquad\qquad\qquad$ $\hat{\mathbf{x}}[0\,|\,0] = E\{\mathbf{x}[0]\}$

Error covariance matrix for estimate $\hat{\mathbf{x}}[0\,|\,0]$: $\quad$ $\mathbf{P}(0\,|\,0) = E\{\mathbf{x}[0]\mathbf{x}^H[0]\}$

Computation: For $n = 1, 2,....$, compute all the following equations in sequence,

$$\hat{\mathbf{x}}[n\,|\,n-1] = \mathbf{A}(n-1,n)\hat{\mathbf{x}}[n-1\,|\,n-1]$$

$$\mathbf{P}(n\,|\,n-1) = \mathbf{A}(n-1,n)\mathbf{P}(n-1\,|\,n-1)\mathbf{A}^H(n-1,n) + \mathbf{Q}_\mathbf{w}(n)$$

Predictor gain:

$$\mathbf{G}(n) = \mathbf{A}(n-1,n)\mathbf{P}(n \mid n-1)\mathbf{C}^{H}(n)\left[\mathbf{C}(n)\mathbf{P}(n \mid n-1)\mathbf{C}^{H}(n) + \mathbf{Q}_{v}(n)\right]^{-1}$$

Predictor: (best prediction at $n+1$ given all observations up to n)

$$\hat{\mathbf{x}}[n+1 \mid n] = \mathbf{A}(n-1,n)\hat{\mathbf{x}}[n \mid n-1] + \mathbf{G}(n)\left[\mathbf{y}[n] - \mathbf{C}(n)\hat{\mathbf{x}}[n \mid n-1]\right]$$

Error covariance matrix:

$$\mathbf{P}(n+1 \mid n) = \left[\mathbf{A}(n-1,n) - \mathbf{G}(n)\mathbf{C}(n)\right]\mathbf{P}(n \mid n-1)\mathbf{A}^{H}(n-1,n) + \mathbf{Q}_{v}(n)$$

6.2.1 Kalman Filter Examples

Example 1: Use the Kalman filter to estimate the following AR(1) process,

$$x[n] = 0.8x[n-1] + w[n]$$
$$y[n] = x[n] + v[n]$$

where $w[n]$ and $v[n]$ are uncorrelated white noise processes with respective variances of $\sigma_w^2 = 0.36$ and $\sigma_v^2 = 1$.

Here, $\mathbf{A}(n) = 0.8$ and $\mathbf{C}(n) = 1$, and the Kalman state equation is,

$$\hat{x}[n] = 0.8\hat{x}[n-1] + K(n)\left[y[n] - 0.8\hat{x}[n-1]\right]$$

Since the state vector is scalar, the equations for computing the Kalman gain are also scalar equations,

Equation 6.18 is, $\quad P(n \mid n-1) = (0.8)^2 P(n-1 \mid n-1) + 0.36$

Equation 6.26 is, $\quad K(n) = P(n \mid n-1)\left[P(n \mid n-1) + 1\right]^{-1}$

Equation 6.27 is, $\quad P(n \mid n) = \left[1 - K(n)\right]P(n \mid n-1)$

With $\hat{x}[0] = E\{x[0]\} = 0$, and $P(0 \mid 0) = E\{|x[0]|^2\} = 1$, the Kalman gain and the error covariances for the first few values of n are as follows,

n	$P(n \mid n-1)$	$K(n)$	$P(n \mid n)$
1	1.0000	0.5000	0.5000
2	0.6800	0.4048	0.4048
3	0.6190	0.3824	0.3824

4	0.6047	0.3768	0.3768
5	0.6012	0.3755	0.3755
6	0.6003	0.3751	0.3751
...			
∞	0.6000	0.3750	0.3750

After a few iterations the Kalman filter settles down into its steady-state solution, Equation 6.22, i.e., $\hat{x}[n] = 0.8\hat{x}[n-1] + 0.375\big[y[n] - 0.8\hat{x}[n-1]\big]$, and final mean square error of $\xi = 0.375$. Notice that this result is the same as the causal Wiener filter solution.

Example 2: From (Bozic 1994) is an example for the radar tracking of a plane. Assume the following states and system (or state) equations to describe the problem,

$x_1[n] = \rho(n)$, aircraft radial range.

$x_2[n] = \dot{\rho}(n)$, aircraft radial velocity.

$x_3[n] = \theta(n)$, aircraft bearing.

$x_4[n] = \dot{\theta}(n)$, aircraft bearing rate or angular velocity.

$$\mathbf{x}[n] = \mathbf{A}(n-1,n)\mathbf{x}[n-1] + \mathbf{w}[n-1]$$

$$\begin{bmatrix} x_1[n] \\ x_2[n] \\ x_3[n] \\ x_4[n] \end{bmatrix} = \begin{bmatrix} 1 & T & 0 & 0 \\ 0 & 1 & 0 & 0 \\ 0 & 0 & 1 & T \\ 0 & 0 & 0 & 1 \end{bmatrix} \begin{bmatrix} x_1[n-1] \\ x_2[n-1] \\ x_3[n-1] \\ x_4[n-1] \end{bmatrix} + \begin{bmatrix} 0 \\ u_1[n-1] \\ 0 \\ u_2[n-1] \end{bmatrix}$$

The noise terms $u_1[n-1]$ and $u_2[n-1]$ represent the change in radial velocity and bearing rate respectively over the time interval T. They are each T times the radial and angular acceleration, are random with zero means, and uncorrelated with each other from time interval to time interval. The radar sensors provide noisy estimates of the range $x_1[n] = \rho(n)$ and bearing $x_3[n] = \theta(n)$ at time intervals T. At time n these two sensors produce respective outputs,

$y_1[n] = x_1[n] + v_1[n]$

$y_2[n] = x_2[n] + v_2[n]$

That is,

$$\mathbf{y}[n] = \mathbf{C}(n)\mathbf{x}[n] + \mathbf{v}[n]$$

$$\begin{bmatrix} y_1[n] \\ y_2[n] \end{bmatrix} = \begin{bmatrix} 1 & 0 & 0 & 0 \\ 0 & 0 & 1 & 0 \end{bmatrix} \begin{bmatrix} x_1[n] \\ x_2[n] \\ x_3[n] \\ x_4[n] \end{bmatrix} + \begin{bmatrix} v_1[n] \\ v_2[n] \end{bmatrix}$$

The additive noise components $\mathbf{v}[n]$ are assumed to be Gaussian with zero means and respective component variances $\sigma_\rho^2(n)$ and $\sigma_\theta^2(n)$. Next, the covariance matrices $\mathbf{R}(n) = \mathbf{Q}_v(n)$ for the system, and $\mathbf{Q}(n) = \mathbf{Q}_w(n)$ for the measurement models are needed. These are derived as follows,

$$\mathbf{Q}_v(n) = \mathbf{R}(n) = E\{\mathbf{v}[n]\mathbf{v}^T[n]\} = \begin{bmatrix} \sigma_\rho^2 & 0 \\ 0 & \sigma_\theta^2 \end{bmatrix}$$

$$\mathbf{Q}_w(n) = \mathbf{Q}(n) = E\{\mathbf{w}[n]\mathbf{w}^T[n]\} = \begin{bmatrix} 0 & 0 & 0 & 0 \\ 0 & \sigma_1^2 & 0 & 0 \\ 0 & 0 & 0 & 0 \\ 0 & 0 & 0 & \sigma_2^2 \end{bmatrix}$$

where $\sigma_1^2(n) = E\{u_1^2[n]\}$, and $\sigma_2^2(n) = E\{u_2^2[n]\}$ are the variances of T times the radial and angular acceleration respectively. Assuming that the probability density function of the acceleration in either direction (ρ or σ) is uniform and equal to $p(u) = \dfrac{1}{2M}$, between limits $\pm M$, the variance is therefore $\sigma_u^2 = \dfrac{M^2}{3}$. More realistic variances are $\sigma_1^2 = T^2 \sigma_u^2$, and $\sigma_2^2 = \dfrac{\sigma_1^2}{R^2}$.

To start the Kalman processing the gain Matrix $\mathbf{K}(k)$ is initialised by specifying the error covariance matrix $\mathbf{P}(k|k)$ in some way. An ad hoc way to do this is to use two measurements of range and bearing at times $k = 1$ and $k = 2$. Take the following estimates,

$$\hat{\mathbf{x}}[2] = \begin{bmatrix} \hat{x}_1[2] = \hat{\rho}(2) = y_1[2] \\ \hat{x}_2[2] = \hat{\dot{\rho}}(2) = \dfrac{1}{T}[y_1[2] - y_1[1]] \\ \hat{x}_3[2] = \hat{\theta}(2) = y_2[2] \\ \hat{x}_4[2] = \hat{\dot{\theta}}(2) = \dfrac{1}{T}[y_2[2] - y_2[1]] \end{bmatrix}$$

$$\mathbf{P}(2\,|\,2) = E\{[\mathbf{x}[2] - \hat{\mathbf{x}}[2]][\mathbf{x}[2] - \hat{\mathbf{x}}[2]]^H\}$$

Therefore,

$$\mathbf{x}[2] - \hat{\mathbf{x}}[2] = \begin{bmatrix} u_1[1] - \dfrac{-v_1[2]}{(v_1[2] - v_1[1])} \\[2pt] u_2[1] - \dfrac{-v_2[2]}{(v_2[2] - v_2[1])} \\ T \end{bmatrix}$$

Noise sources u and v are independent therefore,

$$\mathbf{P}(2\,|\,2) = \begin{bmatrix} \sigma_\rho^2 & \dfrac{\sigma_\rho^2}{T} & 0 & 0 \\[8pt] \dfrac{\sigma_\rho^2}{T} & \dfrac{2\sigma_\rho^2}{T^2} + \sigma_1^2 & 0 & 0 \\[8pt] 0 & 0 & \sigma_\theta^2 & \dfrac{\sigma_\theta^2}{T} \\[8pt] 0 & 0 & \dfrac{\sigma_\theta^2}{T} & \dfrac{2\sigma_\theta^2}{T^2} + \sigma_2^2 \end{bmatrix}$$

For a numerical example assume $R = 160$ km, $T = 15$ seconds and the maximum acceleration $M = 2.1$ ms^{-1}, Let $\sigma_\rho = 1000$ m, and $\sigma_\theta = 0.017$ radians, which define the covariance matrix $\mathbf{R}$. The noise variances in the $\mathbf{Q}$ matrix are computed as $\sigma_1^2 = 330$, and $\sigma_2^2 = 1.3 \times 10^{-8}$. From these values,

$$\mathbf{P}(2\,|\,2) = \begin{bmatrix} 10^6 & 6.7 \times 10^4 & 0 & 0 \\ 6.7 \times 10^4 & 0.9 \times 10^4 & 0 & 0 \\ 0 & 0 & 2.9 \times 10^{-4} & 1.9 \times 10^{-5} \\ 0 & 0 & 1.9 \times 10^{-5} & 2.6 \times 10^{-6} \end{bmatrix}$$

The predictor gain $\mathbf{G}(3)$ is,

$$\mathbf{G}(3) = \mathbf{A}(2,3)\mathbf{P}(3\,|\,2)\mathbf{C}^H(3)\left[\mathbf{C}(3)\mathbf{P}(3\,|\,2)\mathbf{C}^H(3) + \mathbf{R}(3)\right]^{-1}$$

$\mathbf{P}(3\,|\,2)$ can be calculated as,

$$\mathbf{P}(3\,|\,2) = \left[\mathbf{A}(1,2) - \mathbf{G}(2)\mathbf{C}(2)\right]\mathbf{P}(2\,|\,1)\mathbf{A}^H(1,2) + \mathbf{Q}(n)$$

However, $\mathbf{G}(2)$ and $\mathbf{P}(2|1)$ are still not known, but $\mathbf{P}(3|2)$ can still be computed as,

$$\mathbf{P}(3|2) = \mathbf{A}(2,3)\mathbf{P}(2|2)\mathbf{A}^{H}(2,3) + \mathbf{Q}(n) \text{ , i.e.,}$$

$$\mathbf{P}(3|2) = \begin{bmatrix} 5\times10^{6} & 2\times10^{5} & 0 & 0 \\ 2\times10^{5} & 9.3\times10^{3} & 0 & 0 \\ 0 & 0 & 14.5\times10^{-4} & 5.8\times10^{-5} \\ 0 & 0 & 5.8\times10^{-5} & 2.6\times10^{-6} \end{bmatrix}$$

The diagonal values of $\mathbf{P}(3|2)$ give the prediction errors. The first and third elements are the mean square range and bearing prediction errors for $k = 3$ respectively.

The predictor gain $\mathbf{G}(3)$ is now,

$$\mathbf{G}(3) = \begin{bmatrix} 1.33 & 0 \\ 3.3\times10^{-2} & 0 \\ 0 & 1.33 \\ 0 & 3.3\times10^{-2} \end{bmatrix}$$

This process can now be repeated indefinitely to keep estimating the prediction values for the state estimate $\hat{\mathbf{x}}(k+1)$.

6.3 Kalman Filter for Ship Motion

A ship tracking example is a good one to use to develop and further demonstrate the ideas of a Kalman filter because it is an easy to comprehend two-dimensional problem. The ship's state vector $\mathbf{x}$ in its simplest form contains easting and northing $[E, N]^{T}$ position coordinates. It can also include and contain the ship's velocity, acceleration and quantities that describe the interaction between the ship and its environment. The optimum estimator includes all the available observations made to "fix" the ship's position e.g., ranges to shore or seabed stations, satellite observations, gyrocompass heading and Doppler sonar velocities. It also includes the manner in which the ship might be expected to move through the water (ship's dynamics) bearing in mind the various forces (such as wind, current, rudders, thrusters etc.) acting on it. Thus, according to (Cross 1987) the Kalman filter combines information related to both observations and dynamics to produce some smooth optimal track of the ship using LSE as the optimality criterion. The entire LSE minimisation problem can be defined generally by Equation 6.28.

$$\text{LSE minimises } \mathbf{W}^o(\mathbf{x}[n]-\mathbf{x}^o[n])^2+\mathbf{W}^d(\mathbf{x}[n]-\mathbf{x}^d[n])^2 \qquad (6.28)$$

where:

$\mathbf{x}[n]=$ estimate of the state at some time $t = nT$.

$\mathbf{x}^o[n]=$ estimate given by the observations at $t = nT$.

$\mathbf{x}^d[n]=$ estimate given by the modelled dynamics at $t = nT$.

$\hat{\mathbf{x}}[n]=$ LSE of $\mathbf{x}[n]$.

$\mathbf{W}^o$ = weights attached to the observations.

$\mathbf{W}^d$ = weights attached to the dynamics.

T = observation time interval.

The optimal LSE minimises the quantity θ, as defined by Equation 6.29.

$$\theta = \mathbf{v}^T\mathbf{R}\mathbf{v}+\mathbf{w}^T\mathbf{Q}\mathbf{w} \qquad (6.29)$$

where:

$\mathbf{v}$ = residual vector for observations

   (amount by which the estimate fails to satisfy the observations).

$\mathbf{w}$ = amount by which final estimate diverges from prediction by the dynamics.

$\mathbf{R}$ = weight matrix for observations $\Big\}$ inverse of covariance matrix, usually diagonal.
$\mathbf{Q}$ = weight matrix for dynamics

The choice of least squares as the optimum estimator for this problem is justified by the following properties of least squares (Cross 1983),

1.   They are unbiased, i.e., true or average.

2.   They have the smallest variance.

3.   If the errors are Gaussian distributed (or have any continuous symmetrical distribution) they satisfy the maximum likelihood criterion (most probable solution).

Least squares estimates are often called the best linear unbiased estimates and are used almost exclusively for both land and sea surveying.

6.3.1 Kalman Tracking Filter Proper

The general Kalman ship tracking filter can be developed as follows. Firstly, assume that the following information is available,

$n \equiv$ some time where $t = nT$.

$\hat{\mathbf{x}}[n]$ = optimal estimate of ship's vector at time $t = nT$.

$\mathbf{P}(n)$ = covariance matrix of $\hat{\mathbf{x}}(n)$.

$\ell(n)$ = measurements made at time $t = nT$, which are related to the state vector.

The functional relationship $\dot{F}(\mathbf{x}[n])$ between the state vector $\mathbf{x}[n]$ and the measurements $\ell(n)$, made at time $t = nT$, is defined by Equation 6.30. Also, assume that a dynamic ship motion model of the form $F(\mathbf{x},t)=0$ is available.

$$\dot{F}(\mathbf{x}[n]) = \ell(n) \tag{6.30}$$

The problem is to combine all this information to find $\hat{\mathbf{x}}[n \mid n]$ and $\mathbf{P}(n \mid n)$, which the Kalman filter does in two main steps as follows,

1.   Predict the state vector and its covariance matrix using the dynamic model and its associated statistics.

2.   Alter this prediction using the observation model and the statistics of both the observation and dynamic models.

The model $F(\mathbf{x},t)=0$ must be reduced by some means to the discrete form Equation 6.31.

$$\mathbf{x}[n \mid n-1] = \mathbf{A}(n-1,n)\mathbf{x}[n-1 \mid n-1] + \mathbf{w}[n] \tag{6.31}$$

where:

$\mathbf{A}(n-1,n)$ is the transition matrix or dynamics matrix (square matrix), which multiplies the state vector at time $n-1$ to produce the state vector at n.

$\mathbf{x}[n \mid n-1]$ means the value of $\mathbf{x}$ at time n using observational data up to only time $n-1$.

Since the random component $\mathbf{w}[n]$ in Equation 6.31 is unknown the prediction process $\hat{\mathbf{x}}[n \mid n-1]$ would be done by Equation 6.32.

$$\hat{\mathbf{x}}[n \mid n-1] = \mathbf{A}(n-1,n)\hat{\mathbf{x}}[n-1 \mid n-1] \tag{6.32}$$

In practical problems the transition matrix $\mathbf{A}(n-1,n)$ can often be extremely complicated to determine. In simple cases, as will be exemplified in the next Section, the model $F(\mathbf{x},t)=0$ can be expressed directly in the form of Equation 6.32. In the more general case the real physical process of ship motion might be described by a linear differential of the form $\dot{\mathbf{x}} = F(\mathbf{x},\mathbf{w})$. In some special cases the transitional matrix $\mathbf{A}(n-1,n)$ in Equation 6.32 can be derived analytically via

Laplace transforms but usually a numerical integration process is necessary, as defined by Equation 6.33.

$$\dot{\mathbf{A}}(n-1,n) = \mathbf{M}\mathbf{A}(n-1,n) \tag{6.33}$$

where:

$\mathbf{M}$ is a square Jacobian matrix $\partial F/\partial \mathbf{x}$ obtained by differentiating each row of $F(\mathbf{x}, t) = 0$ with respect to each element of the state vector.

The covariance matrix $\mathbf{P}(n\,|\,n-1)$ of the optimum estimate is defined and computed by Equation 6.34 once the transitional matrix $\mathbf{A}(n-1,n)$ is determined.

$$\mathbf{P}(n\,|\,n-1) = \mathbf{A}(n-1,n)\mathbf{P}(n-1\,|\,n-1)\mathbf{A}^{H}(n-1,n) + \mathbf{Q}_{\mathbf{w}}(n) \tag{6.34}$$

where:

$\mathbf{Q}_{\mathbf{w}}$ is the covariance of $\mathbf{w}$, which usually requires a complicated numerical integration to compute. In certain dynamic models as will investigated later $\mathbf{Q}_{\mathbf{w}}$ is a constant and explicitly known.

The next step is to linearize Equation 6.30, $F(\mathbf{x}[n]) = \ell(n)$, to the form defined by Equation 6.35.

$$\mathbf{J}(n)\mathbf{x}[n] = \mathbf{b}[n] + \mathbf{v}[n] \tag{6.35}$$

where:

$\mathbf{J}(n) = $ Jacobian matrix $\partial F(\mathbf{x}[n])/\partial x$.

$\mathbf{b}[n] = $ [Observed − Computed quantities], $\mathbf{b}[n] = F(\mathbf{x}^{o}[n]) - \ell(n)$.

$\mathbf{x}^{o}[n] = $ provisional values of $\mathbf{x}[n]$.

The form of Equation 6.35 has been seen before in the standard LSE problem but in this case it would not be solved by the usual LSE process because there are more parameters than observations. Observation information enters the Kalman filter process along with a gain matrix $\mathbf{K}(n)$ as defined by Equation 6.36.

$$\mathbf{K}(n) = \mathbf{P}(n\,|\,n-1)\mathbf{J}^{H}(n)[\mathbf{J}(n)\mathbf{P}(n\,|\,n-1)\mathbf{J}^{H}(n) + \mathbf{Q}_{\ell}(n)]^{-1} \tag{6.36}$$

where:

$\mathbf{Q}_{\ell} = \mathbf{Q}_{\mathbf{v}}$, the covariance matrix of the observations.

The gain matrix $\mathbf{K}(n)$ correctly combines dynamic model and observation information, consequently Equations 6.37 and 6.38 may be defined accordingly.

$$\hat{\mathbf{x}}[n\,|\,n] = \hat{\mathbf{x}}[n\,|\,n-1] + \mathbf{K}(n)[\mathbf{b}[n] - \mathbf{J}(n)\mathbf{x}[n\,|\,n-1]] \tag{6.37}$$

$$\mathbf{P}(n\,|\,n) = (\mathbf{I} - \mathbf{K}(n)\mathbf{J}(n))\mathbf{P}(n\,|\,n-1) \tag{6.38}$$

Now, there is enough information to define the Kalman filtering process as follows. Given $\hat{\mathbf{x}}[n-1]$ and $\mathbf{P}(n-1)$ initially, then,

1. Compute the predicted state by Equation 6.32,

$$\hat{\mathbf{x}}[n \mid n-1] = \mathbf{A}(n-1,n)\hat{\mathbf{x}}[n-1 \mid n-1]$$

2. Compute the predicted state covariance matrix by Equation 6.34,

$$\mathbf{P}(n \mid n-1) = \mathbf{A}(n-1,n)\mathbf{P}(n-1 \mid n-1)\mathbf{A}^H(n-1,n)+\mathbf{Q}_\mathbf{w}(n)$$

3. Compute the gain matrix $\mathbf{K}(n)$ by Equation 6.36,

$$\mathbf{K}(n) = \mathbf{P}(n \mid n-1)\mathbf{J}^H(n)[\mathbf{J}(n)\mathbf{P}(n \mid n-1)\mathbf{J}^H(n)+\mathbf{Q}_\ell(n)]^{-1}$$

4. Estimate the new filtered state $\hat{\mathbf{x}}[n \mid n]$ by Equation 6.37,

$$\hat{\mathbf{x}}[n \mid n] = \hat{\mathbf{x}}[n \mid n-1]+\mathbf{K}(n)\big[\mathbf{b}[n]-\mathbf{J}(n)\mathbf{x}[n \mid n-1]\big]$$

5. Increment n and go to Step 1.

6.3.2 Simple Example of a Dynamic Ship Model

An example of a possible dynamic ship model $F(\mathbf{x},t)=0$ suitable for navigation at sea can be formed by employing a simple universally useful polynomial model. Assume some unknown process $\mathbf{x}(t) = \mathbf{x}(nT)$ is continuous and a function of position as defined by Equation 6.39.

$$\mathbf{x}(t) = [E,N]^T \tag{6.39}$$

Also, assume for purposes of analysis, that $\delta t = T$. Now, expand $\mathbf{x}(t)$ by Taylor's series as defined by Equation 6.40.

$$\mathbf{x}(t+\delta t) = \mathbf{x}(t)+\dot{\mathbf{x}}(t)\delta t+\ddot{\mathbf{x}}(t)\frac{\delta t^2}{2}+\dddot{\mathbf{x}}(t)\frac{\delta t^3}{6}.....etc. \tag{6.40}$$

If the time derivatives $\dot{\mathbf{x}}(t)$ and $\ddot{\mathbf{x}}(t)$ are assumed to be continuous and all differentials higher than $\dddot{\mathbf{x}}(t)$ are assumed to be negligible Equations 6.41 and 6.42 follow.

$$\dot{\mathbf{x}}(t+\delta t) = \dot{\mathbf{x}}(t)+\ddot{\mathbf{x}}(t)\delta t+\dddot{\mathbf{x}}(t)\frac{\delta t^2}{2} \tag{6.41}$$

$$\ddot{\mathbf{x}}(t+\delta t) = \ddot{\mathbf{x}}(t) + \dddot{\mathbf{x}}(t)\delta t \tag{6.42}$$

Let $t = nT$ and $t = (n+1)T$ be two time epochs separated by time $\delta t = T$ then Equation 6.43 can be defined accordingly.

$$
\begin{bmatrix} \mathbf{x} \\ \dot{\mathbf{x}} \\ \ddot{\mathbf{x}} \end{bmatrix}_{n+1}
=
\begin{bmatrix} 1 & \delta t & \dfrac{\delta t^2}{2} \\ 0 & 1 & \delta t \\ 0 & 0 & 1 \end{bmatrix}
\begin{bmatrix} \mathbf{x} \\ \dot{\mathbf{x}} \\ \ddot{\mathbf{x}} \end{bmatrix}_{n}
+
\begin{bmatrix} \dfrac{\delta t^3}{6} \\ \dfrac{\delta t^2}{2} \\ \delta t \end{bmatrix}
\begin{bmatrix} \dddot{\mathbf{x}} \end{bmatrix}
\tag{6.43}
$$

Specifically for ship positioning define the position, velocity and acceleration vector as follows,

$$
\begin{bmatrix} \mathbf{x}, \dot{\mathbf{x}}, \ddot{\mathbf{x}} \end{bmatrix}^T
=
\begin{bmatrix} E, N, \dot{E}, \dot{N}, \ddot{E}, \ddot{N} \end{bmatrix}^T
$$

Substituting this vector in Equation 6.43 results in Equation 6.44.

$$
\begin{bmatrix} E \\ N \\ \dot{E} \\ \dot{N} \\ \ddot{E} \\ \ddot{N} \end{bmatrix}_{\substack{n+1, \\ \text{or} \\ n}}
=
\begin{bmatrix}
1 & 0 & \delta t & 0 & \dfrac{\delta t^2}{2} & 0 \\
0 & 1 & 0 & \delta t & 0 & \dfrac{\delta t^2}{2} \\
0 & 0 & 1 & 0 & \delta t & 0 \\
0 & 0 & 0 & 1 & 0 & \delta t \\
0 & 0 & 0 & 0 & 1 & 0 \\
0 & 0 & 0 & 0 & 0 & 1
\end{bmatrix}
\begin{bmatrix} E \\ N \\ \dot{E} \\ \dot{N} \\ \ddot{E} \\ \ddot{N} \end{bmatrix}_{\substack{n, \\ \text{or} \\ n-1}}
+
\begin{bmatrix}
\dfrac{\delta t^3}{6} & 0 \\
0 & \dfrac{\delta t^3}{6} \\
\dfrac{\delta t^2}{2} & 0 \\
0 & \dfrac{\delta t^2}{2} \\
\delta t & 0 \\
0 & \delta t
\end{bmatrix}
\begin{bmatrix} \dddot{E} \\ \dddot{N} \end{bmatrix}
\tag{6.44}
$$

Allowing for index adjustments Equation 6.44 is the realisation of Equation 6.31 for the current polynomial based dynamic ship model, i.e.,

$$\mathbf{x}[n \mid n] = \mathbf{A}(n-1, n)\mathbf{x}[n-1] + \mathbf{w}[n].$$

For the ship tracking problem the last part of Equations 6.44 and 6.31 can be equated by defining $\mathbf{w} = \Sigma\mathbf{g}$, where,

$\mathbf{g}$ = random driving noise vector,

and,

$$\Sigma = \begin{bmatrix} \dfrac{\delta t^3}{6} & 0 \\[2mm] 0 & \dfrac{\delta t^3}{6} \\[2mm] \dfrac{\delta t^2}{2} & 0 \\[2mm] 0 & \dfrac{\delta t^2}{2} \\[2mm] \delta t & 0 \\[2mm] 0 & \delta t \end{bmatrix} \text{, a matrix from Equation 6.44.}$$

The random noise vector **g** is the rate of change of the ship's acceleration, which can be assumed to be random and having a covariance matrix of $\mathbf{Q_g}$. Its standard deviation σ depends on sea conditions and could be estimated from the Kalman filtering process itself. The dynamics covariance matrix $\mathbf{Q_w}$ needed to compute the covariance matrix of the predicted state vector, Equation 6.34, is defined by the Equation 6.45.

$$\mathbf{Qw} = \Sigma \; \mathbf{Qg} \; \Sigma^H \tag{6.45}$$

where:

$$\mathbf{Q_g} = \begin{bmatrix} \sigma & 0 \\ 0 & \sigma \end{bmatrix}$$

In detail expression Equation 6.45 is defined by Equation 6.46.

$$\mathbf{Q_w} = \begin{bmatrix} \dfrac{\delta t^3}{6} & 0 \\[2mm] 0 & \dfrac{\delta t^3}{6} \\[2mm] \dfrac{\delta t^2}{2} & 0 \\[2mm] 0 & \dfrac{\delta t^2}{2} \\[2mm] \delta t & 0 \\[2mm] 0 & \delta t \end{bmatrix} \begin{bmatrix} \sigma & 0 \\ 0 & \sigma \end{bmatrix} \begin{bmatrix} \dfrac{\delta t^3}{6} & 0 & \dfrac{\delta t^2}{2} & 0 & \delta t & 0 \\[2mm] 0 & \dfrac{\delta t^3}{6} & 0 & \dfrac{\delta t^2}{2} & 0 & \delta t \end{bmatrix} \tag{6.46}$$

Some typical error and parameter values for this type of ship tracking problem that may be used are as follows,

1.   10 m standard error for each of 3 range positioning systems.

2.   0.1 degrees gyrocompass error of heading.

3.   1% error of velocity of dual axis Doppler velocity sonar systems.

4. $\delta t = T = 10$ seconds between fixes.

There are other more accurate dynamically models utilising the general hydrodynamic equations of a ship's motion that could have been used but this was the simplest case to consider.

6.3.3 Stochastic Models

It is important to assign correct covariance matrices to both the dynamic model and observations. If they are over optimistic, results appear to be of a higher quality than they really are. This is the most serious situation. Overall sizes of covariance matrices can usually be checked by performing statistical tests on $\mathbf{v}^T \mathbf{Q}_\ell^{-1} \mathbf{v}$ and $\mathbf{w}^T \mathbf{Q_w} \mathbf{w}$. The relative sizes of $\mathbf{Q}_\ell$ and $\mathbf{Q_w}$ are also most important.

If $\mathbf{Q}_\ell$ is too optimistic, vis a vis $\mathbf{Q_w}$, then "under-filtering" occurs and the final answer will fit the observations extremely well at the expense of the dynamic model. The result will be an uneven track that will often be seen to be wrong, even though it will never drift from the true track for long. If the opposite is true then "over-filtering" occurs and a very smooth track is produced that may contain rather large position errors. It is feasible to alter covariance matrices manually to, at times, induce soft and hard filtering.

Unless the functional (dynamic and observational) models are known to be correct in practice it is difficult to determine these covariance matrices automatically. A guide to the proper operation of the process is through the prediction of residuals defined by Equation 6.47.

$$\hat{\mathbf{r}}[n] = \mathbf{b}[n] - \mathbf{J}(n)\hat{\mathbf{x}}[n \mid n-1] \tag{6.47}$$

If $\hat{\mathbf{r}}(n)$ is unexpectedly large then either gross observational errors are present or the dynamic model is incorrectly predicting the state vector. Usually the former case causes spasmodic errors whereas the later case will result in a gradual increase in the size of the predicted residuals. If observations and dynamic errors are Gaussian then $\hat{\mathbf{r}}(n)$ should also be Gaussian, where $\mathbf{Q}_{\hat{\mathbf{r}}(n)} = \mathbf{Q}_{\ell(n)} + \mathbf{J}(n)\mathbf{P}(n \mid n)\mathbf{J}^T(n)$, and the quadratic form $\chi[n] = \mathbf{r}^T(n)\mathbf{Q}_{\mathbf{r}(n)}^{-1}\mathbf{r}(n)$ has a Chi-squared distribution, which can be tested in the usual way.

6.3.4 Alternate Solution Methods

The Kalman filter is not the only way to combine observational and dynamic ship model information. It can be done with the general LSE method by writing "pseudo" observation equations for motion and following the standard procedure

(Cross 1982). Bayes filtering can also be used for ship tracking. There is however, only one optimum estimate for the state vector, and proper application of a Kalman, Bayes or LSE filter must lead to identical results for this type of ship tracking problem. What is significant about using a Kalman filter is that it has less computational complexity. Kalman filtering requires only one matrix inversion per recursion, a matrix inversion that is smaller than the required matrix inversions in the other methods.

6.3.5 Advantages of Kalman Filtering

Some advantages of Kalman filtering are,

1.   It enables the convenient combination of a variety of observables in order to solve continuously for a number of state vector elements.

2.   The method can accept measurements in real-time and does not need to wait until enough, for a "fix," have been collected. In fact, it is possible to apply the filter every time a signal observation is made.

3.   The combination of a variety of data and the introduction of a dynamics model greatly increases the reliability of the measurements and for offshore work this is usually more critical than precision. Even completely unreliable data can be made, to some extent, reliable by the inclusion of a dynamics model.

6.3.6 Disadvantage of Kalman Filtering

The main disadvantage of Kalman filtering is,

1.   Its practical application is fraught with dangers and great care must be taken to select the appropriate functional and stochastic models. The modelling errors may cause the filter to diverge. It should not be used blindly without understanding its limitations.

The Kalman filter does not function correctly and may diverge when the Kalman gain $\mathbf{K}(n)$ becomes small. However, the measurements still contain information for the estimates. For small $\mathbf{K}(n)$, the estimator believes the model and for large $\mathbf{K}(n)$ it believes the measurements. The gain $\mathbf{K}(n)$ is proportional to $\dfrac{Q_w}{Q_v}$. As Q_w increases, or Q_v decreases, $\mathbf{K}(n)$ increases, and the filter bandwidth increases.

6.4 Extended Kalman Filter

The Kalman ship track filtering problem described above is involved with the estimation of a state vector in a linear model of a dynamical system. For nonlinear dynamical models it is possible to extend the Kalman filter through an appropriate linearization procedure since the Kalman filter is described in terms of differential (continuous-time) or difference (discrete-time) equations (Haykin 1996). The Extended Kalman Filter (EKF) is actually an approximation that allows the standard Kalman filter to be extended in application to nonlinear state-space models. The basic idea of the EKF is to linearize the state-space model at each time instant around the most recent state estimate. Once each linear model is determined the standard Kalman filter equations are applied.

The Decoupled Extended Kalman Filter (DEKF) can be used to perform the supervised training of a recurrent network, e.g. a recurrent Multi-Layer Perceptron (MLP) neural network structure (Haykin 1999). This is achieved by evolving the system via adaptive filtering to change the recurrent network's weights through training. The DEKF has a superior learning performance over the Backpropagation-of-error learning algorithm (see Chapter 12) because of its information preserving property.

6.5 Exercises

The following Exercises identify some of the basic ideas presented in this Chapter.

6.5.1 Problems

6.1.   Is the Kalman filter useful for filtering nonstationary and nonlinear processes?

6.2.   Develop a Kalman filter to estimate the value of an unknown scalar constant x given measurements that are corrupted by an uncorrelated, zero mean white noise $v[n]$ that has a variance of σ_v^2.

6.3.   Use a Kalman filter to estimate the first-order AR process defined by,

$$x[n] = 0.5x[n-1] + w[n],$$

where $w[n]$ is zero mean white noise with a variance $\sigma_w^2 = 0.64$.

The noisy measurements of $x[n]$ are defined by the equation,

$$y[n] = x[n] + v[n],$$

where $v[n]$ is unit variance zero mean white noise that is uncorrelated with $w[n]$ ($\sigma_v^2 = 1$).

6.4.   Can the exponentially weighted RLS algorithm (refer to Chapter 8) be considered to be a special case of the Kalman filter?

7. Power Spectral Density Analysis

The Power Spectral Density (PSD) is the Fourier transform of the autocorrelation sequence, therefore the problem of estimating the power spectrum is mathematically equivalent to estimating the autocorrelation function. In general power spectral estimation is performed on wide-sense stationary random processes. However, most practical interest is actually in ergodic processes, which are more amenable to analysis. Since there are number of different conventions found in the literature for the definition and normalisation of the PSD special attention must be given to the precise terms used in the development of the power spectral theory to follow.

If $x(t)$ is an ergodic random nonperiodic stationary process for which the condition $\int_{-\infty}^{+\infty}|x(t)|dt < \infty$ is not satisfied it is not possible to apply a Fourier Transform to it. However the autocorrelation function of such a process can be estimated if it is normalised to have a mean of zero. In that case it can be assumed that the autocorrelation function $r_x(\tau \to \infty) = 0$ and $\int_{-\infty}^{+\infty}|r_x(\tau)|d\tau < \infty$, thereby making it possible to apply a Fourier Transform to the autocorrelation function in order to compute the power spectrum.

The two-sided spectral density function of $x(t)$ is a real valued function $P_x(f)$ defined by Equation 7.1.

$$P_x(f) = FT[r_x(\tau)] = \int_{-\infty}^{+\infty} r_x(\tau)e^{-j2\pi f\tau}d\tau \tag{7.1}$$

where:

$r_x(\tau) = \lim_{T\to\infty}\frac{1}{T}\int_0^T x(t)x(t+\tau)dt$, converges as $T \to \infty$.

$r_x(-\tau) = r_x(\tau)$

$r_x(0) \ge |r_x(\tau)|$, for all τ.

$\overline{x(t)} = \sqrt{r_x(\infty)}$, mean of $x(t)$.

$MS(x(t)) = r_x(0)$, mean square of $x(t)$.

The autocorrelation $r_x(k)$ of an ergodic discrete-time process $\{x[n]\}$ can be computed as defined by Equation 7.2.

$$r_x(k) = \lim_{N \to \infty} \left\{ \frac{1}{2N+1} \sum_{n=-N}^{N} x[n+k]x^*[n] \right\} \tag{7.2}$$

However, the problem with Equation 7.2 is that in practice there is never unlimited data available and often it can be very short, especially over periods of stationarity. Another general problem with spectral estimation is that the process data is usually corrupted by noise or an interfering signal. Consequently in practice, the spectrum estimation often involves estimating the Fourier transform of the autocorrelation sequence, theoretically defined by Equation 7.3, from a finite number of noisy discrete measurements $x[n]$.

$$P_x(e^{j\theta}) = \sum_{k=-\infty}^{\infty} r_x(k)e^{-jk\theta} \tag{7.3}$$

If knowledge is available about the process this can help produce a much better power spectral estimate. For example, if it is known that the process consists of one or more sinusoids in Gaussian noise it would be possible to parametrically estimate the power spectrum. Otherwise, it may be possible to extrapolate the data or its autocorrelation in order to improve the estimation algorithm.

Spectral estimation has application in a variety of fields. For example, the Wiener filter is made up from the power spectrums of the signal and noise, which in practice must often be estimated. The power spectrum must also be estimated in signal detection, signal tracking, harmonic analysis, prediction, time series extrapolation and interpolation, spectral smoothing, bandwidth compression, beamforming and direction finding problems (Hayes 1996).

7.1 Power Spectral Density Estimation Techniques

There are two main approaches to spectral estimation, the classical or nonparametric and the nonclassical or parametric approaches. The classical approaches involve taking the Fourier transform of the estimate of the autocorrelation sequence made from a given data set. On the other hand, the nonclassical parametric approaches are based on using a process model either known or guessed *a priori*. Some important classical nonparametric spectral estimation methods include the,

1.   Periodogram method.

2.   Modified periodogram method - windowing.

3.   Bartlett's method - periodogram averaging.

4.   Welch's method.

5.   Blackman-Tukey method.

The Blackman-Tukey method is the most popular of the classical methods as it generally gives the best overall performance. Some other nonparametric spectral estimation methods include the,

1.   Minimum variance method.

2.   Maximum entropy (all poles) method.

Some high-resolution nonclassical parametric spectral estimation approaches include,

1.   Autoregressive methods.

2.   Moving average methods.

3.   Autoregressive moving average methods.

4.   Harmonic methods.

All of these methods for spectral estimation are briefly described and summarised in this Chapter, emulating the general development from (Hayes 1996) but with a slightly different nomenclature, style and emphasis adopted for this book.

7.2 Nonparametric Spectral Density Estimation

The periodogram method was first introduced by Schuster in 1898. It is very easy to compute but it is very limited in its ability to estimate accurate power spectral densities, especially for short data records. Fortunately there are a number of modifications to the basis periodogram method that can improve its statistical properties. These include the modified periodogram method, Bartlett's method, Welch's method, and the Blackman-Tukey method.

7.2.1 Periodogram Power Spectral Density Estimation

The power spectral density of a wide-sense stationary random process is defined as the Fourier transform of its autocorrelation function as given by Equation 7.3. For an ergodic process and an unlimited amount of data the autocorrelation sequence may, in theory, be determined using the time-average defined by Equation 7.2. However, since the process $x[n]$ is actually measured over a finite interval, $n = 0, 1,$

$2, \ldots, N-1$, then the autocorrelation can only be adequately estimated using the finite sum defined by Equation 7.4.

$$\hat{r}_x(k) = \frac{1}{N} \sum_{n=0}^{N-1-k} x[n+k]x^*[n], \qquad k = 0,1,..,N-1 \tag{7.4}$$

The values of $\hat{r}_x(k)$ for $k < 0$ can be defined by the conjugate symmetry $\hat{r}_x(-k) = \hat{r}_x^*(k)$, and the values of $\hat{r}_x(k)$ for $|k| \geq N$ are simply set to zero. The periodogram defined by Equation 7.5 is the estimate of the power spectrum by taking the Fourier transform of the finite sum autocorrelation estimate defined by Equation 7.4.

$$\hat{P}_{per\_x}(e^{j\theta}) = \sum_{k=-N+1}^{N-1} \hat{r}_x(k)e^{-jk\theta} \tag{7.5}$$

where:

θ is an angle in radians.

In practice it is more convenient to express the periodogram directly in terms of the sequence $x[n]$ itself. If $x_N[n]$ is the finite length sequence of length N that is equal to $x[n]$ over the interval $[0, N-1]$ and zero elsewhere then $x_N[n] = w_R[n]x[n]$, where $w_R[n]$ is a rectangular window of length N. Therefore, the estimated autocorrelation sequence may be defined by Equation 7.6 in terms of $x_N[n]$.

$$\hat{r}_x(k) = \frac{1}{N} \sum_{n=-\infty}^{\infty} x_N[n+k]x_N^*[n] = \frac{1}{N} x_N[k] * x_N^*[-k] \tag{7.6}$$

The periodogram is then the Fourier transform of Equation 7.6 as defined by Equation 7.7.

$$\hat{P}_{per\_x}(e^{j\theta}) = \frac{1}{N} X_N(e^{j\theta}) X_N^*(e^{j\theta}) = \frac{1}{N} \left| X_N(e^{j\theta}) \right|^2 \tag{7.7}$$

Here, $X_N(e^{j\theta})$ is the discrete-time Fourier transform of the N-point data sequence $x_N[n]$ as defined by Equation 7.8.

$$X_N(e^{j\theta}) = \sum_{n=-\infty}^{\infty} x_N[n]e^{-jn\theta} = \sum_{n=0}^{N-1} x[n]e^{-jn\theta} \tag{7.8}$$

From this it can be seen that the periodogram is proportional to the squared magnitude of the discrete-time Fourier transform of $x_N[n]$, which can easily be computed as defined by Equation 7.9.

$$x_N[n] \overset{DFT}{\to} X_N(k) \to \frac{1}{N} \left| X_N(k) \right|^2 = \hat{P}_{per\_x}(e^{j2\pi k/N}) \tag{7.9}$$

The periodogram has a convenient interpretation in terms of parallel filter banks. It is as though there are N bandpass filters in parallel as defined by Equations 7.10 and 7.11.

$$h_i[n] = \frac{1}{N}e^{jn\theta_i}w_R[n] = \begin{cases} \frac{1}{N}e^{jn\theta_i}, & 0 \le n < N \\ 0, & \text{otherwise} \end{cases} \tag{7.10}$$

$$H_i(e^{j\theta}) = \sum_{n=0}^{N-1} h_i[n]e^{-jn\theta} = e^{-j(\theta-\theta_i)(N-1)/2}\frac{\sin[N(\theta-\theta_i)/2]}{N\sin(\theta-\theta_i)/2} \tag{7.11}$$

One single ith bandpass filter centred at frequency θ_i with a bandwidth of $\Delta\theta = 2\pi/N$ is illustrated in Figure 7.1.

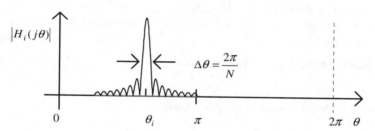

Figure 7.1. Magnitude of one Bandpass filter in the Periodogram's Filter Bank

If the wide-sense stationary process $x[n]$ is filtered with the bandpass filter h_i the filter output is defined by Equation 7.12.

$$y_i[n] = x[n] * h_i[n] = \frac{1}{N}\sum_{k=n-N+1}^{n} x[k]e^{j(n-k)\theta_i} \tag{7.12}$$

However, this filter introduces a small leakage error because it accumulates small amounts of energy from overlap with other filter frequency band tails. The magnitude of the filter at θ_i is equal to unity, i.e., $\left|H_i(e^{j\theta})\right|_{\theta=\theta_i} = 1$, therefore the power spectrums of signals $x[n]$ and $y[n]$ are equal at θ_i. Also, since the bandwidth of each filter is narrow the power spectrum of $x[n]$ may be assumed to be approximately constant over the passband of the filter. The periodogram, defined by Equation 7.13, can therefore be viewed as the estimate of the power spectrum that is formed by using a parallel filter bank of these filters, and being derived from a one point sample average of the power in the filtered process $y_i[n]$.

$$\hat{P}_{per\_x}(e^{j\theta_i}) = N\left|y_i[N-1]\right|^2 = \frac{1}{N}\left|\sum_{k=0}^{N-1} x[k]e^{-jk\theta_i}\right|^2 \tag{7.13}$$

For the periodogram to be a consistent estimate of the power spectrum it is necessary that it be mean square convergent as defined by Equation 7.14.

$$\lim_{N \to \infty} = E\left\{ \left[\hat{P}_{per\_x}(e^{j\theta}) - P_x(e^{j\theta}) \right]^2 \right\} = 0 \tag{7.14}$$

To be mean square convergent it must be asymptotically unbiased as defined by Equation 7.15 and have a variance that goes to zero as N goes to infinity, as defined by Equation 7.16.

$$\lim_{N \to \infty} = E\{ \hat{P}_{per\_x}(e^{j\theta}) \} = P_x(e^{j\theta}) \tag{7.15}$$

$$\lim_{N \to \infty} = Var\{ \hat{P}_{per\_x}(e^{j\theta}) \} = 0 \tag{7.16}$$

Unfortunately, it turns out that the periodogram is not a consistent estimate of the power spectrum because, although it is asymptotically unbiased, the variance does not go to zero as the record length increases to infinity. The variance depends on the process, but for white Gaussian noise it is proportional to the square of the power spectrum, i.e., $Var\{ \hat{P}_{per\_x}(e^{j\theta}) \} = P_x^2(e^{j\theta})$, and it therefore does not decrease as the amount of the data increases.

A summary of the properties of the periodogram is as follows,

Spectral Estimate: $\hat{P}_{per\_x}(e^{j\theta}) = \dfrac{1}{N} \left| \displaystyle\sum_{n=0}^{N-1} x[n]e^{-jn\theta} \right|^2$

Bias: $E\{ \hat{P}_{per\_x}(e^{j\theta}) \} = \dfrac{1}{2\pi} P_x(e^{j\theta}) * W_B(e^{j\theta})$

where:

$W_B(e^{j\omega})$ is the frequency response of the Bartlett (triangular) window $w_B[k]$.

$$w_B[k] = \begin{cases} \dfrac{N - |k|}{N}, & |k| \le N \\ 0, & |k| > N \end{cases}$$

Resolution: $\Delta\theta = 0.89 \dfrac{2\pi}{N}$

Variance: $Var\{ \hat{P}_{per\_x}(e^{j\theta}) \} \approx P_x^2(e^{j\theta})$

7.2.2 Modified Periodogram - Data Windowing

The periodogram is proportional to the squared magnitude of the Fourier transform of the rectangular windowed signal $x_N[n] = w_r[n]x[n]$ as defined by Equation 7.17.

$$\hat{P}_{per\_x}(e^{j\theta}) = \frac{1}{N}\left|X_N(e^{j\theta})\right|^2 = \frac{1}{N}\left|\sum_{k=-\infty}^{\infty}x[n]w_R[n]e^{-jn\theta}\right|^2 \qquad (7.17)$$

Since the rectangular window has relatively high side lobes they contribute to the leakage error and this limits the dynamic range of the power spectral estimate by masking weak spectral components. This problem can be alleviated by replacing the rectangular window with other windows having better leakage characteristics. Ideally a window spectrum should approximate an impulse function. The window's main lobe should be as narrow as possible and the maximum sidelobe should be as small as possible relative to the main lobe. The problem is that in practice these cannot be optimised simultaneously. A reduction in sidelobes results in a broadening of the window's mainlobe in the frequency domain and vice versa. The mainlobe is directly related to the filter bank bandwidth, which therefore affects the frequency resolution.

There are many different types of suitable windows. Some common ones expressed in a form compatible with the present windowing requirements for spectral estimation based on sample sequence lengths of N are,

Hamming : $w[n] = 0.54 - 0.46\cos(\pi n / N)$

Hanning : $w[n] = 0.50 - 0.50\cos(\pi n / N)$

Blackman : $w[n] = 0.42 - 0.50\cos(\pi n / N) + 0.08\cos(2\pi n / N)$

where :

N is the sample sequence length.

$w[n] = 0$ outside of the sequence length internal $[0, N-1]$.

Two of the most often used windows for PSD estimation are,

Bartlett : $w[n] = \dfrac{N - |n|}{N}$

Welch : $w[n] = \left(\dfrac{N - |n|}{N}\right)^2$

The Bartett window is easy to implement but the Welch window is recommended for PSD estimation because it produces one of the best periodogram based methods.

The modified periodogram is the periodogram of a process that is windowed using an arbitrary window $w[n]$ as defined by Equation 7.18.

$$\hat{P}_M(e^{j\theta}) = \frac{1}{NU}\left|\sum_{k=-\infty}^{\infty} x[n]w[n]e^{-jn\theta}\right|^2 \tag{7.18}$$

where:

N is the length of the window.

U is a constant chosen to make $\hat{P}_M(e^{j\theta})$ asymptotically unbiased.

The constant U is defined by Equation 7.19.

$$U = \frac{1}{N}\sum_{n=0}^{N}\left|w[n]\right|^2 \tag{7.19}$$

The variance of $\hat{P}_M(e^{j\theta})$ is approximately the same as that for the periodogram, i.e., $Var\{\hat{P}_M(e^{j\theta})\} \approx P_x^2(e^{j\theta})$. Although the window provides no benefit in respect to variance reduction it does provide a trade-off between spectral resolution (mainlobe width) and spectral masking (sidelobe amplitude). The spectral resolution of $\hat{P}_M(e^{j\theta})$ is defined to be the 3 dB bandwidth of the window, i.e., Resolution $[\hat{P}_M(e^{j\theta})] = (\Delta\theta)_{3dB}$. Table 7.1 compares the side-lobe levels and resolutions for a number of common windows.

Table 7.1

Window	Side-lobe Level (dB)	3dB BW $(\Delta\theta)_{3dB}$
Rectangular	-13	$0.89\,(2\pi/N)$
Bartlett	-27	$1.28\,(2\pi/N)$
Hanning	-32	$1.44\,(2\pi/N)$
Hamming	-43	$1.30\,(2\pi/N)$
Blackman	-58	$1.68\,(2\pi/N)$

A summary of the properties of the modified periodogram is as follows,

Spectral Estimate: $\hat{P}_M(e^{j\theta}) = \dfrac{1}{NU}\left|\sum_{n=-\infty}^{\infty} w[n]x[n]e^{-jn\theta}\right|^2$, $U = \dfrac{1}{N}\sum_{n=0}^{N-1}\left|w[n]\right|^2$

Bias: $E\{\hat{P}_M(e^{j\theta})\} = \dfrac{1}{2\pi NU}P_x(e^{j\theta}) * \left|W(e^{j\theta})\right|^2$

Resolution: Window dependent.

Variance: $Var\{\hat{P}_M(e^{j\theta})\} \approx P_x^2(e^{j\theta})$

7.2.3 Bartlett's Method - Periodogram Averaging

Bartlett's method of spectral estimation involves periodogram averaging which, unlike the periodogram on its own, produces a consistent power spectrum. The periodogram is asymptotically unbiased so the expected value of the periodogram converges to $P_x(e^{j\theta})$ as the data record of length N goes to infinity. A consistent estimate of the mean $E\{\hat{P}_{per\_x}(e^{j\theta})\}$ can be found by averaging periodograms since each periodogram is uncorrelated with the others. Let $x_i[n]$ for $i = 1,2,...., K$, be K uncorrelated records of a random process $x[n]$ over the interval $0 \le n < L$. If the periodogram of $x_i[n]$ is defined by Equation 7.20 then the average of K periodograms is defined by Equation 7.21.

$$\hat{P}^i_{per\_x}(e^{j\theta}) = \frac{1}{L}\left|\sum_{k=0}^{L-1} x_i[n]e^{-jn\theta}\right|^2 \tag{7.20}$$

$$\hat{P}_{per\_x}(e^{j\theta}) = \frac{1}{K}\sum_{i=1}^{K} \hat{P}^i_{per\_x}(e^{j\theta}) \tag{7.21}$$

Since the data records are uncorrelated the variance can be defined by Equation 7.22 as simply the average of the K periodogram variances.

$$Var\{\hat{P}_{per\_x}(e^{j\theta})\} = \frac{1}{K}Var\{\hat{P}^i_{per\_x}(e^{j\theta})\} \approx \frac{1}{K}P^2_x(e^{j\theta}) \tag{7.22}$$

Clearly, the variance goes to zero as $K \to \infty$, therefore it can be said that $\hat{P}_{per\_x}(e^{j\theta})$ is a consistent estimate of the power spectrum as K and L go to infinity. Since in practice there is only a single record of length N Bartlett proposed that $x[n]$ be partitioned into K nonoverlapping subsequences of length L where $N = KL$ (Refer to Figure 7.2). The Bartlett power spectral density estimate is therefore defined by Equation 7.23.

$$\hat{P}_B(e^{j\theta}) = \frac{1}{N}\sum_{i=0}^{K-1}\left|\sum_{n=0}^{L-1} x[n+iL]w[n]e^{-jn\theta}\right|^2 \tag{7.23}$$

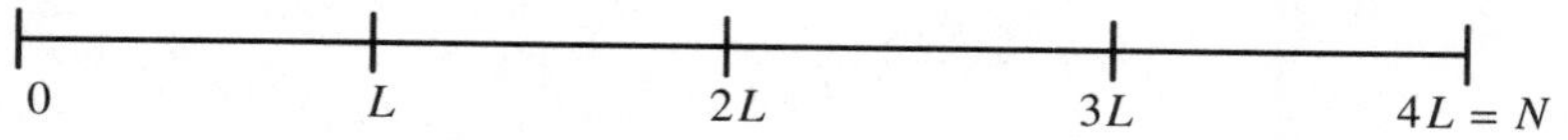

$K = 4$, and $N = 4L$

Figure 7.2. Nonoverlapping Subsequences

A summary of the properties of Bartlett's method is as follows,

Spectral Estimate: $\hat{P}_B(e^{j\theta}) = \dfrac{1}{N} \sum\limits_{i=0}^{K-1} \left| \sum\limits_{n=0}^{L-1} x[n+iL]w[n]e^{-jn\theta} \right|^2$

Bias: $E\{\hat{P}_B(e^{j\theta})\} = \dfrac{1}{2\pi} P_x(e^{j\theta}) * W_B(e^{j\theta})$

Resolution: $\Delta\theta = 0.89K\dfrac{2\pi}{N}$

Variance: $Var\{\hat{P}_B(e^{j\theta})\} \approx \dfrac{1}{K} P_x^2(e^{j\theta})$

7.2.4 Welch's Method

Welch's contribution to the problem is two fold. Firstly, he proposed that the subsequences in the Bartlett method be overlapped and secondly he suggested that a data window be applied to each sequence. In effect, this is averaging of overlapping modified periodograms. If the successive sequences of length L are offset by D points then the ith subsequence is defined by Equation 7.24.

$$x_i[n] = x[n+iD], \qquad n = 0,1, L-1 \tag{7.24}$$

The amount of overlap between successive sequences is $L - D$ points. If K sequences cover all the N data points then $N = L + (K - 1) D$. Refer to Figure 7.3 for an example of $K = 8$..

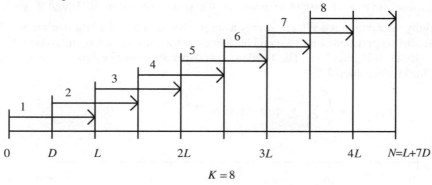

$K = 8$

Figure 7.3. Overlapping Subsequences

An important case is when there is a 50% overlap, i.e., $D = L/2$. In this case $K = 2(N/L) - 1$ sections of length L. This maintains the same resolution as Bartlett's method while doubling the number of averaged modified periodograms, and thereby reducing the variance. With a 50% overlap it is also possible to form $K = (N/L) - 1$ subsequences of length $2L$. This increases the resolution while maintaining the same variance as Bartlett's method. By allowing the subsequences to overlap it is possible to increase the number and/or length of the sequences that are averaged. However, this does trade a reduction in variance for a reduction in resolution.

Welch's method may be written in two ways as defined by Equation 7.25.

$$\hat{P}_W(e^{j\theta}) = \frac{1}{KLU} \sum_{i=0}^{K-1} \left| \sum_{n=0}^{L-1} x[n+iD]w[n]e^{-jn\theta} \right|^2$$

$$= \frac{1}{K} \sum_{i=0}^{K-1} \hat{P}_M^i(e^{j\theta}) \tag{7.25}$$

It follows that the expected value of Welch's estimate is defined by Equation 7.26.

$$E\{\hat{P}_W(e^{j\theta})\} = E\{\hat{P}_M(e^{j\theta})\} = \frac{1}{2\pi LU} P_x(e^{j\theta}) * \left| W_B(e^{j\theta}) \right|^2 \tag{7.26}$$

The variance is more difficult to compute because the overlapping subsequences cannot be assumed to be uncorrelated. It has been shown that with a Bartlett window and 50% overlap the variance is approximately as defined by Equation 7.27.

$$Var\{\hat{P}_W(e^{j\theta})\} \approx \frac{9}{8K} P_x^2(e^{j\theta}) \tag{7.27}$$

Although the variance according to Welch's method is 9/8 times larger than for Bartlett's method, for a fixed amount of data N and a given resolution (sequence length L), twice as many sections may be averaged with a 50% overlap. The variance for Welch's method is therefore approximately 9/16 times lower than for Bartlett's method.

A summary of the properties of Welch's method is as follows,

Spectral Estimate:
$$\hat{P}_W(e^{j\theta}) = \frac{1}{KLU} \sum_{i=0}^{K-1} \left| \sum_{n=0}^{L-1} x[n+iD]w[n]e^{-jn\theta} \right|^2$$

$$U = \frac{1}{L} \sum_{i=0}^{L-1} |w[n]|^2$$

Bias:
$$E\{\hat{P}_W(e^{j\theta})\} = \frac{1}{2\pi LU} P_x(e^{j\theta}) * \left| W_B(e^{j\theta}) \right|^2$$

Resolution: Window dependent.

Variance: $Var\{\hat{P}_W(e^{j\theta})\} \approx \dfrac{9L}{16N} P_x^2(e^{j\theta}),$

assuming a Bartlett window and 50% overlap.

7.2.5 Blackman-Tukey Method

The Blackman-Tukey method of spectral estimation reduces the statistical variability of the periodogram by smoothing. This is done by applying a window to the autocorrelation estimate $\hat{r}_x(k)$ to reduce the error variance caused by having a finite data record. The variance of the autocorrelation estimate is greatest close to the ends so the window serves to reduce this effect by tapering the values close to the ends. The Blackman-Tukey power spectrum estimate is defined by Equation 7.28.

$$\hat{P}_{BT}(e^{j\theta}) = \sum_{k=-M}^{M} \hat{r}_x(k)w[k]e^{-jk\theta},$$ (7.28)

for window extending $-M$ to M, where $|M| < N-1$

For $M < N$ and a rectangular window the power spectrum will have a smaller variance but at the expense of resolution since a smaller number of autocorrelation estimates are used. The Blackman-Tukey estimate smooths the periodogram by convolving with the Fourier transform of the autocorrelation window $W(e^{j\theta})$, i.e.,

$\hat{P}_{BT}(e^{j\theta}) = \dfrac{1}{2\pi}\hat{P}_{per\_x}(e^{j\theta}) * W(e^{j\theta})$. To ensure that $\hat{P}_{BT}(e^{j\theta})$ is guaranteed to be nonnegative $W(e^{j\theta})$ must also be nonnegative and real-valued, which therefore requires $w[k]$ to be conjugate symmetric.

A summary of the properties of the Blackman-Tukey method is as follows,

Spectral Estimate: $\hat{P}_{BT}(e^{j\theta}) = \sum_{k=-M}^{M} \hat{r}_x(k)w[k]^{-jk\theta}$

Bias: $E\{\hat{P}_{BT}(e^{j\theta})\} \approx \dfrac{1}{2\pi}P_x(e^{j\theta}) * W(e^{j\theta})$

Resolution: Window dependent.

Variance: $Var\{\hat{P}_{BT}(e^{j\theta})\} \approx P_x^2(e^{j\theta})\dfrac{1}{N}\sum_{k=-M}^{M}w^2[k]$

7.2.6 Performance Comparisons of Nonparametric Methods

In each nonparametric method there is trade-off between variance and resolution. The methods can therefore be compared by looking at their normalised variance v and figure of merit μ computed as the product of the normalised variance and the resolution, i.e., $\mu = v\Delta\theta$. The figure of merit for each is approximately similar and it is inversely proportional to the length of the sequence N. It is clear that the performance of the nonparametric methods is limited by the availability of data. Table 7.2 shows the comparisons amongst the four nonparametric methods described in the preceding Sections.

Table 7.2

Method	Normalised Variance v	Resolution $\Delta\theta$	Figure of Merit $\mu = v\Delta\theta$
Periodogram	1	$0.89 \left(2\pi/N\right)$	$0.89 \left(2\pi/N\right)$
Bartlett	$1/K$	$0.89K \left(2\pi/N\right)$	$0.89 \left(2\pi/N\right)$
Welch	$9/(8K)$	$1.28K \left(2\pi/L\right)$	$0.72 \left(2\pi/N\right)$
Blackman-Tukey	$2M/(3N)$	$0.64K \left(2\pi/M\right)$	$0.43 \left(2\pi/N\right)$

7.2.7 Minimum Variance Method

In the Minimum Variance method of spectral estimation the power spectrum is estimated by filtering the process with a bank of narrowband bandpass filters. If $x[n]$ is taken to be zero mean wide-sense stationary having a power spectrum of $P_x(e^{j\theta})$ this signal can be filtered with a bank of ideal bandpass filters $h_i[n]$ where the filters are defined by Equation 7.29.

$$\left|H_i(e^{j\theta})\right| = \begin{cases} 1, & \left|\theta - \theta_i\right| < \Delta/2 \\ 0, & \text{otherwise} \end{cases} \tag{7.29}$$

The power spectrum of the filtered signal $y_i[n]$ is $P_{y_i}(e^{j\theta}) = P_x(e^{j\theta})\left|H_i(e^{j\theta})\right|^2$. The power in this output $y_i[n]$ is computed by Equation 7.30.

$$E\{\left|y_i(n)\right|^2\} = \frac{1}{2\pi}\int_{-\pi}^{+\pi}P_i(e^{j\theta})d\theta = \frac{1}{2\pi}\int_{-\pi}^{+\pi}P_x(e^{j\theta})\left|H_i(e^{j\theta})\right|^2 d\theta$$

$$= \frac{1}{2\pi}\int_{\theta_i-\Delta/2}^{\theta_i+\Delta/2}P_x(e^{j\theta})d\theta \tag{7.30}$$

$$\approx P_x(e^{j\theta})\frac{\Delta}{2\pi}, \quad \text{if } \Delta \text{ is small enough.}$$

Since ideal bandpass filters cannot be designed it is necessary to try to design an optimum bandpass filter for each bank that will reject as much of the out of band signal power as possible. The minimum variance technique is based on this approach and involves the following three steps,

1.   Design a bank of filters $h_i[n]$ centred at θ_i such that each filter passes the signal component at θ_i with no distortion and rejects the maximum amount of out of band signal power.

2.   Filter the signal $x[n]$ with each filter in the bank and estimate the amount of power passing through each bank.

3.   Set the estimate of the power spectrum of $x[n]$, $\hat{P}_x(e^{j\theta_i})$, to be equal to the power estimate in Step 2 and divide it by the filter bandwidth.

For Step 1 a complex valued linear phase FIR bandpass filter design is chosen, centred at frequency θ_i and having a unity gain at θ_i, as defined by Equation 7.31.

$$H_i(e^{j\theta}) = \sum_{n=0}^{q} h_i[n]e^{-jn\theta_i} = 1 \tag{7.31}$$

If vectors $\mathbf{h}_i$ and $\mathbf{e}_i$ are designed to represent the filter coefficients and complex exponentials $e^{jk\theta_i}$ respectively then they are defined as follows,

$$\mathbf{h}_i = [h_i[0], h_i[1], h_i[2], ..., h_i[q]]^T,$$

and,

$$\mathbf{e}_i = \left[1, e^{j\theta_1}, e^{j2\theta_2}, ..., e^{jq\theta_q}\right]^T$$

Equation 7.31 can now be written more compactly in vector form as defined by Equation 7.32.

$$\mathbf{h}_i^H \mathbf{e}_i = \mathbf{e}_i^H \mathbf{h}_i = 1 \tag{7.32}$$

The power in the filter output $y_i[n]$ can be expressed in the terms of the autocorrelation matrix $\mathbf{R}_x$ as defined by Equation 7.33.

$$E\{|y_i[n]|^2\} = \mathbf{h}_i^H \mathbf{R}_x \mathbf{h}_i \tag{7.33}$$

Now, the filter design problem becomes one of minimising Equation 7.33 subject to the linear constraint defined by Equation 7.32. The solution to this is the required optimum filter as defined by Equation 7.34.

$$\mathbf{h}_i = \frac{\mathbf{R}_x^{-1} \mathbf{e}_i}{\mathbf{e}_i^H \mathbf{R}_x^{-1} \mathbf{e}_i} \tag{7.34}$$

The minimum value of $E\{|y_i[n]|^2\}$ that gives the power in $y_i[n]$ is used as the power estimate $\hat{\sigma}_x^2(\theta_i)$ and is defined by Equation 7.35.

$$\hat{\sigma}_x^2(\theta_i) = \min_{\mathbf{h}_i} E\{|y_i[n]|^2\} = \frac{1}{\mathbf{e}_i^H \mathbf{R}_x^{-1} \mathbf{e}_i} \tag{7.35}$$

Although these equations were derived for a specific frequency θ_i they are valid for all frequencies since the originally chosen frequency was arbitrary. Therefore, the optimum filter $\mathbf{h}$ for estimating the power in $x[n]$ at frequency θ and the power estimate $\hat{\sigma}_x^2(\theta)$ are defined by Equations 7.36 and 7.37 respectively.

$$\mathbf{h} = \frac{\mathbf{R}_x^{-1} \mathbf{e}}{\mathbf{e}^H \mathbf{R}_x^{-1} \mathbf{e}} \tag{7.36}$$

$$\hat{\sigma}_x^2(\theta) = \frac{1}{\mathbf{e}^H \mathbf{R}_x^{-1} \mathbf{e}} \tag{7.37}$$

where:

$$\mathbf{e} = \left[1, e^{j\theta}, e^{j2\theta}, \ldots, e^{jq\theta}\right]^T$$

The next thing to do is to estimate the power spectrum by dividing the power estimate by the bandwidth of the bandpass filter. The bandwidth Δ is defined as the value that produces the correct power spectral density for white noise. The minimum variance of the power in white noise is $E\{|y_i[n]|^2\} = \sigma_x^2/(q+1)$. Therefore, from Equation 7.30 the power spectral estimate can be defined by Equation 7.38.

$$\hat{P}_x(e^{j\theta_i}) = \frac{E\{|y_i[n]|^2\}}{\Delta/2\pi} = \frac{\sigma_x^2}{(q+1)} \frac{2\pi}{\Delta} \tag{7.38}$$

From this, it is evident that $\Delta = 2\pi/(q+1)$ because that makes $\hat{P}_x(e^{j\theta_i}) = \sigma_x^2$ as required. Therefore the general minimum variance power spectrum estimate $\hat{P}_{MV}(e^{j\theta})$ becomes as defined by Equation 7.39.

$$\hat{P}_{MV}(e^{j\theta}) = \frac{q+1}{\mathbf{e}^H \mathbf{R}_x^{-1} \mathbf{e}} \tag{7.39}$$

The minimum variance method requires the inversion of the autocorrelation matrix $\mathbf{R}_x$. Since $\mathbf{R}_x$ is Toeplitz the inversion may be easily found using either the Levinson recursion or the Cholesky decomposition.

The last remaining issue is to decide what the FIR filter order q should be. In theory the higher order the better but in practice the order must be $q \leq N$, where N

is the data record length. For a fixed data record the autocorrelation matrix must be estimated by computing $r_x(k)$ for lags $k = 0,1,...,N-1$. Since the autocorrelation estimates close to N have a large variance then q has to be chosen to be much smaller than N.

7.2.8 Maximum Entropy (All Poles) Method

One of the limitations with the classical approach to spectrum estimation is that the autocorrelation sequence can only be estimated for lags $|k| < N$, where N is the available data record length. There are many processes of interest, including narrowband processes, that have autocorrelations that decay slowly with k. Consequently, autocorrelation estimates of these types of processes will suffer if it is necessary to set $r_x(k) = 0$, for $|k| \geq N$. What is needed is an effective method of extrapolating the autocorrelation sequence for $|k| \geq N$. One way to achieve this extrapolation is by the maximum entropy method.

Given the autocorrelation sequence $r_x(k)$ for a wide-sense stationary process for lags $|k| \leq p$ the problem is how to extrapolate the values of $r_x(k)$ for lags $|k| > p$. This modified power spectrum can be defined by Equation 7.40.

$$P_x(e^{j\theta}) = \sum_{k=-p}^{p} r_x(k)e^{-jk\theta} + \sum_{|k|>p} r_e(k)e^{-jk\theta} \qquad (7.40)$$

where:

$r_e(k)$ are the extrapolated values of $r_x(k)$.

With the addition of the extrapolation the power spectrum $\hat{P}_x(e^{j\theta})$ must correspond to a valid power spectrum that is real valued and nonnegative for all θ. To achieve a unique extrapolation it is necessary to impose additional constraints. Burg developed the maximum entropy method by imposing the constraint on the extrapolation to be a maximum entropy (a measure of randomness) extrapolation. This is equivalent to finding the sequence $r_e(k)$ that makes $x[n]$ as white or as random as possible and thus introduces the least amount of structure on $x[n]$, i.e., $\hat{P}_x(e^{j\theta})$ is "as flat as possible."

The entropy $H(x)$ for a random Gaussian process with a power spectrum of $P_x(e^{j\theta})$ is defined by Equation 7.41.

$$H(x) = \frac{1}{2\pi} \int_{-\pi}^{+\pi} \ln P_x(e^{j\theta})d\theta \qquad (7.41)$$

Given the autocorrelation sequence $r_x(k)$ for a wide-sense stationary random Gaussian process for lags $|k| \leq p$ the maximum entropy power spectrum is the one that maximises Equation 7.41 subject to the constraint that the inverse discrete-time Fourier transform of $P_x(e^{j\theta})$ equals $r_x(k)$ for lags $|k| \le p$ as defined by Equation 7.42.

$$\frac{1}{2\pi} \int_{-\pi}^{+\pi} P_x(e^{j\theta}) e^{jk\theta} d\theta = r_x(k), \quad |k| \le p \tag{7.42}$$

The values of $r_e(k)$ that maximise the entropy Equation 7.41 can be found by differentiating $H(x)$ with respect to $r_e^*(k)$ and setting it equal to zero as defined by Equation 7.43.

$$\frac{\partial H(x)}{\partial r_e^*(k)} = \frac{1}{2\pi} \int_{-\pi}^{+\pi} \frac{1}{P_x(e^{j\theta})} \frac{\partial P_x(e^{j\theta})}{\partial r_e^*(k)} d\theta = 0$$

$$= \frac{1}{2\pi} \int_{-\pi}^{+\pi} \frac{e^{jk\theta}}{P_x(e^{j\theta})} d\theta = 0, \quad |k| > p \tag{7.43}$$

Defining $Q_x(e^{j\theta}) = 1/P_x(e^{j\theta})$ and substituting it into Equation 7.43 reveals that the inverse discrete-time Fourier transform of $Q_x(e^{j\theta})$ is a finite length sequence that is equal to zero for $|k| > p$, as defined by Equation 7.44.

$$q_x[k] = \frac{1}{2\pi} \int_{-\pi}^{+\pi} Q_x(e^{j\theta}) e^{jk\theta} d\theta = 0, \quad |k| > p \tag{7.44}$$

Notice that $Q_x(e^{j\theta})$ can be defined by Equation 7.45 and that the maximum entropy power spectrum estimate for a Gaussian process is an all-pole power spectrum $\hat{P}_{ME}(e^{j\theta})$ defined by Equation 7.46.

$$Q_x(e^{j\theta}) = \frac{1}{P_x(e^{j\theta})} = \sum_{k=-p}^{p} q_x[k] e^{-jk\theta} \tag{7.45}$$

$$\hat{P}_{ME}(e^{j\theta}) = \frac{1}{\sum_{k=-p}^{p} q_x[k] e^{-jk\theta}} \tag{7.46}$$

Equation 7.46 can be re-expressed as Equation 7.47 by using the spectral factorisation theorem.

$$\hat{P}_{ME}(e^{j\theta}) = \frac{|b_0|^2}{A(e^{j\theta}) A^*(e^{j\theta})} = \frac{|b_0|^2}{\left|1 + \sum_{k=1}^{p} a_k e^{-jk\theta}\right|^2} = \frac{|b_0|^2}{\left|e^H a\right|^2}, \tag{7.47}$$

where:

$$e = \left[1, e^{j\theta}, ..., e^{jp\theta}\right]^T$$
$$a = \left[1, a_0, ..., a_p\right]^T$$

The required coefficients b_0 and a_k must be chosen in such a way that the inverse discrete-time Fourier transform of $\hat{P}_{ME}(e^{j\theta})$ produces an autocorrelation sequence that matches the given values for $r_x(k)$ for lags $|k| \le p$. If the coefficients a_k are the solution to the autocorrelation normal Equations 7.48 re-expressed as 7.49 and if $\varepsilon_p = |b_0|^2 = r_x(0) + \sum_{k=1}^{p} a_k r_x^*(k)$, then the autocorrelation constraint set in Equation 7.42 will be satisfied.

$$\begin{bmatrix} r_x(0) & r_x^*(1) & r_x^*(2) & .. & r_x^*(p) \\ r_x(1) & r_x(0) & r_x^*(1) & .. & r_x^*(p-1) \\ r_x(2) & r_x(1) & r_x(0) & .. & r_x^*(p-2) \\ \vdots & \vdots & \vdots & \ddots & \vdots \\ r_x(p) & r_x(p-1) & r_x(p-2) & .. & r_x(0) \end{bmatrix} \begin{bmatrix} 1 \\ a_1 \\ a_2 \\ \vdots \\ a_p \end{bmatrix} = \varepsilon_p \begin{bmatrix} 1 \\ 0 \\ 0 \\ \vdots \\ 0 \end{bmatrix} \tag{7.48}$$

$$\mathbf{R}_p \mathbf{a} = \varepsilon_p \mathbf{u}_1$$

i.e.,

$$\mathbf{a} = \varepsilon_p \mathbf{R}_x^{-1} \mathbf{u}_1 \tag{7.49}$$

Thus, the final $\hat{P}_{ME}(e^{j\theta})$ is defined by Equation 7.50.

$$\hat{P}_{ME}(e^{j\theta}) = \frac{\varepsilon_p}{\left|\mathbf{e}^H \mathbf{a}\right|^2} \tag{7.50}$$

where:

$$e = \left[1, e^{j\theta}, ..., e^{jp\theta}\right]^T$$
$$a = \left[1, a_0, ..., a_p\right]^T$$

Since $\hat{P}_{ME}(e^{j\theta})$ is an all-pole power spectrum, then $r_x(k)$ satisfies the *Yule-Walker* equations defined by Equation 7.51.

$$r_x(m) = -\sum_{k=1}^{p} a_k r_x(k-m), \text{ for } m > 0 \tag{7.51}$$

Consequently, the maximum entropy method extrapolates the autocorrelation sequence according to the recursion defined by Equation 7.51.

7.3 Parametric Spectral Density Estimation

Parametric spectral density methods potentially offer high-resolution with small variance given shorter data records. However to achieve this it is necessary to have some knowledge about the process that can be incorporated into the spectrum estimation algorithm design. The knowledge may be some *a priori* knowledge about the process or it may be knowledge of how it was generated or it may be experimental knowledge indicating that a particular model works well for that type of process. The most common types of models that are used include autoregressive, moving average, autoregressive moving average and harmonic (complex exponentials in noise) models. It is, however, important when choosing the model that it is appropriate for the process, otherwise it could give very misleading results. If the model is inappropriate it may well be the case that a nonparametric estimate will be more accurate and correct.

The forms of the spectrums for autoregressive, moving average, and autoregressive moving average processes are represented by Equations 7.52, 7.53 and 7.54 respectively.

$$P_x(e^{j\theta}) = \frac{|b_0|^2}{\left|1 + \sum_{k=1}^{p} a_k e^{-jk\theta}\right|^2} \tag{7.52}$$

$$P_x(e^{j\theta}) = \left|\sum_{k=0}^{q} b_k e^{-jk\theta}\right|^2 = \sum_{k=0}^{q} r_x(k) e^{-jk\theta} \tag{7.53}$$

where:
Through the *Yule-Walker* equations,

$$r_x(k) = \sum_{m=0}^{q-k} b_{m+k} b_m^*, \ k = 0,1,..,q, \text{ with, } r_x(-k) = r_x^*(k), \text{ and } r_x(k) = 0, \text{ for } |k| > q$$

$$P_x(e^{j\theta}) = \frac{\left|\sum_{k=0}^{q} b_k e^{-jk\theta}\right|^2}{\left|1 + \sum_{k=1}^{p} a_k e^{-jk\theta}\right|^2} \tag{7.54}$$

7.3.1 Autoregressive Methods

The autoregressive power spectrum estimate of an autoregressive process represented by an all-pole filter driven by unit variance white noise is defined by Equation 7.55.

$$\hat{P}_{AR}(e^{j\theta}) = \frac{\left|\hat{b}_0\right|^2}{\left|1 + \sum_{k=1}^{p} \hat{a}_k e^{-jk\theta}\right|^2} \tag{7.55}$$

Some autoregressive approaches to spectrum estimation include the *Yule-Walker*, Covariance, Burg, and Least Squares approaches. Each one uses a different technique to estimate the all-pole model parameters but once they are estimated the power spectrum is generated in the same way.

7.3.1.1 Yule-Walker Approach

The *Yule-Walker* method is actually equivalent to the maximum entropy method with the only difference being in the assumption made about the process $x[n]$. In the *Yule-Walker* method it is assumed that the process is autoregressive, whereas in the maximum entropy method the process is assume to be Gaussian.

In the autocorrelation method of all-pole modelling, the autoregressive coefficient estimates $\hat{a}_k$ and $\hat{b}_0$ are computed by solving the autocorrelation normal Equations 7.56 and 7.57.

$$\begin{bmatrix} \hat{r}_x(0) & \hat{r}_x^*(1) & \hat{r}_x^*(2) & .. & \hat{r}_x^*(p) \\ \hat{r}_x(1) & \hat{r}_x(0) & \hat{r}_x^*(1) & .. & \hat{r}_x^*(p-1) \\ \hat{r}_x(2) & \hat{r}_x(1) & \hat{r}_x(0) & .. & \hat{r}_x^*(p-2) \\ : & : & : & \ddots & : \\ \hat{r}_x(p) & \hat{r}_x(p-1) & \hat{r}_x(p-2) & .. & \hat{r}_x(0) \end{bmatrix} \begin{bmatrix} 1 \\ \hat{a}_1 \\ \hat{a}_2 \\ : \\ \hat{a}_p \end{bmatrix} = \hat{\varepsilon}_p \begin{bmatrix} 1 \\ 0 \\ 0 \\ : \\ 0 \end{bmatrix} \tag{7.56}$$

$$\hat{\varepsilon}_p = \left|\hat{b}_0\right|^2 = \hat{r}_x(0) + \sum_{k=1}^{p} \hat{a}_k \hat{r}_x^*(k) \tag{7.57}$$

The biased autocorrelation estimate is computed by Equation 7.58.

$$\hat{r}_x(k) = \frac{1}{N} \sum_{k=0}^{N-1-k} x[n+k]x^*[k], \quad k = 0,1,..,p \tag{7.58}$$

The autocorrelation matrix is Toelplitz, therefore it is possible to solve these equations efficiently by using the Levinson-Durbin recursion. When the autoregressive coefficients, computed from the autocorrelation normal equations, are incorporated into Equation 7.55 to estimate to the spectrum, this is called the *Yule-Walker* method.

The autocorrelation method applies a rectangular window to the data in the autocorrelation sequence estimation. For small data records this results in a lower resolution spectrum estimate than provided by other approaches such as the covariance and Burg methods. If the autocovariance method includes over modelling, i.e., p is too large, the spectral line splitting artefact can occur. Spectral line splitting is when a single spectral peak is modelled by two and distinct peaks.

7.3.1.2 Covariance, Least Squares and Burg Methods

In the covariance method of all-pole modelling, the autoregressive coefficient estimates $\hat{a}_k$ are computed by solving Equations 7.59.

$$\begin{bmatrix} \hat{r}_x(1,1) & \hat{r}_x(2,1) & \hat{r}_x(3,1) & .. & \hat{r}_x(p,1) \\ \hat{r}_x(1,2) & \hat{r}_x(2,2) & \hat{r}_x(3,2) & .. & \hat{r}_x(p,2) \\ \hat{r}_x(1,3) & \hat{r}_x(2,3) & \hat{r}_x(3,3) & .. & \hat{r}_x(p,3) \\ \vdots & \vdots & \vdots & \ddots & \vdots \\ \hat{r}_x(1,p) & \hat{r}_x(2,p) & \hat{r}_x(3,p) & .. & \hat{r}_x(p,p) \end{bmatrix} \begin{bmatrix} \hat{a}_1 \\ \hat{a}_2 \\ \hat{a}_3 \\ \vdots \\ \hat{a}_p \end{bmatrix} = - \begin{bmatrix} \hat{r}_x(0,1) \\ \hat{r}_x(0,2) \\ \hat{r}_x(0,3) \\ \vdots \\ \hat{r}_x(0,p) \end{bmatrix} \qquad (7.59)$$

where:

$$\hat{r}_x(k,l) = \sum_{n=p}^{N-1} x[n-l]x^*[n-k] \qquad (7.60)$$

Unlike the equations in the autocorrelation method these covariance Equations 7.59 and 7.60 are not Toeplitz. However, they do have the advantage that no windowing is necessary to form the autocorrelation sequence estimate $\hat{r}_x(k,l)$. This means that, for short data records the covariance method produces higher resolution spectrum estimates than the autocorrelation method, but when $N \gg p$ the differences are negligible.

The modified covariance method, also known as the forward-backward method and the least squares method, is similar to the covariance method. The covariance method finds the autoregressive model that minimises the sum of the squares of the forward prediction error, whereas the modified covariance method finds the autoregressive model that minimises the sum of the squares of both the forward and backward prediction errors. In the modified covariance method the autocovariance estimation Equation 7.60 is simply replaced by Equation 7.61. This autocorrelation matrix is not Toeplitz either.

$$\hat{r}_x(k,m) = \sum_{n=p}^{N-1} [x[n-m]x^*[n-k] + x[n-p+m]x^*[n-p+k]] \qquad (7.61)$$

The modified covariance method gives statistically stable spectrum estimates with high resolution. Due to additive noise the modified covariance method tends to shift spectral peaks away from their true locations for sinusoids in white noise. However, the shift appears to be less severe and less sensitive to phase than with other autoregressive estimation methods. Also, the modified covariance method is not subject to the spectral line splitting artefact of the autocovariance method.

The Burg algorithm, like the modified covariance method, finds a set of all-pole model parameters that minimise the sum of the squares of the forward and backward prediction errors. However, it is less accurate than the modified covariance method because, for the sake of stability, it performs the minimisation sequentially with respect to the reflection coefficients. The Burg algorithm is more accurate than the autocorrelation method. However for sinusoids in white noise it does suffer spectral line splitting and the peak locations are highly dependent on the sinusoid phases.

7.3.1.3 Model Order Selection for the Autoregressive Methods

The selection of the model order p of an autoregressive process is very important to the success of the method. If the order is too small the estimated spectrum will be smoothed, resulting in poor resolution. If the order is too large the spectrum estimate may include spurious peaks and may also lead to line splitting. It is tempting to simply increase the order progressively until the modelling error is a minimum. However, the model error is a monotonically nonincreasing function of the model order so this will not be reliable. Still, it is possible to overcome this problem by incorporating a penalty function that increases with the model order.

7.3.2 Moving Average Method

A moving average process is generated by filtering unit variance white noise $w[n]$ with a FIR filter as defined by Equation 7.62 having a power spectrum defined by Equation 7.63.

$$x[n] = \sum_{k=0}^{q} b_k w[n-k] \tag{7.62}$$

Because the autocorrelation sequence of a moving average process is of finite length then the moving average spectrum estimate $\hat{P}_{MA}(e^{j\theta})$ may be conveniently defined by Equation 7.63.

$$\hat{P}_{MA}(e^{j\theta}) = \sum_{k=0}^{q} \hat{r}_x(k)e^{-jk\theta} \tag{7.63}$$

where:
$\hat{r}_x(k)$ is a suitable estimate.

Equation 7.63 is equivalent to the Blackman-Tukey spectrum estimate using a rectangular window. However the difference is that Equation 7.63 assumes that the process actually is a moving average process of order q, in which case the true autocorrelation is zero for $|k| > q$ and $\hat{P}_{MA}(e^{j\theta})$ is therefore unbiased.

Another approach to moving average spectrum estimation is to estimate the parameters b_k from $x[n]$ and then substitute the estimates into Equation 7.53 as defined by Equation 7.64.

$$\hat{P}_{MA}(e^{j\theta}) = \left| \sum_{k=0}^{q} \hat{b}_k e^{-jk\theta} \right|^2 \tag{7.64}$$

7.3.3 Autoregressive Moving Average Method

An autoregressive moving average process has a power spectrum defined by Equation 7.54. The autoregressive parameters of the model may be estimated from the modified *Yule-Walker* equations either directly or by using a least squares approach. Then, the moving average parameters can be estimated using a moving average modelling technique such as Durbin's method.

7.3.4 Harmonic Methods

When the process can be modelled by a sum of complex exponentials represented as the signal $s[n]$ in white noise $w[n]$ defined by Equation 7.65, then it is possible to use harmonic methods of spectrum estimation that take this model into account.

$$x[n] = s[n] + w[n] = \sum_{i=1}^{p} A_i e^{jn\theta_i} + w[n] \tag{7.65}$$

where:

$A_i = |A_i| e^{j\phi_i}$, i.e., complex.

ϕ_i are uncorrelated random variables over $[-\pi, \pi]$.

In this case the power spectrum of $x[n]$ is composed of a set of p impulses with area $2\pi |A_i|$ at frequencies θ_i, for $i = 1, 2, .., p$ plus the power spectrum of the white noise $w[n]$. Consequently, it is possible to perform an eigendecomposition of the autocorrelation matrix of $x[n]$ into a sinusoidal signal subspace and noise subspace. A frequency estimation function can then be used to estimate the frequencies from the signal subspace.

7.3.4.1 Eigendecomposition of the Autocorrelation Matrix
A wide-sense stationary process consisting of p complex exponentials in white noise is defined by Equation 7.65 and has a $M \times M$ autocorrelation sequence defined by Equation 7.66.

$$r_x(k) = \sum_{i=1}^{p} P_i e^{jk\theta_i} + \sigma_w^2 \delta(k) \tag{7.66}$$

where:

$P_i = |A_i|^2$, $\sigma_w^2 =$ variance of the white noise.

The autocorrelation matrix is defined by Equation 7.67.

$$\mathbf{R}_x = \mathbf{R}_s + \mathbf{R}_w = \sum_{i=1}^{p} P_i \mathbf{e}_i \mathbf{e}_i^H + \sigma_w^2 \mathbf{I} = \mathbf{EPE}^H + \sigma_w^2 \mathbf{I} \tag{7.67}$$

where:

$$\mathbf{e}_i = \left[1, e^{j\theta_i}, e^{j2\theta_i}, ..., e^{j(M-1)\theta_i} \right]^T, \quad i = 1, 2, ..., p.$$

$\mathbf{E} = [\mathbf{e}_1, ..., \mathbf{e}_p], \quad M \times p$ matrix.

$\mathbf{P} = \text{diag}\{P_1, ..., P_p\}, \quad p \times p$ diagonal matrix.

The eigenvalues of $\mathbf{R}_x$ are $\lambda_i = \lambda_i^s + \sigma_w^2$, where λ_i^s are the eigenvalues of $\mathbf{R}_s$. $\mathbf{R}_x$ is a matrix of rank p, therefore the first p eigenvalues of $\mathbf{R}_s$ will be greater than σ_w^2 and the last M-p eigenvalues will be equal to σ_w^2. Consequently, the eigenvalues and eigenvectors of $\mathbf{R}_x$ may be divided into two groups, the signal eigenvectors $\mathbf{v}_1, ..., \mathbf{v}_p$ that have eigenvalues greater than σ_w^2 and the noise eigenvectors $\mathbf{v}_{p+1}, ..., \mathbf{v}_M$ that have eigenvalues equal to σ_w^2. If the eigenvectors have been normalised the spectral theorem can be used to decompose $\mathbf{R}_x$ according to Equation 7.68.

$$\mathbf{R}_x = \mathbf{R}_s + \mathbf{R}_w = \sum_{i=1}^{p} (\lambda_i^s + \sigma_w^2) \mathbf{v}_i \mathbf{v}_i^H + \sum_{i=p+1}^{M} \sigma_w^2 \mathbf{v}_i \mathbf{v}_i^H$$
$$= \mathbf{V}_s \mathbf{V}_s^H + \mathbf{V}_{ww} \mathbf{V}_w^H \tag{7.68}$$

where:

$\mathbf{V}_s = [\mathbf{v}_1, ..., \mathbf{v}_p], \quad M \times p$ matrix.

$\mathbf{V}_w = [\mathbf{v}_{p+1}, ..., \mathbf{v}_M], \quad M \times (M - p)$ matrix.

$\mathbf{V}_{ss} = \text{diag}\{(\lambda_1^s + \sigma_w^2), ..., (\lambda_p^s + \sigma_w^2)\}, \quad p \times p$ matrix.

$\mathbf{V}_{ww} = \text{diag}\{\sigma_w^2, ..., \sigma_w^2\}, \quad (M - p) \times (M - p)$ matrix.

The signal and white noise spaces are orthogonal as defined by Equation 7.69.

$$\mathbf{e}_i^H \mathbf{v}_k = 0, \quad i = 1, ..., p, \text{ and } k = p + 1, ..., M \tag{7.69}$$

Therefore, the frequencies can be estimated using various techniques based on the frequency estimation function defined by Equation 7.70.

$$\hat{P}(e^{j\theta}) = \frac{1}{\sum_{i=p+1}^{M} \alpha_i \left| \mathbf{e}^H \mathbf{v}_i \right|^2} \tag{7.70}$$

where:

α_i are appropriately chosen constants.

Two methods that use Equation 7.70 are the Pisarenko Harmonic Decomposition (PHD) and the Multiple SIgnal Classification (MUSIC) methods.

7.3.4.1.1 Pisarenko's Method

Pisarenko's method is mainly of theoretical interest only because it is sensitive to noise. This method assumes that the number of complex exponentials p is known and that $p+1$ values of the autocorrelation sequence are either known or have been estimated. For a $(p+1) \times (p+1)$ autocorrelation matrix the dimension of the white noise subspace is one and is spanned by the eigenvector $\mathbf{v}_{min}$ corresponding to the minimum eigenvalue, $\lambda_{min} = \sigma_w^2$. The eigenvector $\mathbf{v}_{min}$ is orthogonal to each of the signal vectors $\mathbf{e}_i$ as defined by Equation 7.71.

$$\mathbf{e}_i^H \mathbf{v}_{min} = \sum_{k=0}^{p} v_{min}[k] e^{-jk\theta_i} = 0, \quad i = 1,..,p \tag{7.71}$$

Therefore $V_{min}(e^{j\theta}) = \sum_{k=0}^{p} v_{min}[k] e^{-jk\theta}$ is equal to zero at each of the p complex frequencies. This means that the z-Transform of the white noise eigenvector (the eigenfilter), Equation 7.72, has p zeros on the unit circle from which the complex exponentials can be extracted.

$$V_{min}(e^{j\theta}) = \sum_{k=0}^{p} v_{min}[k] z^{-k} = \prod_{k=1}^{p} \left(1 - e^{j\theta_k} z^{-1}\right) \tag{7.72}$$

The frequency estimation function is defined by Equation 7.73.

$$\hat{P}_{PHD}(e^{j\theta}) = \frac{1}{\left| \mathbf{e}^H \mathbf{v}_{min} \right|^2} \tag{7.73}$$

Equation 7.73 is called a pseudospectrum since it does not contain any information about the power in the complex exponentials nor a component due to noise. To complete the power spectrum estimation it is necessary to find the powers P_i from the eigenvalues of $\mathbf{R}_x$. The signal subspace $\mathbf{v}_1,..,\mathbf{v}_p$ vectors have been normalised, therefore $\mathbf{v}_i^H \mathbf{v}_i = 1$ and eigenvalues are defined by Equation 7.74.

$$\mathbf{R}_x \mathbf{v}_i = \lambda_i \mathbf{v}_i, \quad i = 1,2,..,p \tag{7.74}$$

Multiplying on the left of Equation 7.74 with $\mathbf{v}_i^H$ gives Equation 7.75.

$$\mathbf{v}_i^H \mathbf{R}_x \mathbf{v}_i = \lambda_i \mathbf{v}_i^H \mathbf{v}_i = \lambda_i, \quad i = 1,2,..,p$$

Substituting the expression for $\mathbf{R}_x$ defined in Equation 7.67 into Equation 7.75 results in Equation 7.76.

$$\mathbf{v}_i^H \mathbf{R}_x \mathbf{v}_i = \mathbf{v}_i^H \left\{ \sum_{k=1}^{p} P_k \mathbf{e}_k \mathbf{e}_k^H + \sigma_w^2 \mathbf{I} \right\} \mathbf{v}_i = \lambda_i, \quad i=1,2,..,p, \text{i.e.,}$$

(7.76)

$$\sum_{k=1}^{p} P_k \left| \mathbf{e}_k^H \mathbf{v}_i \right|^2 = \lambda_i - \sigma_w^2, \quad i=1,2,..,p$$

where:

$$\left| \mathbf{e}_k^H \mathbf{v}_i \right|^2 = \left| V_i(e^{j\theta_k}) \right|^2$$

$$V_i(e^{j\theta_k}) = \sum_{m=0}^{p} v_i[m] e^{-jm\theta}$$

Equation 7.76 can be redefined as Equation 7.77, which is a set of p linear equations in p unknowns P_k. This can then be solved for the required powers P_k.

$$\sum_{k=1}^{p} P_k \left| V_i(e^{j\theta_k}) \right|^2 = \lambda_i - \sigma_w^2, \quad i=1,2,..,p$$

(7.77)

7.3.4.1.2 MUSIC

MUSIC is an improvement to the Pisarenko Harmonic Decomposition. The autocorrelation matrix $\mathbf{R}_x$ is a $M \times M$ matrix of $x[n]$ with $M > p+1$. If the eigenvalues of $\mathbf{R}_x$ are arranged in decreasing order, $\lambda_1 \geq \lambda_2 \geq \lambda_3 \geq ... \geq \lambda_M$, their corresponding eigenvectors, $\mathbf{v}_1, \mathbf{v}_2, \mathbf{v}_3,.., \mathbf{v}_M$, can be divided into two groups, the p signal eigenvectors and the M-p white noise eigenvectors. The white noise eigenvalues will only be approximately equal to σ_w^2 if an inexact autocorrelation is used. Since the smallest $M - p$ eigenvalues should all be the same it is possible to derive a better estimate for them by taking the average of these smallest $M - p$ eigenvalues.

To estimate the frequencies of the complex exponentials is a little more involved. The eigenvectors of $\mathbf{R}_x$ have a length of M and the white noise subspace eigenfilters defined by Equation 7.78 will therefore have $M - 1$ roots (zeros).

$$V_i(z) = \sum_{k=0}^{M-1} v_i[k] z^{-k}, \quad i = p+1,..,M$$

(7.78)

Ideally only p of these roots would be expected to lie on the unit circle at the frequencies of the complex exponentials. The eigenspectrum is defined by Equation 7.79 and it will exhibit sharp peaks at the p frequencies of the complex exponentials.

$$\left| V_i(e^{j\theta}) \right|^2 = \frac{1}{\left| \sum_{k=0}^{M-1} v_i[k] e^{-jk\theta} \right|^2}$$

(7.79)

The remaining $M - p - 1$ zeros may lie anywhere including close to the unit circle, which would give rise to spurious peaks. In the MUSIC algorithm the effects of these spurious peaks are reduced by averaging, using the frequency estimation function defined by Equation 7.80.

$$\hat{P}_{MUSIC}(e^{j\theta}) = \frac{1}{\sum\limits_{i=p+1}^{M} \left| e^{H} \mathbf{v}_i \right|^2} \tag{7.80}$$

The frequencies of the complex exponentials are taken as the locations of the p largest peaks in Equation 7.80. Once the peaks have been found the energy in each peak is found by solving Equation 7.77 as before.

7.4 Exercises

The following Exercises identify some of the basic ideas presented in this Chapter.

7.4.1 Problems

7.1.   What are the two main approaches to spectral estimation and in what way do they differ?

7.2.   What is the power spectrum of white noise having a variance of σ_x^2?

7.3.   Assume that a random process can be described by two equal amplitude sinusoids in unit random variance white noise as defined by the following equation,

$$x[n] = A\sin(n\theta_1 + \phi_1) + A\sin(n\theta_2 + \phi_2) + v[n]$$

Also assume that $\Delta\theta = |\theta_1 - \theta_2| = 0.02\pi$ radians and $F_s = 1000$ Hz. What is the minimum value of data length N such that the two sinusoids can be resolved via the nonparametric periodogram spectral estimation method? Compute the frequency resolution for this problem?

7.4.   Does the variance of the periodogram spectral estimate of white noise reduce as the data length N increases?

7.5.   For a total data length N what can you say about the relationship between the resolution and variance as function of K, the number of nonoverlapping data sections, for Bartlett's spectral estimation method?

7.6.   By looking at the performance comparisons of the various nonparametric spectral estimation methods what general conclusions can be drawn?

7.7.   Compute the optimum filter for estimating the Minimum Variance (MV) power spectrum of white noise having a variance of σ_x^2.

7.8.   Compute the MV spectral estimate of a random phase complex exponential process in white noise, defined by,

$$x[n] = |A_1| e^{j\phi} e^{jn\theta_1} + w[n]$$

where :

   ϕ is a random variable uniformly distributed over $[-\pi, \pi]$.

   $w[n]$ has a variance of σ_w^2.

Use Woodbury's identity to find the required inverse of the autocorrelation matrix.

7.9.   Compute the qth order Maximum Entropy (ME) spectral estimate for Problem 7.8.

PART IV. ADAPTIVE FILTER THEORY

The optimum Wiener solution for an adaptive FIR filter is represented by the Wiener-Hopf solution. This solution is strictly only applicable for stationary processes and it requires knowledge of the input autocorrelation matrix and the crosscorrelation between the input and desired response. In real applications the processes are more likely to be nonstationary and it may therefore not be possible to know the autocorrelation and crosscorrelation functions explicitly. Since the Wiener-Hopf solution is not practicable in many situations other solution methods must be used for adaptive filters. One of those methods is the iterative method called the method of steepest gradient descent (Principe *et al* 2000). A most popular one is the Least Mean Squares (LMS) algorithm which is actually a robust simplification of the steepest gradient descent method. It is also possible to solve the adaptive FIR filter problem using Recursive Least Squares (RLS) estimation, which is a special case of the Kalman filter. An adaptive linear shift-invariant filter model can be usefully viewed as a dynamical system that continually tries to converge to the Wiener solution as data flows through it.

The FIR or transversal filter structure is a good one to use for adaptive filters for a number of reasons. Firstly, the mean square error for this filter is a quadratic function of the tap-weights. This means that the error surface is a paraboloid, which has a single minimum that is easy to find. Secondly, the transversal filter is guaranteed to be stable. Adaptive Infinite Impulse Response (IIR) filters can often provide better performance for a given filter order. However, they have potential instability problems that may affect both the convergence time as well as the general numerical sensitivity of the filter. Despite their problems there are many applications where adaptive IIR filters may nevertheless be preferred, for example in echo cancellation, where the IIR filter structure offers the best system model.

FIR filters have the drawback that they often require many, perhaps thousands, of filter coefficients (tap-weights) to achieve desired levels of performance in real applications. One method of reducing the large amounts of computation required for these types of adaptive systems is to perform the computations in the frequency domain. Here, a block updating strategy is introduced where the filter coefficients are updated every so many samples and Fast Fourier Transform (FFT) routines are used to reduce the amount of computation required to implement convolutions and correlations.

Many adaptive engineering problems are inherently nonlinear and therefore are better addressed by nonlinear solutions. One approach is to use adaptive polynomial filters like the adaptive Volterra filters, which are a generalisation of adaptive linear FIR filters. The Volterra filter is a nonlinear filter but it depends linearly on the coefficients of the filter itself, and its behaviour can be described in the frequency domain by means of a type of multi-dimensional convolution. Because of this linear coefficient dependence "optimum linear filter theory" can be easily extended to "optimum nonlinear Volterra filter theory."

Theoretical approaches related to adaptive filters are also relevant to adaptive control systems as they incorporate many of the same ideas. The main difference between the two is mostly a matter of configuration and application rather than underlying theory and principles of operation. Adaptive control systems and adaptive filters are usually treated as separate fields so a review of adaptive control principles is provided to show some of the connections and similarities between them.

From a pragmatic view point adaptive control can be seen as a special type of nonlinear feedback control in which the states of the process are separated into two categories related to the rate of change involved. In this view the slowly changing states are seen as the parameters and the fast ones are the ordinary feedback states. The two main methods for adaptive control are the direct and indirect methods depending on how the parameters are adjusted. The direct methods have adjustment rules which tell how the regulator parameters should be updated. Indirect methods on the other hand update the process parameters and then the regulator parameters are obtained from the solution of a design problem. One most important direct method is the Model-Reference Adaptive System (MRAS) and one important indirect method is the Self-Tuning Regulator (STR). Although different in detail these two methods are nevertheless closely related in principle.

8. Adaptive Finite Impulse Response Filters

Adaptive signal processing has undergone a large increase in interest, especially over more recent years, due mainly to the advancing developments in VLSI circuit design. These advances have allowed for much faster real-time digital signal processing. Seismic signals (10^2 Hz), speech and acoustic signals (10^2 - 10^5 Hz) and electromagnetic signals (at 10^6 Hz and above) are now all reasonable candidates for real-time adaptive signal processing. Adaptive signal processing systems are mainly time-varying digital systems. The adaptive notion derives from the desire to emulate living systems, which adapt to their changing environments. Adaptive signal processing has its roots in adaptive control and the mathematics of iterative processes. Early developments in adaptive control occurred in the 1940s and 1950s. More recently the work of Bernard Widrow and his colleagues beginning around 1960 has given us the most popular adaptive algorithm called the Widrow-Hoff LMS algorithm (Widrow and Hoff 1960, Widrow and Sterns 1985, Widrow *et al* 1975, Widrow and Winter 1988), most commonly applied to the Adaptive Linear Combiner (ALC) structure.

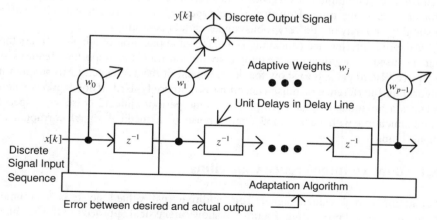

Figure 8.1. Adaptive Linear Combiner

The ALC as shown in Figure 8.1 is also known by the following names,

1. ADALINE - with the addition of a threshold element on the output.

2.   Adaptive transversal filter.

3.   Adaptive Finite Impulse Response (FIR) filter.

Adaptive linear filters are useful in situations where the measured input conditions are uncertain, or where they change with time. Under these circumstances the system achieves good performance by altering the filter parameters (coefficients or tap-weights) $\{w_j\}$ based on real valued input data $\{x[k]\}$ or the estimated statistical properties of the data. A system that searches for improved performance guided by a computational algorithm for adjustment of the parameters or weights is called an adaptive system, which by nature is therefore time-varying. An adaptive process can be open-loop or closed-loop as defined below,

1.   An open-loop process first makes measurements on the input data, learns the statistics of the data, and then applies this knowledge to solve for $\{w_j\}$ to optimise performance.

2.   A closed-loop operates in an iterative manner and updates w_j with the arrival of new data and current signal processor performance feedback. The optimum set of values of $\{w_j\}$ is approached sequentially.

The ALC shown in Figure 8.1 is a closed-loop process, which is a simple, robust, and commonly used adaptive filter structure that is typically adapted by the LMS algorithm. The FIR filter structure realises only zeros whereas other structures realise both poles and zeros. The more complex filters based on lattice structures can achieve more rapid convergence under certain conditions, but at the expense of longer processing time. However these types of lattice filters can also suffer instability and may not be very robust under certain conditions.

Adaptive interference cancelling is a classical application of adaptive filters that can be used as a convenient example to demonstrate the important features of adaptive FIR filters and to show the development of the associated LMS adaptation algorithm. Interference or noise cancelling is commonly used in applications where there is a broadband signal corrupted by some periodic inband noise, e.g., speech communication with background engine noise or a medical signal corrupted by mains hum.

8.1 Adaptive Interference Cancelling

Separating a real valued signal from additive real valued noise is a common problem in signal processing. Figure 8.2 shows a classical approach to this problem using optimum filtering. The optimum filter tries to pass the discrete signal $s[k]$ without distortion while trying to stop the noise $n[k]$. In general this cannot be done perfectly. Even the best filter distorts the signal somewhat, and some noise still gets through.

Signal + Noise
$s[k] + n[k]$

Optimum Filter

$y[k] = \hat{s}[k]$

Figure 8.2. Optimum Noise Filter

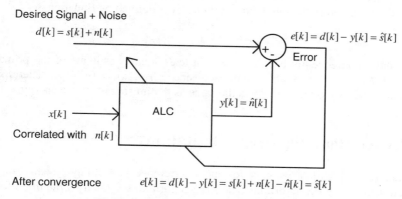

Desired Signal + Noise
$d[k] = s[k] + n[k]$

$e[k] = d[k] - y[k] = \hat{s}[k]$
Error

$x[k]$

ALC

$y[k] = \hat{n}[k]$

Correlated with $n[k]$

After convergence $e[k] = d[k] - y[k] = s[k] + n[k] - \hat{n}[k] = \hat{s}[k]$

Figure 8.3. Adaptive Noise Cancelling System

Figure 8.3 shows the adaptive filter solution, which is viable when there is an additional reference noise input $x[k]$ available that is correlated with the original corrupting noise $n[k]$. The filter filters the reference noise $x[k]$ and produces an estimate of the actual noise $n[k]$ ($y[k] = \hat{n}[k]$) and subtracts it from the primary input $s[k] + n[k]$ to compute an estimate of the signal $\hat{s}[k]$. Here the signal $d[k] = s[k] + n[k]$ acts as the desired response and the system output $e[k]$ acts as the error for the adaptive filter algorithm. When the adaptive filter converges the error $e[k]$ becomes the estimate of the signal $\hat{s}[k]$. Adaptive noise cancelling generally performs much better than the classical filtering approach since the noise is subtracted rather than filtered as such. Furthermore, little or no prior knowledge of $s[k]$, $n[k]$ and $x[k]$ or their interrelationship is needed.

In order to show how the adaptive noise canceller works the assumptions must be made that $s[k]$, $n[k]$, $x[k]$ and $y[k]$ are statistically stationary and have zero means; that $s[k]$ is uncorrelated with $n[k]$ and $x[k]$, and that $x[k]$ is correlated with $n[k]$. In this case, dropping the time index k, the output is defined by Equation 8.1.

$$e = s + n - y \tag{8.1}$$

The power or energy of this signal is computed by squaring it as defined by Equation 8.2.

$$e^2 = s^2 + (n - y)^2 + 2s(n - y) \tag{8.2}$$

Taking expectations of both sides results in the reduction defined by Equation 8.3.

$$E\{e^2\} = E\{s^2\}+E\{(n-y)^2\}+2E\{s(n-y)\}$$
$$= \{s^2\}+E\{(n-y)^2\}$$

(8.3)

(Note: the expectation of a signal squared is the same as the variance for a zero mean signal)

Adapting the filter to minimise the error energy $E\{e^2\}$ will not affect the signal energy $E\{s^2\}$ therefore the minimum error energy is defined by Equation 8.4.

$$E_{min}\{e^2\}= E\{s^2\}+E_{min}\{(n-y)^2\}$$

(8.4)

The signal y consequently becomes the best least squares estimate of the primary noise n. $E\{(e-s)^2\}$ is also minimised since, $(e-s)=(n-y)$. Therefore, minimising the total output energy is the same as minimising the noise energy.

8.2 Least Mean Squares Adaptation

The adaptive FIR filter weights for the interference cancelling system can be adapted using the LMS algorithm. The LMS algorithm is actually a modified form of the so called Wiener-Hopf equations used to adapt the FIR filter weights toward the minimum by gradient descent. The Wiener-Hopf equations define the optimum Wiener solution, economically expressed in matrix form. The LMS algorithm simplifies the computation by estimating the gradient from instantaneous values of the correlation matrix of the tap inputs and the cross-correlation vector between the desired response and the tap-weights.

Figure 8.4 shows a more practical interference cancelling model, which includes a delay in the desired signal path. The delay is not required to show the theoretical development of the LMS algorithm but it is needed in practical implementations, as will be discussed later. The optimum Wiener solution is developed first, followed by the practical LMS algorithm modification.

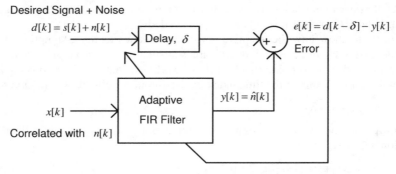

Figure 8.4. Adaptive Noise Cancelling System with Delay

8.2.1 Optimum Wiener Solution

The input to the adaptive FIR filter is the p-dimensional vector $\mathbf{x}[k]=[x[k], x[k-1], ..., x[k-p+1]]^T$. This vector $\mathbf{x}[k]$ can be easily formed by passing the discrete signal $x[k]$ through a delay line having p taps and then feeding the output of each tap into a summing junction as shown in Figure 8.5. Since all the signals in this application are zero mean there is no need to introduce a Direct Current (DC) offset in the system to remove their means but this can be easily done if needed (refer to Chapter 10 on "Adaptive Volterra Filters" for details).

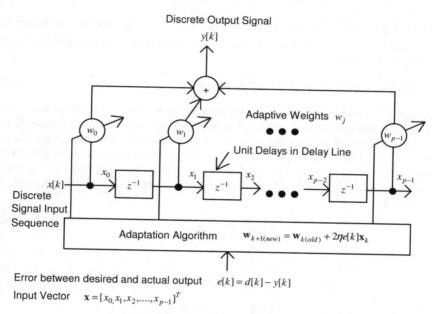

Figure 8.5. LMS Adaptive FIR Filter

The two important system vectors are the input vector $\mathbf{x}$ and the tap-weight vector $\mathbf{w}$ as follows,

$$\mathbf{x} = \begin{bmatrix} x_0 \\ x_1 \\ : \\ x_{p-1} \end{bmatrix} = \begin{bmatrix} x[k] \\ x[k-1] \\ : \\ x[k-p+1] \end{bmatrix}, \quad \mathbf{w} = \begin{bmatrix} w_0 \\ w_1 \\ : \\ w_{p-1} \end{bmatrix}.$$

Assume that the delay $\delta = 0$ for the time being. The system matrix equations are then defined by Equations 8.5.

$$y[k] = \sum_{m=0}^{p-1} w_m x[k-m] = \mathbf{x}^T \mathbf{w} = \mathbf{w}^T \mathbf{x}$$

$$e[k] = d[k] - y[k] = d[k] - \mathbf{x}^T \mathbf{w} = d[k] - \mathbf{w}^T \mathbf{x} \qquad (8.5)$$

$$e^2[k] = d^2[k] - 2d[k]\mathbf{x}^T \mathbf{w} + \mathbf{w}^T \mathbf{x}\mathbf{x}^T \mathbf{w}$$

The mean square error (expectation of the error power or energy) is defined by Equation 8.6.

$$E\{e^2[k]\} = E\{d^2[k]\} - 2E\{d[k]\mathbf{x}[k]^T\}\,\mathbf{w}[k] + \mathbf{w}^T[k]E\{\mathbf{x}[k]\mathbf{x}^T[k]\}\mathbf{w}[k] \qquad (8.6)$$

Let $\mathbf{r}_{dx}^T = E\{d[k]\mathbf{x}^T[k]\}$ and $\mathbf{R}_x = E\{\mathbf{x}[k]\mathbf{x}^T[k]\}$. Since the correlation matrix $\mathbf{R}_x$ is a symmetric positive definite matrix then the expectation of the error energy is defined by Equation 8.7.

$$E\{e^2[k]\} = E\{d^2[k]\} - 2\mathbf{r}_{dx}^T \mathbf{w} + \mathbf{w}^T \mathbf{R}_x \mathbf{w} \qquad (8.7)$$

Equation 8.7 is a quadratic function of the weights $\mathbf{w}$, a concave hyperparaboloid, which has a single minimum solution known as the optimum Wiener solution. Refer to Figure 8.6 for a two weight example depiction of this function. The Wiener solution is said to be optimum in the mean square sense, and it can be said to be truly optimum for second-order stationary noise statistics. To find this optimum solution the error Equation 8.7 is differentiated and equated to zero as defined by Equation 8.8.

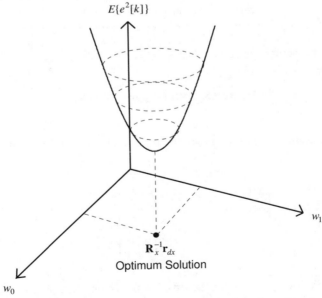

Figure 8.6. Concave Hyperparaboloid Error Function

$$\nabla_{E\{e^2[k]\}} = \frac{\partial E\{e^2[k]\}}{\partial \mathbf{w}} = -2\mathbf{r}_{dx} + 2\mathbf{R}_x\mathbf{w} = 0 \tag{8.8}$$

This is the gradient of the mean square error function. The optimum solution to Equation 8.8 is defined by the Wiener-Hopf Equation 8.9.

$$\mathbf{w}_0 = \mathbf{R}_x^{-1}\mathbf{r}_{dx} \tag{8.9}$$

where:

$\mathbf{w}_0$ is the optimum weight vector, also called the Wiener weight vector.

8.2.2 The Method of Steepest Gradient Descent Solution

Often the solution of the Wiener-Hopf equations is not practicable and other solutions must be used. One of those methods is the method of steepest gradient descent. It is an iterative method in which an initial estimate of $\mathbf{w}[k]$ is progressively refined until the minimum is reached. The iterative equation for the estimate $\hat{\mathbf{w}}[k+1]$ is defined by Equation 8.10.

$$\hat{\mathbf{w}}[k+1] = \hat{\mathbf{w}}[k] - \eta\nabla_{E\{e^2[k]\}} \tag{8.10}$$

where:

η is a small positive gain factor that controls stability and rate of convergence.

The weights are updated progressively a little at a time in the opposite direction to the steepest gradient of the error function. The true gradient of the mean square error function $\nabla_{E\{e^2[k]\}}$ can be estimated by substituting the current estimate of $\mathbf{w}[k]$ into the gradient equation as defined by Equation 8.11. This process is iterated until the gradient goes to zero, where the minimum solution is found.

$$\hat{\nabla}_{E\{e^2[k]\}} = -2\mathbf{r}_{dx} + 2\mathbf{R}_x\hat{\mathbf{w}}[k] \tag{8.11}$$

If $\mu = 2\eta$ the iteration Equation 8.10 is more economically expressed as Equation 8.12.

$$\hat{\mathbf{w}}[k+1] = \hat{\mathbf{w}}[k] + \mu(\mathbf{r}_{dx} - \mathbf{R}_x\hat{\mathbf{w}}[k]) \tag{8.12}$$

The stability and convergence rate of the steepest descent algorithm are determined by the value of μ and the eigenvalues of the correlation matrix $\mathbf{R}_x$. Since $\mathbf{R}_x$ is a correlation matrix it is positive definite and symmetric, which means there is an orthogonal matrix $\mathbf{Q}$ and a diagonal matrix Λ such that Equation 8.13 holds.

$$\mathbf{R}_x = \mathbf{Q}\Lambda\Lambda^T \tag{8.13}$$

A new zero mean variable $\mathbf{v}[k]$ can be introduced to make an affine change of coordinates to that defined by Equation 8.14.

$$v[k] = Q^T (\hat{w}[k] - w_o) \tag{8.14}$$

where:

w_0 is the Wiener weight vector, i.e., $w_0 = R_x^{-1} r_{dx}$, $r_{dx} = R_x w_0$.

The iteration process now becomes as defined by Equations 8.15 or 8.16.

$$Qv[k+1] + w_0 = Qv[k] + w_0 + \mu(r_{dx} - Q\Lambda\Lambda[k] - R_x w_0) \tag{8.15}$$

$$v[k+1] = (I - \mu\Lambda)v[k] \tag{8.16}$$

For vector $v[k]$,

$$v[k] = \begin{bmatrix} v_0[k] \\ v_1[k] \\ : \\ v_{p-1}[k] \end{bmatrix},$$

a set of linear difference equations can be developed as defined by Equation 8.17.

$$v_m[k+1] = (1 - \mu\lambda_m)v_m[k] \tag{8.17}$$

where:

λ_m is the mth element along the diagonal of Λ.

The solution to the set of Equations 8.17 is defined by the set of Equations 8.18.

$$v_m[k] = (1 - \mu\lambda_m)^k v_m[0] \tag{8.18}$$

These solution Equations 8.18 represent a geometric sequence that is bounded if

$-1 < (1 - \mu\lambda_m) < 1$, or if $0 < \mu < \dfrac{2}{\lambda_{max}}$.

Here λ_{max} is the maximum eigenvalue of R_x, which sets the condition for stability. Since R_x is positive definite all the eigenvalues are positive. When the stability condition is satisfied, the Equations 8.18 imply that the approach to the limit takes place exponentially.

The time constant τ_m represents the number of iterations required for v_m to decay to $1/e$ (≈ 0.3679) of its initial value of $v_m(0)$, i.e., the convergence rate τ_m must satisfy Equation 8.19.

$$1 - \mu\lambda_m = \exp\left(-\frac{1}{\tau_m}\right)$$

$$\tau_m = -\frac{1}{\ln(1 - \mu\lambda_m)}$$

(8.19)

For the case of $\mu \ll 1$ the convergence rate τ_m can be simplified to Equation 8.20, where it can be seen that the rate of steepest descent is limited by the smallest eigenvalue of $\mathbf{R}_x$, $\lambda_m = \lambda_{\min}$.

$$\tau_m \approx \frac{1}{\mu\lambda_m}$$

(8.20)

8.2.3 The LMS Algorithm Solution

The LMS adaptive algorithm is a practical method for finding close approximate solutions to the Wiener-Hopf equations which is not dependent on *a priori* knowledge of the autocorrelation of the input process and the crosscorrelation between the input and the desired output. It is an implementation of the method of steepest gradient descent as defined by Equation 8.21.

$$\mathbf{w}[k+1] = \mathbf{w}[k] - \eta\nabla_{E\{e^2[k]\}}$$

(8.21)

where:

η is a small positive gain factor that controls stability and rate of convergence.

The true gradient of the mean square error function $\nabla_{E\{e^2[k]\}}$ can be estimated by an instantaneous gradient by assuming that $e^2[k]$, the square of a single error sample, is an adequate estimate of the mean square error. The estimate of $\nabla_{E\{e^2[k]\}}$ is then defined by Equation 8.22.

$$\hat{\nabla}_{E\{e^2[k]\}} = \frac{\partial e^2[k]}{\partial \mathbf{w}} = 2e[k]\frac{\partial e[k]}{\partial \mathbf{w}} = 2e[k]\frac{\partial(d[k] - y[k])}{\partial \mathbf{w}}$$

$$\hat{\nabla}_{E\{e^2[k]\}} = 2e[k]\frac{\partial(d[k] - \mathbf{w}^T\mathbf{x}[k])}{\partial \mathbf{w}}$$

(8.22)

Equation 8.22 can be simplified to Equation 8.23 by differentiation.

$$\hat{\nabla}_{E\{e^2[k]\}} = -2e[k]\mathbf{x}[k]$$

(8.23)

From this the Widrow-Hoff LMS algorithm can be defined by Equation 8.24.

$$\mathbf{w}[k+1] = \mathbf{w}[k] + 2\eta\, e[k]\, \mathbf{x}[k] \qquad\qquad (8.24)$$

where:

η = learning rate parameter.

$e[k]$ = error (desired output - actual output).

$\mathbf{x}[k]$ = $[\, x_0,\, x_1,\ldots,\, x_{p-1}\,]^T$, the tap vector at instance k.

$\mathbf{w}[k]$ = $[\, w_0, w_1,\ldots,\, w_{p-1}\,]^T$, the tap-weight vector at instance k.

The LMS algorithm uses gradients of the mean square error function but it does not require squaring, averaging or differentiation and the expected value of $\mathbf{w}$ converges to the Wiener weight vector. Starting with an arbitrary (random) initial weight vector the algorithm will converge in the mean as new vectors $\mathbf{x}[k]$ are progressively presented to it in turn. It will stay stable as long as η is greater than zero but less than the reciprocal of the largest eigenvalue λ_{max} of the correlation matrix $\mathbf{R}_x$ as defined by Equation 8.25.

$$\frac{1}{\lambda_{max}} > \eta > 0 \qquad\qquad (8.25)$$

The largest eigenvalue λ_{max} is less than or equal to the trace of $\mathbf{R}_x$ as defined by Equation 8.26.

$$\lambda_{max} \le \sum_{k=0}^{p} \lambda_k = \mathrm{tr}\!\left[\mathbf{R}_x\right] = r_x(0) + r_x(1) + .. + r_x(p) \qquad\qquad (8.26)$$

In practice the step size η should be at least 10 times smaller than the maximum to achieve an adequate convergence. Learning or convergence occurs in an approximately exponential manner.

The derivation above strictly assumes an infinite two sided (noncasual) tapped delay line. This can be closely approximated with a finite delay line with a suitable delay δ for casual signals as shown in Figure 8.4. The value for δ is not overly critical but a value of approximately $\dfrac{p}{2}$ produces best results.

A summary of the LMS Adaptive Algorithm for a pth-order FIR Adaptive Filter is,

Parameters: $p - 1$ = FIR filter order (p is the vector size).

                        $\mu = 2\eta$ = step size.

                        $\mathbf{x}[k]$ = real valued input vector.

                        $\mathbf{w}[k]$ = filter coefficient vector.

Initialisation: $\mathbf{w}[0] = \mathbf{0}$ or very small random values.

Computation: For $k = 0, 1, 2,\ldots..$

                        i. $y[k] = \mathbf{w}^T \mathbf{x}[k]$

    ii. $e[k] = d[k] - y[k]$

    iii. $\mathbf{w}[k+1] = \mathbf{w}[k] + \mu \, e[k] \, \mathbf{x}[k]$

8.2.4 Stability of the LMS Algorithm

The method of steepest gradient descent and the LMS algorithm use gradient descent to try to minimise the mean square error in different ways. The method of steepest descent attempts to minimise the error at each step by repeatedly stepping in the negative direction of the gradient until the gradient goes to zero. The LMS algorithm takes single steps in the negative direction of the gradient at each step and depends on long-term averaging behaviour for eventual minimisation of the mean square error.

The proper choice of the step size $\mu = 2\eta$ is critical to the performance of the LMS algorithm. There are two forms of convergence which must be achieved,

1.   Convergence in the mean - the expected value of $\mathbf{w}[k]$ should approach the Wiener solution as the k tends to infinity.

2.   Convergence in the mean square - the final, steady-state value of the mean square error is finite.

The sequence of input vectors $\mathbf{x}[k]$ are assumed to be statistically independent. Both the input vector $\mathbf{x}[k]$ and the desired response $d[k]$ are assumed to have Gaussian distributions and to be statistically independent of all previous desired responses. The filter weights at time $k+1$, $\mathbf{w}[k+1]$, depend only on the previous input vectors, $\mathbf{x}[k]$, $\mathbf{x}[k-1]$,...., $\mathbf{x}[1]$, the previous desired responses, $d[k]$, $d[k-1]$,...., $d[1]$, and the initial value of the filter weights $\mathbf{w}[0]$. It therefore follows that $\mathbf{w}[k+1]$ is independent of both $\mathbf{x}[k+1]$ and $d[k+1]$.

Convergence in the mean requires that $\mathbf{w}[k]$ tends to the optimum Wiener vector $\mathbf{w}_0$. If it is assumed that,

$$\mathbf{e}[k] = \mathbf{w}[k] - \mathbf{w}_0$$

equivalent to,

$$\lim_{k \to \infty} E\{\mathbf{e}[k]\} = 0$$

then Equation 8.27 can be shown to hold.

$$E\{\mathbf{e}[k+1]\} = (\mathbf{I} - \mu \mathbf{R}_x) E\{\mathbf{e}[k]\} \tag{8.27}$$

This is similar to the condition for stability of the method of steepest descent. It can be shown by a similar argument previously developed that the condition for convergence in the mean is defined by Equation 8.26.

$$0 < \eta < \frac{1}{\lambda_{\max}} \tag{8.28}$$

where:

$\lambda_{\max}$ is largest eigenvalue of matrix $\mathbf{R}_x$.

Convergence in the mean square can be investigated by looking at the correlation matrix of the error $\mathbf{e}[k]$ as defined by Equation 8.29.

$$\mathbf{K}[k] = E\{\mathbf{e}[k]^T \mathbf{e}[k]\} \tag{8.29}$$

It is possible to derive a difference equation for the change in this matrix from one time step to the next as defined by Equation 8.30.

$$\mathbf{K}(k+1) = \mathbf{K}(k) - \mu(\mathbf{R}_x\mathbf{K}(k) + \mathbf{K}(k)\mathbf{R}_x) + \mu^2 \mathrm{tr}[\mathbf{R}_x\mathbf{K}(k)] +$$
$$\mu^2 \mathbf{R}_x\mathbf{K}(k)\mathbf{R}_x + \mu^2 J_{\min}\mathbf{R}_x \tag{8.30}$$

where:

$\mu = 2\eta$

tr[ ] denotes the trace of the matrix.

$J_{\min}$ is the value of the mean square error given by the Wiener solution $\mathbf{w}_0$.

It can be also be shown that the error function $J(k)$ can de defined by Equation 8.31.

$$J(k) = J_{\min} + \mathrm{tr}[\mathbf{R}_x\mathbf{K}(k)] \tag{8.31}$$

The trace is always positive, since the matrices in the brackets are positive definite, so the mean square error at time k is always greater than the mean square of the Wiener solution. The limiting value of the mean square error $J(\infty)$ is defined by Equation 8.32.

$$J(\infty) = \frac{J_{\min}}{1 - \dfrac{\sum\limits_{i=0}^{p} \mu\lambda_i}{2 - \mu\lambda_i}} \tag{8.32}$$

where:

λ_i are the eigenvalues of $\mathbf{R}_x$.

From these equations it can be shown that $J(k)$ converges to a steady-state value $J(\infty)$ if and only if,

$$0 < \mu < \frac{2}{\lambda_{\max}},$$

where $\lambda_{\max}$ is the maximum eigenvalue of $\mathbf{R}_x$, and, $\sum\limits_{i=0}^{p} \dfrac{\mu\lambda_i}{2 - \mu\lambda_i} < 1$.

Although the assumption that the sequence of input vectors is statistically independent is strictly false, in practical problems the equations above are nevertheless usually reliable. Also, when μ is small the LMS algorithm can track changes in the optimum set of filter weights in nonstationary situations if the changes take place slowly enough.

8.2.5 The Normalised LMS Algorithm

The filter weight update equation for the LMS algorithm is defined by Equation 8.33.

$$\mathbf{w}[k+1] = \mathbf{w}[k] + 2\eta\, e\, \mathbf{x}[k] \tag{8.33}$$

However, the problem with this update equation is that if the input vector magnitude is large the filter weights also change by a larger amount. Consequently, it could be desirable to normalise this vector in some way. The normalised LMS algorithm can be reformulated in terms of a constrained optimisation problem as follows. Given the input vector $\mathbf{x}[k]$, the desired response $d[k]$ and the current filter weights $\mathbf{w}[k]$, find the updated filter weights $\mathbf{w}[k+1]$ that minimise the squared Euclidean norm of the difference $\mathbf{w}[k+1] - \mathbf{w}[k]$ subject to the constraint $d[k] = \mathbf{w}[k+1]\mathbf{x}[k]$. This problem can be solved using Lagrange multipliers and thus providing the weight update Equation 8.34.

$$\mathbf{w}[k+1] = \mathbf{w}[k] + \frac{2\eta e[k]}{\|\mathbf{x}[k]\|^2}\mathbf{x}[k] \tag{8.34}$$

This is the update equation for the normalised LMS algorithm. It is convergent in the mean square sense if $0 < \eta < 1$. However, this update equation can still lead to numerical problems if $\|\mathbf{x}[k]\|$ is small, so the alternative form defined by Equation 8.35 can be used in practice to avoid this.

$$\mathbf{w}[k+1] = \mathbf{w}[k] + \frac{2\eta e[k]}{(a + \|\mathbf{x}[k]\|^2)}\mathbf{x}[k] \tag{8.35}$$

where:
   $a > 0$

8.3 Recursive Least Squares Estimation

Recursive Least Squares (RLS) estimation can be seen as a special case of Kalman filtering. It is actually an extension of LSE where the estimate of the coefficients of an optimum filter are updated using a combination of the previous set of coefficients and a new observation. In effect this uses all the past values of the time series to construct the filter as in the Wiener filter. When the statistical information related to the process is unknown it must be estimated from the data. One effective way to do this without having knowledge of ensemble averages (expectations) is to use error measures like the least squares error that do not depend on expectations, and that may be computed directly from the data. One important thing to note about LSE methods is that they produce solutions that are dependent on the data themselves rather than the data's statistics. In contrast to LSE, minimising the mean square error produces the same solution for data having the same statistics.

RLS estimation can be used in conjunction with any ARMA filter model but it is commonly applied to the MA or FIR filter model in the adaptive filtering context. In this case the pth-order FIR filter coefficients are time-varying and are adapted using a least squares recursive algorithm. Consider a zero mean complex valued time series $x[k]$, $i = 1, 2, ...$, and a corresponding set of desired filter responses $d[k]$. Vector samples from the time series are denoted as,

$$\mathbf{x}[k] = \left[x[k], x[k-1], ..., x[k-p+1]\right]^T$$

and the coefficients of a time-varying FIR filter at time n are denoted as,

$$\mathbf{w}[n] = \left[w_0[n], w_1[n], ..., w_{p-1}[n]\right]^T$$

The error between the desired response at time k and the filter output for an input of $\mathbf{x}[k]$ with the filter coefficients at time n is defined by Equation 8.36

$$e[n,k] = d[k] - \mathbf{w}^T[n]\mathbf{x}[k] \tag{8.36}$$

The quality of the filter coefficients at time n may be measured by considering the weighted sum of the differences between the desired output at all earlier times and the output that would have been produced by filtering the time series with the current values of the filter coefficients. The error term $\xi(n)$ for this can be defined by Equation 8.37 (Widrow and Sterns 1985).

$$\xi(n) = \sum_{k=0}^{n} \beta(n,k)\left|e[n,k]\right|^2 = \sum_{k=0}^{n} \beta(n,k)\left|d[k] - \mathbf{w}^T[n]\mathbf{x}[k]\right|^2 \tag{8.37}$$

Weighting factors, $0 < \beta(n,k) \le 1$, for $k = 0,1,2,..,n$, are used to discount the effects of earlier errors so as to enable the filter to track changes in nonstationary time series.

8.3.1 The Exponentially Weighted Recursive Least Squares Algorithm

Exponential weighting (or forgetting) factors of the form $\beta(n,k) = \lambda^{n-k}$ are commonly used, where λ is less than but close to 1 (in the special case when $\lambda=1$ this is referred to as a growing window). In the general exponential weighting case the error function $\xi(n)$ is defined by Equation 838.

$$\xi(n) = \sum_{k=0}^{n} \lambda^{n-k} \left| d[k] - \mathbf{w}^T[n]\mathbf{x}[k] \right|^2 \tag{8.38}$$

The prewindowed autocorrelation $\Phi(n)$ of the input vector $\mathbf{x}[k]$ at time n can be specified by a recursive function as defined by Equation 8.39.

$$\begin{aligned}
\Phi(n) &= \sum_{k=0}^{n} \lambda^{n-k} \mathbf{x}^*[k]\mathbf{x}^T[k] \\
&= \left[\sum_{k=0}^{n-1} \lambda^{n-k} \mathbf{x}^*[k]\mathbf{x}^T[k] \right] + \lambda^{n-n} \mathbf{x}^*[n]\mathbf{x}^T[n] \\
&= \lambda \left[\sum_{k=0}^{n-1} \lambda^{n-1-k} \mathbf{x}^*[k]\mathbf{x}^T[k] \right] + \mathbf{x}^*[n]\mathbf{x}^T[n] \\
&= \lambda \Phi(n-1) + \mathbf{x}^*[n]\mathbf{x}^T[n]
\end{aligned} \tag{8.39}$$

The prewindowed crosscorrelation $\theta(n)$ between the input vector and the desired series at time n can in a similar fashion be specified by a recursive function as defined by Equation 8.40.

$$\begin{aligned}
\theta(n) &= \sum_{k=0}^{n} \lambda^{n-k} d[k]\mathbf{x}^*[k] \\
&= \lambda \theta(n-1) + d[n]\mathbf{x}^*[n]
\end{aligned} \tag{8.40}$$

The optimum filter coefficients that minimise the weighted error $\xi(n)$ can be determined by setting the derivative of $\xi(n)$ with respect to $w_m^*[n]$ to zero for $m = 0,1,\ldots,p\text{-}1$, resulting in Equation 8.41.

$$\Phi(n)\hat{\mathbf{w}}[n] = \theta(n) \tag{8.41}$$

To solve Equation 8.41 for the estimate of the filter coefficients $\hat{\mathbf{w}}[n]$ it is necessary to find the inverse of $\Phi(n)$. This can be done recursively using the Matrix Inversion Lemma, also known as Woodbury's identity. Let $\mathbf{A}$ and $\mathbf{B}$ be positive definite $M \times M$ matrices, let $\mathbf{C}$ be a $M \times N$ matrix, and let $\mathbf{D}$ be a positive definite $N \times N$ matrix. If it is true that $\mathbf{A} = \mathbf{B}^{-1} + \mathbf{C}\mathbf{D}^{-1}\mathbf{C}^T$ then Equation 8.42 follows.

$$\mathbf{A}^{-1} = \mathbf{B} - \mathbf{B}\mathbf{C}(\mathbf{D} + \mathbf{C}^T\mathbf{B}\mathbf{C})^{-1}\mathbf{C}^T\mathbf{B} \tag{8.42}$$

The recursive expression for the exponentially weighted correlation matrix is in the form to which the Matrix Inversion Lemma can be applied if the following equivalences are made,

$$\mathbf{A} = \Phi(n)$$
$$\mathbf{B}^{-1} = \lambda\Phi(n-1)$$

$$C = \mathbf{x}[n]$$
$$D = 1$$

The recursive expression for the inverse of the correlation matrix is therefore defined by Equation 8.43.

$$\Phi^{-1}(n) = \lambda^{-1}\Phi^{-1}(n-1) - \frac{\lambda^{-2}\Phi^{-1}(n-1)\mathbf{x}^{*}[n]\mathbf{x}^{T}[n]\Phi^{-1}(n-1)}{1 + \lambda^{-1}\mathbf{x}^{T}[n]\Phi^{-1}(n-1)\mathbf{x}^{*}[n]} \qquad (8.43)$$

If $\mathbf{P}(n)$ is substituted for $\Phi^{-1}(n)$ and $\mathbf{k}(n) = \dfrac{\lambda^{-1}\mathbf{P}(n-1)\mathbf{x}^{*}[n]}{1 + \lambda^{-1}\mathbf{x}^{T}[n]\mathbf{P}(n-1)\mathbf{x}^{*}[n]}$, where $\mathbf{k}(n)$

is called the gain vector, it is possible to write a recursive expression for $\mathbf{P}(n)$ as defined by Equation 8.44.

$$\mathbf{P}(n) = \lambda^{-1}\mathbf{P}(n-1) - \lambda^{-1}\mathbf{k}(n)\mathbf{x}^{T}[n]\mathbf{P}(n-1) \qquad (8.44)$$

Equation 8.44 is a form of the Riccati equation. With some algebraic manipulation of Equation 8.44 it can be shown that the gain vector $\mathbf{k}(n)$ is defined by Equation 8.45.

$$\mathbf{k}(n) = \mathbf{P}(n)\mathbf{x}^{*}[n] = \Phi^{-1}(n)\mathbf{x}^{*}[n] \qquad (8.45)$$

Given all these equations it is now possible to develop the recursive expression for the update of the filter weights as defined by Equations 8.46 and 8.47.

$$\begin{aligned}
\hat{\mathbf{w}}(n) &= \Phi^{-1}(n)\theta(n) \\
&= \mathbf{P}(n)\theta(n) \\
&= \lambda\mathbf{P}(n)\theta(n-1) + d[n]\mathbf{P}(n)\mathbf{x}^{*}[n] \\
&= \mathbf{P}(n-1)\theta(n-1) - \mathbf{k}(n)\mathbf{x}^{T}[n]\mathbf{P}(n-1)\theta(n-1) + d[n]\mathbf{P}(n)\mathbf{x}^{*}[n] \\
&= \Phi^{-1}(n-1)\theta(n-1) - \mathbf{k}(n)\mathbf{x}^{T}[n]\Phi(n-1)\theta(n-1) + \\
&\quad d[n]\mathbf{P}(n)\mathbf{x}^{*}[n]
\end{aligned} \qquad (8.46)$$

$$\begin{aligned}
\hat{\mathbf{w}}[n] &= \hat{\mathbf{w}}[n-1] - \mathbf{k}(n)\mathbf{x}^{*}[n]\hat{\mathbf{w}}[n-1] + d[n]\mathbf{P}(n)\mathbf{x}^{*}[n] \\
&= \hat{\mathbf{w}}[n-1] - \mathbf{k}(n)\mathbf{x}^{*}[n]\hat{\mathbf{w}}[n-1] + d[n]\mathbf{k}(n) \\
&= \hat{\mathbf{w}}[n-1] + (d[n] - \mathbf{x}^{T}[n]\hat{\mathbf{w}}[n-1])\mathbf{k}(n) \\
&= \hat{\mathbf{w}}[n-1] + \alpha(n)\mathbf{k}(n)
\end{aligned} \qquad (8.47)$$

where:

$\alpha(n) = d[n] - \mathbf{x}^{T}[n]\hat{\mathbf{w}}[n-1] = d[n] - \hat{\mathbf{w}}^{T}[n-1]\mathbf{x}^{*}[n]$ is the innovation or *a priori* estimation error at time n.

When the innovation error $\alpha(n)$ is small, the current set of filter weights are close to their optimum values (in the least squares sense), and only a small correction needs to be applied to the weights, and vice versa. The recursion begins by setting the filter coefficients to zero and setting $\mathbf{P}(0)$ to a multiple of the identity matrix to ensure that the correlation remains nonsingular.

A summary of the RLS algorithm for a FIR Adaptive Filter is as follows,

    i.   Set: $\mathbf{w}^T[n-1] = \mathbf{0},$

                    $\mathbf{P}(0) = \delta^{-1}\mathbf{I}$, where δ is a small positive constant.

    ii.   Compute: $\mathbf{k}(n) = \dfrac{\lambda^{-1}\mathbf{P}(n-1)\mathbf{x}^*[n]}{1+\lambda^{-1}\mathbf{x}^T[n]\mathbf{P}(n-1)\mathbf{x}^*[n]}$

                            $\alpha(n) = d[n] - \hat{\mathbf{w}}^T[n-1]\mathbf{x}^*[n]$

                            $\hat{\mathbf{w}}[n] = \hat{\mathbf{w}}[n-1] + \alpha(n)\mathbf{k}(n)$

                            $\mathbf{P}(n) = \lambda^{-1}[\mathbf{P}(n-1) - \mathbf{k}(n)\mathbf{x}^T[n]\mathbf{P}(n-1)]$

    iii.   Repeat: Step ii until $\hat{\mathbf{w}}[n]$ converges or continue indefinitely.

8.3.2 Recursive Least Squares Algorithm Convergence

The convergence of $\hat{\mathbf{w}}[n]$ in the mean, in the mean square, and the convergence of the *a priori* estimation error $\alpha(n)$ is of interest to us. Assume that the desired response and the input vector are related by a multiple linear regression model of the form $d[n] = e_r[n] + \mathbf{w}_r^T[n]\mathbf{x}[n]$, where $\hat{\mathbf{w}}_r[n]$ is the regression parameter vector and $e_r[n]$ is the measurement error at time n. $\hat{\mathbf{w}}_r[n]$ is a constant vector, and the measurement error $e_r[n]$ is assumed to be white zero mean noise with variance σ_w^2. For the purpose of the following general derivations λ is assumed to be equal to unity, which is close enough to typical values that are chosen in practice.

8.3.2.1 Convergence of the Filter Coefficients in the Mean
Initialising the RLS recursion by setting $\Phi(0) = \delta\mathbf{I}$ introduces a bias into the estimate of the filter coefficients that it produces. If $\mathbf{b}(n)$ is a bias at time n, then $E\{\hat{\mathbf{w}}[n]\} = \mathbf{w}_r + \mathbf{b}(n)$. It can be shown that $\mathbf{b}(n) = -\delta\Phi^{-1}(n)\mathbf{w}_r$, and this tends to zero as n tends to infinity, provided that the data sequence $\mathbf{x}[n]$ is ergodic. This shows convergence in the mean.

8.3.2.2 Convergence of the Filter Coefficients in the Mean Square

If $e_w[n] = \hat{w}[n] - w_r$, it can be shown that the correlation matrix of this error sequence is $K(n) = E\{e_w[n]e_w^T[n]\} = \sigma_w^2 \Phi^{-1}(n)$. It can also be shown that $\|K(n)\| \approx \sigma_w^2 / n\lambda_{\min}$, where $\lambda_{\min}$ is the smallest eigenvalue of the ensemble-averaged correlation matrix of the $x[n]$. This means that the convergence properties depend on the conditioning of this matrix. Since the norm $\|K(n)\|$ is inversely proportional to time then there is convergence of the filter coefficients in the mean square to the regression parameter.

8.3.2.3 Convergence of the RLS Algorithm in the Mean Square

The *a priori* estimation error at time n, $\alpha(n) = d[n] - \hat{w}^T[n-1]x[n]$, can be used to define the mean square error of the RLS algorithm at time n, $\xi(n) = E\{|\alpha(n)|^2\}$. It can be shown that $\xi(n) \approx \sigma_w^2 + M\sigma_w^2 / n$ for large n, so the mean square error approaches the variance of the measurement error as n tends to infinity.

8.3.3 The RLS Algorithm as a Kalman Filter

The exponentially weighted RLS algorithm is actually a special case of the Kalman filter. The state space model for the set of filter coefficients is a random walk. Let $w[n]$ be a vector of filter coefficients at time n and set $w[n+1] = w[n] + v[n]$, where $v[n]$, the process noise vector, is assumed to have zero mean and autocorrelation matrix $Q(n)$. Here the state transition matrix is the identity matrix. The measurement model is taken to be $d[n] = x^T[n]w[n] + e[n]$, where the measurement error $e[n]$ is assumed to have a mean of zero and a mean square valued $\xi_{\min}$. In this case the Kalman gain is defined by Equation 8.48, and the estimate of the filter coefficients at time n is defined by Equation 8.49.

$$K(n) = \tilde{P}(n)x[n][x^T[n]\tilde{P}(n)x[n] + \xi_{\min}]^{-1} \tag{8.48}$$

$$\hat{w}[n] = \hat{w}[n] + K(n)(d[n] - x^T[n]\hat{w}[n]) \tag{8.49}$$

The error covariance matrix for the estimate is defined by Equation 8.50, and the prior estimate and its covariance are defined by Equations 8.50 and 8.51 respectively.

$$P(n) = [I - K(n)x^T[n]]\tilde{P}(n) \tag{8.50}$$

$$\hat{w}[n+1] = \hat{w}[n], \text{ and } \tilde{P}(n+1) = P(n) + Q(n) \tag{8.51}$$

Note that symbol $\mathbf{P}(n)$ here is not the same as the $\mathbf{P}(n)$ in the description of the RLS recursion formula. If $\mathbf{Q}(n) = (\lambda^{-1} - 1)\mathbf{P}(n)$ then the Kalman filter is the same as the RLS recursion formula.

8.4 Exercises

The following Exercises identify some of the basic ideas presented in this Chapter.

8.4.1 Problems

8.1.   Assume $x[n]$ is a second-order autoregressive process defined by the difference equation,

$$x[n] = 1.2728x[n-1] - 0.81x[n-2] + v[n],$$

where $v[n]$ is unit variance white noise the optimum causal predictor for $x[n]$ is given by, $\hat{x}[n] = 1.2728x[n-1] - 0.81x[n-2]$.

Show how you could go about computing the predictor coefficients 1.2728 and -0.81 using the LMS algorithm.

8.2.   In Problem 8.1 if the autocorrelation sequence $r_x(k)$ is given as $r_x(0) = 5.7523$ and $r_x(1) = 4.0450$ find the maximum bound for the step size in the LMS algorithm.

8.3.   Use the normalised LMS algorithm to derive the FIR filter coefficient update equations for the second-order AR(2) linear prediction equation of Problem 8.1.

8.4.   If the process and the FIR filter coefficients are complex valued show that the adaptive LMS update equation is $\mathbf{w}[k+1] = \mathbf{w}[k] + \mu e[k] \mathbf{x}^*[k]$.

8.5.   Compute and sketch a graph of the expectation of the squared error as a function of the single real-valued filter coefficient for a simple zero-order MA(0) process .

8.6.   For the exponentially weighted RLS algorithm prove that $\Phi(n)\mathbf{w}[n] = \theta(n)$ by setting the derivative of the weighted error $\xi(n)$ with respect to $\mathbf{w}*[n]$ to zero, given that,

$$\Phi(n) = \sum_{k=0}^{n} \lambda^{n-k}\mathbf{x}^*[k]\mathbf{x}^T[k] \text{ and, } \theta(n) = \sum_{k=0}^{n} \lambda^{n-k}d[k]\mathbf{x}^*[k]$$

9. Frequency Domain Adaptive Filters

All filter equations can be solved either in the time or frequency domains. However time domain solutions for real-time filter operation require the convolution function, which can be very computationally intensive for large impulse response sequences. One way to reduce this computational burden is to perform the convolution computation in the frequency domain, where the convolution function becomes a much simpler complex multiplication function. All that is needed to exploit this fact is the existence of an efficient means of transforming to the frequency domain and back again. Fortunately, the Fast Fourier Transform (FFT) provides such a transform that can be used to dramatically reduce the overall computational burden, especially for large data block samples. The advent of the FFT has made it possible to produce very efficient adaptive filters. The only drawback is that there are inevitable errors introduced due to the fact that continuous signals must be broken up into finite blocks to be able to apply the FFT to them. Nevertheless, with judicious design these errors can be kept to a minimum and fortunately they have few detrimental effects for many practical problems. The main way to achieve continuous linear convolutions (required by real filters) from circular convolutions (typical of digital processing algorithms) is by a process of overlapping and selection. Two main ways of doing this are by the overlap-save and the overlap-add methods.

9.1 Frequency Domain Processing

Many of the adaptive filtering methods for FIR filters have the drawback that they often require many, perhaps thousands, of filter coefficients to achieve desired levels of performance in real applications. One method of reducing the large amounts of computation required for these algorithms is to use a block updating strategy, where the filter coefficients are updated at each time block of points instead of at each successive time n. Then Fast Fourier Transform (FFT) routines are used to reduce the amount of computation required to implement convolutions by doing them in the frequency domain. Figure 9.1 shows a representative general frequency domain processing arrangement. The input sequences $x[n]$ and $d[n]$ can be acquired at any time from the input delay lines and transformed from the time domain to the frequency domain, where the processing is performed and then the result $y[n]$ is transformed back to the time domain when it is needed. Often all the processing can be done and the results taken directly in the frequency domain, thereby obviating the need even for the final time domain transformation.

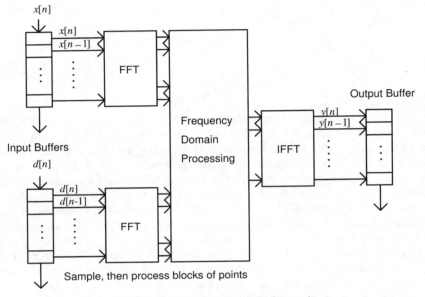

Figure 9.1. Frequency Domain Processing

9.1.1 Time Domain Block Adaptive Filtering

Before looking at the frequency domain adaptive filtering approach it is instructive to review the standard and block LMS algorithms in the time domain. The notation for the time domain is as follows. The real input signal vector at time n is defined as,

$$\mathbf{x}[n] = \begin{bmatrix} x[n] \\ x[n-1] \\ \vdots \\ x[n-p+1] \end{bmatrix}$$

The filter coefficient vector at time n is defined as,

$$\mathbf{w}[n] = \begin{bmatrix} w_0[n] \\ w_1[n-1] \\ \vdots \\ w_{p-1}[n-p+1] \end{bmatrix}$$

The filter output at time n is defined as,

$$y[n] = \mathbf{x}^T[n]\mathbf{w}[n]$$

If the desired response at time n is $d[n]$ the error $e[n]$ is $e[n] = d[n] - y[n]$. The coefficient update formula for the standard LMS algorithm is $\mathbf{w}[n+1] = \mathbf{w}[n] + 2\eta e[n]\mathbf{x}[n]$, where η is the step size parameter, which controls the convergence rate. These computations are computed continuously at every successive time n.

If $L \geq 1$ is the block size the update formula for the Block LMS (BLMS) algorithm is updated on the basis of a block of input data at a time as defined by Equation 9.1.

$$\mathbf{w}[n+L] = \mathbf{w}[n] + 2\eta \sum_{m=0}^{L-1} e[n+m]\mathbf{x}[n+m] \qquad (9.1)$$

where:

$e[n+m] = d[n+m] - \mathbf{x}^T[n+m]\mathbf{w}[n]$ is now the error for the BLMS.

A new time index k for the filter coefficient update representing L increments in n, allows the BLMS update formula to be defined by Equation 9.2.

$$\mathbf{w}[k+1] = \mathbf{w}[k] + 2\eta \sum_{m=0}^{L-1} e[kL+m]\mathbf{x}[kL+m] \qquad (9.2)$$

Using the BLMS requires that the maximum value of the step size be scaled down by a factor of L as compared to the LMS algorithm in order to preserve stability. However, this scaling can make the BLMS algorithm converge more slowly than the simpler LMS algorithm.

9.1.2 Frequency Domain Adaptive Filtering

Adaptive algorithms, where the filtering is done in the frequency domain, have a recursion similar to the block update of the BLMS. Filtering in the frequency domain is made possible by the fact that time domain filtering operations of linear convolution (and linear correlation) can be done simply by complex multiplication in the frequency domain by the Convolution Theorem for the Discrete Fourier Transform (DFT). This Convolution theorem actually gives circular convolutions. Therefore to perform the required linear convolutions some simple corrections are needed. However, by moving the computations into the frequency domain a computational gain can be achieved if a FFT algorithm is used to implement the DFT more efficiently.

The Frequency Domain Adaptive Filter (FDAF) is formed by taking discrete Fourier transforms of time domain lengths of M points (Shynk 1992). Once converted to the frequency domain real variables are transformed to complex variables, which means that complex matrix arithmetic must be used. The time domain coefficients become complex variables in the frequency domain and are denoted by uppercase nonbolded letters. The notation for the frequency domain is as follows. The input signal matrix at time k, $\mathbf{X}(k)$, is defined as,

$$\mathbf{X}(k) = \begin{bmatrix} X_0(k) & 0 & .. & 0 \\ 0 & X_1(k) & .. & 0 \\ \vdots & \vdots & \ddots & \vdots \\ 0 & 0 & .. & X_{M-1}(k) \end{bmatrix}$$

The FIR filter coefficient vector at time k, $\mathbf{W}(k)$, is defined as,

$$\mathbf{W}(k) = \begin{bmatrix} W_0(k) \\ W_1(k) \\ \vdots \\ W_{p-1}(k) \end{bmatrix}$$

The adaptive filter output at time k, $\mathbf{Y}(k)$, is then defined by Equation 9.3.

$$\mathbf{Y}(k) = \mathbf{X}(k)\mathbf{W}(k) \tag{9.3}$$

where:
   The desired response vector at time k is $\mathbf{D}(k)$.
   The error vector at time k is $\mathbf{E}(k)$.

The error vector is defined as the difference between a desired response vector,

$$\mathbf{D}(k) = \begin{bmatrix} D_o(k) \\ D_1(k) \\ \vdots \\ D_{p-1}(k) \end{bmatrix},$$

and the corresponding filter output vector $\mathbf{Y}(k)$. $\mathbf{D}(k)$ can be defined directly in the frequency domain, or it may be alternatively computed from a block sequence of desired values in the time domain.

   The filter coefficient update formula for the FDAF can now defined by Equation 9.4.

$$\mathbf{W}(k+1) = \mathbf{W}(k) + 2\mathbf{G}\mathbf{M}(k)\mathbf{X}^H(k)\mathbf{E}(k) \tag{9.4}$$

The **G** matrix in Equation 9.4 is the matrix that contains the gradient $\mathbf{X}^H(k)\mathbf{E}(k)$ in order to convert from circular to linear convolutions, and the matrix $\mathbf{M}(k)$, the diagonal matrix of scalar step sizes is,

$$\mathbf{M}(k) = \begin{bmatrix} \eta_0(k) & 0 & .. & 0 \\ 0 & \eta_1(k) & .. & 0 \\ \vdots & \vdots & \ddots & \vdots \\ 0 & 0 & .. & \eta_{M-1}(k) \end{bmatrix}$$

The step sizes may be chosen to vary inversely with the signal power in the band, corresponding to a block frequency parameter m as defined below for Equation 9.5. If this is done it is necessary to estimate these power values on an ongoing basis.

To compute the required frequency domain vectors a $M \times M$ DFT matrix **F**, with inverse $\mathbf{F}^{-1} = \mathbf{F}^H/M$ is required. This matrix **F** is defined by Equation 9.5.

$$\mathbf{F} = \begin{bmatrix} F_{00} & F_{01} & .. & F_{0(M-1)} \\ F_{10} & F_{11} & .. & F_{1(M-1)} \\ \vdots & \vdots & \ddots & \vdots \\ F_{(M-1)0} & F_{(M-1)1} & .. & F_{(M-1)(M-1)} \end{bmatrix} \tag{9.5}$$

where:

$$F_{ml} = e^{\frac{-j2\pi ml}{M}}, \qquad m,l = 0,1,..,M-1$$

Premultiplying a time domain vector by **F** will compute its DFT and premultiplying a frequency domain vector by $\mathbf{F}^{-1}$ will compute a vector of time domain samples. In practice the DFT is implemented with a FFT algorithm in order to achieve the desired computational advantage, but here it is shown in full for equation clarity.

By the Convolution Theorem for the DFT, the product of two DFT sequences is actually the DFT of their circular convolution, where the linear and circular convolutions have terms in common. Nevertheless, it is still possible to derive linear convolutions from circular convolutions by a process of overlapping and selection. There are two methods of doing this known as the overlap-save and the overlap-add methods. In order to generate p correct output samples the DFT length must be greater that $2p - 1$. Therefore DFTs with length $M = 2p$ can be used where the optimal block size is $L = p$.

9.1.2.1 The Overlap-save Method
In the overlap-save method, vectors $\mathbf{W}(k)$ and $\mathbf{x}[k]$ have dimensions of $2p$. This is achieved by defining vectors as follows,

$$\mathbf{W}(k) = \mathbf{F} \begin{bmatrix} w_0[k] \\ w_1[k] \\ \vdots \\ w_{p-1}[k] \\ 0 \\ \vdots \\ 0 \end{bmatrix}$$

$\mathbf{X}(k)$ is a frequency domain diagonal vector matrix whose elements are components of vector,

$$\mathbf{F} \begin{bmatrix} \mathbf{x}(kp - p) \\ \mathbf{x}(kp) \end{bmatrix} = \mathbf{F} \begin{bmatrix} x[kp - p] \\ x[kp - p + 1] \\ \vdots \\ x[kp - 1] \\ x[kp] \\ x[kp + 1] \\ \vdots \\ x[kp + p - 1] \end{bmatrix}$$

The frequency domain filter output $\mathbf{Y}(k)$ is defined by Equation 9.6 and in the time domain the output is the vector,

$$\mathbf{y}[k] = \begin{bmatrix} y[kp] \\ y[kp + 1] \\ \vdots \\ y[kp + p - 1] \end{bmatrix},$$

whose elements are the last components of $\mathbf{F}^{-1}\mathbf{Y}(k)$.

$$\mathbf{Y}(k) = \mathbf{X}(k)\mathbf{W}(k) \tag{9.6}$$

If the desired output in the time domain at time k is the vector $\mathbf{d}[k]$ as follows,

$$\mathbf{d}(k) = \begin{bmatrix} d(kp) \\ d(kp + 1) \\ \vdots \\ d(kp + p - 1) \end{bmatrix},$$

then the error in the time domain is $\mathbf{e}(k) = \mathbf{d}(k) - \mathbf{y}(k)$, and the transform of this vector with p zeros prepended provides the error in the frequency domain as follows,

$$E(k) = \mathbf{F} \begin{bmatrix} 0 \\ \vdots \\ 0 \\ \mathbf{e}(k) \end{bmatrix}$$

If $\Delta\mathbf{w}[k]$ is a vector consisting of the first p components of the vector $\mathbf{F}^{-1}\mathbf{X}^H(k)\mathbf{E}(k)$, it can be shown that the update formula for the BLMS algorithm in the frequency domain is defined by Equation 9.7

$$\mathbf{W}(k+1) = \mathbf{W}(k) + 2\eta\mathbf{F} \begin{bmatrix} \Delta\mathbf{w}[k] \\ 0 \\ \vdots \\ 0 \end{bmatrix} \tag{9.7}$$

Now, if a $2p \times 2p$ matrix $\mathbf{g}$ is set to $\mathbf{g} = \begin{bmatrix} \mathbf{I} & \mathbf{0} \\ \mathbf{0} & \mathbf{0} \end{bmatrix}$ with four $p \times p$ submatrices then the update formula can be defined by Equation 9.8.

$$\mathbf{W}(k+1) = \mathbf{W}(k) + 2\eta\mathbf{F}\mathbf{g}\mathbf{F}^{-1}\mathbf{X}^H(k)\mathbf{E}(k) \tag{9.8}$$

If Equation 9.8 is compared with the general form of the FDAF, $\mathbf{W}(k+1) = \mathbf{W}(k) + 2\mathbf{G}\mathbf{M}\mathbf{X}^H(k)\mathbf{E}(k)$, it can be seen that $\mathbf{G} = \mathbf{F}\mathbf{g}\mathbf{F}^{-1}$. If $\mathbf{k} = \begin{bmatrix} \mathbf{0} & \mathbf{I} \end{bmatrix}$ is defined then $\mathbf{y}[k] = \mathbf{k}\mathbf{F}^{-1}\mathbf{Y}(k)$ and $E(k) = \mathbf{F}\mathbf{k}^T\mathbf{e}[k]$.

The overlap-save algorithm can therefore be summarised as follows,

Set $\mathbf{W}(0) = \begin{bmatrix} 0 \\ 0 \\ \vdots \\ 0 \end{bmatrix}$, and $P_m(0) = \delta_m,$ $m = 0,1,...,2p-1$, where δ_m are small

positive random numbers that are the first estimates of $P_m(0)$, the measures of power used to set the step sizes.

For each block of p input samples, the input matrix $\mathbf{X}(k)$ is made to be a diagonal matrix whose elements are the Fourier transform of the $2p$ time domain points $\mathbf{x}[kp-p],....,\mathbf{x}[kp+p-1]$. Compute the following in sequence,

$$\mathbf{Y}(k) = \mathbf{X}(k)\mathbf{W}(k)$$

$$\mathbf{y}[k] = \mathbf{kF}^{-1}\mathbf{Y}(k)$$

$$\mathbf{e}[k] = \mathbf{d}[k] - \mathbf{y}[k]$$

$$\mathbf{E}(k) = \mathbf{Fk}^T\mathbf{e}[k]$$

$$P_m(k) = \lambda P_m(k-1) + (1-\lambda)\left|X_m(k)\right|^2, \quad m = 0,..,2p-1$$

$$\mathbf{M}(k) = \begin{bmatrix} \eta/P_0(k) & 0 & .. & 0 \\ 0 & \eta/P_1(k) & .. & 0 \\ : & : & \ddots & : \\ 0 & 0 & .. & \eta/P_{2p-1}(k) \end{bmatrix}$$

$$\mathbf{W}(k+1) = \mathbf{W}(k) + 2\mathbf{FgF}^{-1}\mathbf{M}(k)\mathbf{X}^H(k)\mathbf{E}(k)$$

where:
  $\lambda \approx 1$ is a weighting parameter and η is the step size parameter.

For efficiency sake, as indicated previously, the multiplications by $\mathbf{F}$ and $\mathbf{F}^{-1}$ should be carried out using FFT routines and not by using straight matrix multiplication as shown in the derivation.

9.1.2.2 The Overlap-add Method

The overlap-add method is an alternative way of partitioning the data and reassembling the results to obtain a linear convolution. There are two differences between the overlap-save and the overlap-add methods. Firstly, in the overlap-add method, the frequency domain data vector is defined as $\mathbf{X}(k) = \mathbf{X}'(k) + \mathbf{JX}'(k-1)$, where $\mathbf{X}'(k)$ is the diagonal matrix whose elements are components of the Fourier transform of the $2p$ points, $x[kp],...., x[kp+p-1], 0,...., 0$ and $\mathbf{J}$ is the $2p \times 2p$ diagonal matrix with $J_{mm} = (-1)^m$ for $m = 0,...., 2p-1$. Secondly, instead of $\mathbf{k} = \begin{bmatrix} \mathbf{0} & \mathbf{I} \end{bmatrix}$, $\overline{\mathbf{k}} = \begin{bmatrix} \mathbf{I} & \mathbf{0} \end{bmatrix}$ is used.

The overlap-add algorithm can therefore be summarised as follows,

$$\text{Set } \mathbf{W}(0) = \begin{bmatrix} 0 \\ 0 \\ : \\ 0 \end{bmatrix}, \text{ and, } P_m(0) = \delta_{m,} \quad m = 0,1,...,2p-1$$

For each block of p input samples, the input matrix $\mathbf{X}'(k)$ is made to be a diagonal matrix whose elements are the Fourier transform of the $2p$ time domain points $x[kp],...., x[kp+p-1], 0,...., 0$. Compute the following in sequence,

$$\mathbf{X}(k) = \mathbf{X}'(k) + \mathbf{JX}'(k-1)$$

$$\mathbf{Y}(k) = \mathbf{X}(k)\mathbf{W}(k)$$

$$\mathbf{y}[k] = \overline{\mathbf{k}}\mathbf{F}^{-1}\mathbf{Y}(k)$$

$$\mathbf{e}[k] = \mathbf{d}[k] - \mathbf{y}[k]$$

$$\mathbf{E}(k) = \mathbf{F}\overline{\mathbf{k}}^{T}\mathbf{e}[k]$$

$$P_m(k) = \lambda P_m(k-1) + (1-\lambda)\left|X_m(k)\right|^2, \quad m = 0,..,2p-1$$

$$\mathbf{M}(k) = \begin{bmatrix} \eta/P_0(k) & 0 & .. & 0 \\ 0 & \eta/P_1(k) & .. & 0 \\ : & : & \ddots & : \\ 0 & 0 & .. & \eta/P_{2p-1}(k) \end{bmatrix}$$

$$\mathbf{W}(k+1) = \mathbf{W}(k) + 2\mathbf{F}\mathbf{g}\mathbf{F}^{-1}\mathbf{M}(k)\mathbf{X}^{H}(k)\mathbf{E}(k)$$

where:

$\lambda \approx 1$ is a weighting parameter and η is the step size parameter.

9.1.2.3 The Circular Convolution Method

The overlap-save and the overlap-add methods are designed to compensate for the fact that using the DFT to compute convolutions gives circular convolution in a situation where linear convolutions are required. It is possible to reduce the computational complexity at the expense of degraded performance by omitting this compensation. This is equivalent to making $\mathbf{G} = \mathbf{I}$ in the weight update formula and making certain other simplifications.

For circular convolutions the $\mathbf{F}$ matrix can be a $p \times p$ matrix of a p-point DFT. The p-point DFTs are computed once for each block of p samples, with no overlap. The weight vector is $\mathbf{W}(k) = \mathbf{F}\mathbf{w}[k]$ and $\mathbf{X}(k)$ is the diagonal matrix whose elements are the components of the Fourier transform of the p-points, $\mathbf{x}[kp],...., \mathbf{x}[kp+p-1]$. As before, $\mathbf{Y}(k) = \mathbf{X}(k)\mathbf{W}(k)$ and the time domain output vector is $\mathbf{y}[k] = \mathbf{F}^{-1}\mathbf{Y}(k)$. The error is computed in the frequency domain as the difference between $\mathbf{Y}(k)$ and the Fourier transform of the vector of desired time domain output values $\mathbf{d}[k]$.

The circular convolution algorithm can therefore be summarised as follows,

$$\text{Set } \mathbf{W}(0) = \begin{bmatrix} 0 \\ 0 \\ : \\ 0 \end{bmatrix}, \text{ and } P_m(0) = \delta_m, \quad m = 0,1,..., p$$

For each block of p input samples, the input matrix $\mathbf{X}(k)$ is made to be a diagonal matrix whose elements are the Fourier transform of the p time domain points $\mathbf{x}[kp],...., \mathbf{x}[kp+p-1]$. Compute the following in sequence,

$$\mathbf{D}(k) = \mathbf{F}\mathbf{d}[k]$$

$$\mathbf{Y}(k) = \mathbf{X}(k)\mathbf{W}(k)$$

$$\mathbf{E}(k) = \mathbf{D}(k) - \mathbf{Y}(k)$$

$$P_m(k) = \lambda P_m(k-1) + (1-\lambda)|X_m(k)|^2, \quad m = 0,..,p-1$$

$$\mathbf{M}(k) = \begin{bmatrix} \eta(k)/P_0(k) & 0 & .. & 0 \\ 0 & \eta(k)/P_1(k) & .. & 0 \\ \vdots & \vdots & \ddots & \vdots \\ 0 & 0 & .. & \eta(k)/P_{p-1}(k) \end{bmatrix}$$

$$\mathbf{W}(k+1) = \mathbf{W}(k) + 2\mathbf{M}(k)\mathbf{X}^H(k)\mathbf{E}(k)$$

9.1.2.4 Computational Complexity

The time domain LMS algorithm requires $2p^2$ real multiplications for every p output samples. On the other hand computational savings are made with the overlap-save and overlap-add methods, where both have the same computational complexity, each requiring $10p\log_2(2p) + 16p$ real multiplications. More efficiency is gained through use of the circular convolution method, which requires only $3p\log_2(p) + 8p$ multiplications.

9.2 Exercises

The following Exercises identify some of the basic ideas presented in this Chapter.

9.2.1 Problems

9.1.   Compute how many multiplication operations are saved in the coefficient update equation for a time domain BLMS algorithm per block of L input points. Is it worth the trouble?

9.2.   Show how linear convolutions on sequences can be done by using frequency domain operations.

9.3.   What computational savings can be had by using a radix-2 decimation-in-time FFT to perform the DFTs required in Problem 9.2?

9.4.   For frequency domain filtering define the inverse of the $p \times p$ DFT matrix $\mathbf{F}$.

9.5.   What is the main purpose of the so called overlap-save and the overlap-add methods in frequency domain filtering?

9.6   What is the disadvantage of the circular convolution method?

10. Adaptive Volterra Filters

Adaptive Polynomial filters are a nonlinear generalisation of adaptive linear filters that are based on nonrecursive or recursive linear difference equations. Polynomial filters are often referred to as Volterra filters when are based on the Volterra series that was first studied by Vito Volterra in 1880. However, the first use of the Volterra series to model the input-output relationship of a nonlinear system was by Norbert Wiener (Wiener 1958). The nonlinear Volterra filter has the property that it depends linearly on the coefficients of the filter itself, therefore the principles of "optimum linear filter theory" can be naturally extended to "optimum nonlinear Volterra filter theory."

10.1 Nonlinear Filters

While linear filters have been used widely in engineering, many problems are inherently nonlinear and may better be addressed by nonlinear solutions. Nonlinear filters are particularly useful for applications with signal dependent or multiplicative noises and for non-Gaussian signal statistics. There are many types of nonlinear filters including,

1. Homomorphic Filters.

2. Morphological Filters.

3. Order Statistics Filters.

4. Nonlinear Median Filters.

5. Artificial Neural Networks.

6. Polynomial and Volterra Filters.

Homomorphic filters are often used in image enhancement, seismic signal processing, models of human visual systems, and for the removal of multiplicative noise and are amongst the oldest nonlinear filters. In morphological filters the geometric features of signals are used in such applications as edge detection and shape recognition. Order statistics filters are based on the order of the data samples in the signal and are computationally simple and robust. The median filter is a common example of an order statistics filter often used in image processing for the removal of impulse noise without destroying edge detail. Artificial Neural Network (ANN) filters represent a general nonlinear filtering model that can be applied to a very wide range of nonlinear problems. Polynomial filters based the on a truncated Volterra series expansion represent one of the most important methods of approach to nonlinear problems. They, like ANNs, are more general than the other models mentioned above. Polynomial filters can arise from a nonlinear generalisation of linear FIR filters. They can also be based on recursive nonlinear difference equations (Mathews 1991). Polynomial filters that are based on the truncated Volterra series are referred to as Volterra filters. Volterra filters, like neural network based filters, have found application in many problems including,

1.   Modelling of nonlinear systems.

2.   Noise and echo cancellation.

3.   Nonlinear communication channel equalisation.

4.   Signal detection and estimation.

5.   Texture discrimination.

6.   Spatial and temporal prediction.

7.   Nonlinear interpolation of image sequences.

8.   Edge preserving smoothing enhancement of noisy images.

9.   Removal of impulse noise in images without edge blurring.

Volterra filters are based on the Volterra series, which can be described as a Taylor series with memory. The Volterra filter depends linearly on the coefficients of the filter itself and its behaviour can be described in the frequency domain by means of a type of multi-dimensional convolution. Because of this linear coefficient dependence "optimum linear filter theory" can be routinely extended to "optimum nonlinear Volterra filter theory." Some of the main adaptation principles used for adaptive Volterra filters are,

1.   Least Means Squares (LMS).

2.   Recursive Least Squares (RLS).

3.   Fast Kalman.

10.2 The Volterra Series Expansion

If $x[n]$ is a real input sequence to a discrete-time causal nonlinear system that produces the real output sequence $y[n]$ the Volterra series expansion for $y[n]$ in terms of $x[n]$ is defined by Equation 10.1 (Mathews 1991).

$$y[n] = h_0 +$$

$$\sum_{m_1=0}^{\infty} h_1[m_1]x[n-m_1] +$$

$$\sum_{m_1=0}^{\infty}\sum_{m_2=0}^{\infty} h_2[m_1,m_2]x[n-m_1]x[n-m_2] + ... + \tag{10.1}$$

$$\sum_{m_1=0}^{\infty}\sum_{m_2=0}^{\infty} ... \sum_{m_N=0}^{\infty} h_N[m_1,m_2,..,m_N]x[n-m_1]x[n-m_2]...x[n-m_N] + ...,$$

where:

$h_N[m_1, m_2,...., m_N]$ are the elements of the N-th order kernel of the system.

There is no loss of generality in assuming that the kernels are symmetric, i.e., the value of $h_N[m_1, m_2,...., m_N]$ is left unchanged by all the possible permutations of the indices $m_1, m_2,...., m_N$.

In practice the full Volterra series cannot be used because it is strictly infinite. However, it can be applied by truncating it in two ways: limiting the order N of the series, and limiting the number of terms in each summation to some finite number p.

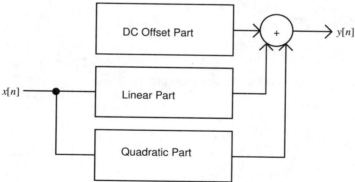

Figure 10.1. A Second-order Volterra Filter

10.3 A LMS Adaptive Second-order Volterra Filter

A most common and useful subset of the Volterra filter is the second-order Volterra filter (Lau *et al* 1992, Leung *et al* 1992), shown in Figure 10.1. It uses the first three terms of the Volterra series and embodies only a quadratic nonlinear part.

This second-order Volterra filter consists of a Direct Current (DC) offset part, a linear part and a quadratic part as defined by Equation 10.2.

$$y[n] = h_0 + \sum_{m_1=0}^{p-1} h_1[m_1]x[n-m_1] + \sum_{m_1=0}^{p-1}\sum_{m_2=0}^{p-1} h_2[m_1,m_2]x[n-m_1]x[n-m_2]$$

(10.2)

or simply,

$$y[n] = h_0 + \mathbf{a}^T[n]\mathbf{x}[n] + \mathbf{x}^T[n]\mathbf{B}[n]\mathbf{x}[n]$$

where:

h_0 is the scalar constant to make $y[n]$ an unbiased estimate (removes the DC component), while $\mathbf{a}[n]$ and $\mathbf{B}[n]$ are the linear and quadratic kernels, respectively.

The input vector $\mathbf{x}[n]$ is the p-dimensional input vector,

$$\mathbf{x}[n] = \begin{bmatrix} x[n] \\ x[n-1] \\ \vdots \\ x[n-p+1] \end{bmatrix}$$

The weight vector $\mathbf{a} \equiv \mathbf{a}[n]$ of the linear part is,

$$\mathbf{a} = \begin{bmatrix} a_0 \\ a_1 \\ \vdots \\ a_{p-1} \end{bmatrix} = \begin{bmatrix} h_1[0;n] \\ h_1[1;n] \\ \vdots \\ h_1[p-1;n] \end{bmatrix}$$

The weight matrix $\mathbf{B} \equiv \mathbf{B}[n]$ of the quadratic part is,

$$\mathbf{B} = \begin{bmatrix} b_{00} & b_{01} & .. & b_{0p-1} \\ b_{10} & b_{11} & .. & b_{1p-1} \\ \vdots & \vdots & \ddots & \vdots \\ b_{p-10} & b_{p-11} & .. & b_{p-1p-1} \end{bmatrix}$$

$$\mathbf{B} = \begin{bmatrix} h_2[0,0;n] & h_2[0,1;n] & .. & h_2[0,p-1;n] \\ h_2[1,0;n] & h_2[1,1;n] & .. & h_2[1,p-1;n] \\ \vdots & \vdots & \ddots & \vdots \\ h_2[p-1,0;n] & h_2[p-1,1;n] & .. & h_2[p-1,p-1;n] \end{bmatrix}$$

In the nonlinear equalisation problem, the Volterra filter is designed to optimally minimise the difference between the desired signal $d[n]$ and the Volterra filter output $y[n]$. Using the LMS algorithm the filter parameters are initially chosen arbitrarily and subsequently updated per each input sample as follows,

$$h_0[n+1] = h_0[n] + \mu_0 e[n]$$

$$\mathbf{a}[n+1] = \mathbf{a}[n] + \mu_a e[n]\mathbf{x}[n]$$

$$\mathbf{B}[n+1] = \mathbf{B}[n] + \mu_b e[n]\mathbf{x}[n]\mathbf{x}^T[n]$$

$$e[n] = d[n] - h_0[n] - \mathbf{a}^T[n]\mathbf{x}[n] - \mathbf{x}^T[n]\mathbf{B}[n]\mathbf{x}[n]$$

where:

μ_0, μ_a, μ_b are chosen small positive step sizes (DC, linear, quadratic parts).

$e[n] = d[n] - y[n]$ is the estimation error and $d[n]$ is the desired output value.

The constants μ_0, μ_a and μ_b are small positive values that control the speed of convergence and the steady-state and tracking properties of the filter. The gradients of the mean square error function $E\{e^2[n]\}$ with respect to h_0, $\mathbf{a}$ and $\mathbf{B}$ for real data are respectively,

$$\nabla_{h_o} = -2E\{e[n]\}$$

$$\nabla_{\mathbf{a}} = -2E\{e[n]\mathbf{x}[n]\}$$

$$\nabla_{\mathbf{B}} = -2E\{e[n]\mathbf{x}[n]\mathbf{x}^T[n]\}$$

The computational complexity of the LMS algorithm is approximately of the order $O\{(3p^2 + 9p)/2\}$, where $O\{.\}$ denotes "order of."

10.4 A LMS Adaptive Quadratic Filter

For many applications the input signal is assumed to be zero mean, which allows for the removal of the DC offset part that its associated with the update equation. In this case an alternative more compact matrix notation for the second-order Volterra or quadratic filter is often used as defined by Equation 10.3.

$$y[n] = \mathbf{h}^T[n]\mathbf{x}[n] \tag{10.3}$$

The vectors $\mathbf{h}[n]$ and $\mathbf{x}[n]$ in Equation 10.3 are defined as follows,

$$\mathbf{h}[n] = \begin{bmatrix} h_1[0;n] \\ \vdots \\ h_1[p-1;n] \\ h_2[0,0;n] \\ \vdots \\ h_2[0,p-1;n] \\ h_2[1,0;n] \\ \vdots \\ h_2[p-1,p-1;n] \end{bmatrix}, \text{ and, } \mathbf{x}[n] = \begin{bmatrix} x[n] \\ x[n-1] \\ \vdots \\ x[n-p+1] \\ x^2[n] \\ x[n]x[n-1] \\ \vdots \\ x[n]x[n-p+1] \\ x[n-1]x[n] \\ x^2[n-1] \\ \vdots \\ x^2[n-p+1] \end{bmatrix}$$

Given that the error is $e[n] = d[n] - \mathbf{h}^T[n]\mathbf{x}[n]$, the LMS update equation for this form is as defined by Equation 10.4.

$$\mathbf{h}[n+1] = \mathbf{h}[n] + e[n]\mathbf{M}\mathbf{x}[n] \tag{10.4}$$

where:

$\mathbf{M}$ is a diagonal gain matrix with μ_a in the first p diagonal places and μ_b in the remaining diagonal places.

In Equation 10.4 μ_a and μ_b are chosen to satisfy $0 < \mu_a, \mu_b < \dfrac{2}{\lambda_{max}}$, where λ_{max} is the maximum eigenvalue of the autocorrelation matrix of the vector $\mathbf{x}[n]$, i.e., $E\{\mathbf{x}[n]\mathbf{x}[n]^T\}$.

10.5 A RLS Adaptive Quadratic Filter

It is possible to derive a compact exponentially weighted RLS algorithm for the quadratic Volterra filter in a way that is similar to the derivation of the compact linear case. Taking the compact notation used in Equation 10.3 the weighted error function $\xi(n)$ is defined by Equation 10.5.

$$\xi(n) = \sum_{k=0}^{n} \lambda^{n-k} (d[k] - \mathbf{h}^T[n]\mathbf{x}[k]) \tag{10.5}$$

where:

$\lambda \approx 1$ is the weight factor.

$\xi(n)$ is minimised via Equation 10.6.

$$\mathbf{h}[n] = \mathbf{C}^{-1}(n)\mathbf{P}(n) \qquad (10.6)$$

where:

$$\mathbf{C}(n) = \sum_{k=0}^{n} \lambda^{n-k} \mathbf{x}[k]\mathbf{x}^{T}[k]$$

$$\mathbf{P}(n) = \sum_{k=0}^{n} \lambda^{n-k} d[k]\mathbf{x}[k]$$

Matrices $\mathbf{C}(n)$ and $\mathbf{P}(n)$ can be updated recursively by using Equations 10.7 and 10.8 respectively.

$$\mathbf{C}(n) = \lambda\mathbf{C}(n-1) + \mathbf{x}[n]\mathbf{x}^{T}[n] \qquad (10.7)$$

$$\mathbf{P}(n) = \lambda\mathbf{P}(n-1) + d[n]\mathbf{x}[n] \qquad (10.8)$$

The explicit inversion of the matrix $\mathbf{C}(n)$ is avoided by using the matrix inversion lemma as follows. Initialise $\mathbf{h}[0]$ to be the zero vector and set $\mathbf{C}^{-1}(0) = \delta^{-1}\mathbf{I}$, where δ is a small positive value. For $n=1,2,....$, compute the following in sequence,

$$\mathbf{k}(n) = \frac{\lambda^{-1}\mathbf{C}^{-1}(n-1)\mathbf{x}[n]}{1 + \lambda^{-1}\mathbf{x}^{T}[n]\mathbf{C}^{-1}(n-1)\mathbf{x}[n]}$$

$$\varepsilon[n] = d[n] - \mathbf{h}^{T}[n-1]\mathbf{x}[n]$$

$$\mathbf{h}[n] = \mathbf{h}[n-1] + \varepsilon[n]\mathbf{k}(n)$$

$$\mathbf{C}^{-1}(n) = \lambda^{-1}[\mathbf{C}^{-1}(n-1) - \mathbf{k}(n)\mathbf{x}^{T}[n]\mathbf{C}^{-1}(n-1)]$$

$$e[n] = d[n] - \mathbf{h}^{T}[n]\mathbf{x}[n]$$

In both the compact LMS and RLS algorithms described above it is possible to reduce the size of the vectors represented by Equation 10.3. There is a redundancy in the Equation 10.3 due to the fact that the Volterra kernels are symmetric. Consequently, it is possible to ignore the quadratic kernels either above or below the diagonal of the quadratic kernel matrix without serious consequences. The vectors $\mathbf{h}[n]$ and $\mathbf{x}[n]$ in Equation 10.3 can therefore be redefined as follows,

$$\mathbf{h}[n] = \begin{bmatrix} h_1[0;n] \\ \vdots \\ h_1[p-1;n] \\ h_2[0,0;n] \\ \vdots \\ h_2[0,p-1;n] \\ h_2(1,1;n) \\ h_2(2,1;n) \\ \vdots \\ h_2[p-1,p-1;n] \end{bmatrix}$$

and,

$$\mathbf{x}[n] = \begin{bmatrix} x[n] \\ x[n-1] \\ \vdots \\ x[n-p+1] \\ x^2[n] \\ x[n]x(n-1) \\ \vdots \\ x[n]x[n-p+1] \\ x^2[n-1] \\ x[n-1]x[n-2] \\ \vdots \\ x^2[n-p+1] \end{bmatrix}$$

All the remaining equations for the LMS and RLS algorithms are as before.

10.6 Exercises

The following Exercises identify some of the basic ideas presented in this Chapter.

10.6.1 Problems

10.1.   What is a key feature of the Volterra filter that allows for the use of optimum filter theory?

10.2.   What is the main disadvantage of the adaptive Volterra filter?

10.3. Compute the gradients of the mean square error function, $E\{e^2[n]\}$, with respect to h_0, **a** and **B**, for the second-order Volterra filter in the case when the data are complex.

10.4. What would the LMS weight update equations be for Problem 10.3?

11. Adaptive Control Systems

Classical control theory has mostly been concerned with the design of feedback systems for time-invariant plants with known transfer functions (Levine 1996). However, the assumptions of known mathematical models and time-invariance are not valid for many modern control problems. For example, in robotics the dynamic models vary with robot attitude and load variations. Chemical reactor transfer functions vary as a function of reagent mix, catalyst and time. These types of problems might be solved using a classical approach by designing a robust fixed controller that ensures stability for all possible plant dynamics. However, this approach may often be at the expense of suboptimal control behaviour. The other approach is to use adaptive control algorithms that can learn from the plant input-output behaviour and thereby develop on-line self-tuning controllers to improve the closed loop performance. There are typically two main themes found in relation to learning, or adaptive controllers. Systems may have unknown but constant dynamics or the dynamics may be time-varying.

Many of the past algorithms and approaches in adaptive control have often been somewhat ad hoc, lacking good systematic methods. They used and applied methods gathered from a wide range of areas including nonlinear system theory, stability theory, singular perturbations and averaging theory, stochastic control theory, parameter estimation theory, and optimisation theory. Nevertheless, useful adaptive control techniques are beginning to emerge after a long period of research and experimentation. There is still much more work required on stability issues but some important theoretical results have already been established. The field is now sufficiently mature to have a number of adaptive regulator products appearing in the market place.

In adaptive systems design it is desirable to find the simplest possible parameter adjustment rules. However, these rules must generally be nonlinear rules. There are two main methods for adaptive control depending on the parameter adjustment rules, the direct and indirect methods. The direct methods have adjustment rules that tell how the regulator parameters should be updated. Indirect methods, on the other hand, update the process parameters and then the regulator parameters are obtained from the solution of a design problem. One most important direct method is the Model-Reference Adaptive System (MRAS) and one important indirect method is the Self-Tuning Regulator (STR). Although different in detail these two methods are closely related in principle.

11.1 Main Theoretical Issues

Because adaptive control systems are inherently nonlinear in their operation they are complex and difficult to analyse. Because of this complexity it is necessary to invoke various theories to achieve adequate design and analysis. These theories include nonlinear systems, stability, recursive parameter estimation, system identification, optimal control, and stochastic control theories.

Although it is possible to establish that a particular nonlinear system solution is stable the same solution applied to other cases may not be stable. Since an adaptive system is seeking and finding new solutions on an ongoing basis it is only in very special cases that it is possible to speak of a "stable adaptive system" in a global sense. Often the best that can be done is to find the stable equilibrium solutions and then determining the local behaviours by linearization techniques.

A typical problem in adaptive control is to design a parameter adjustment rule that is guaranteed to result in a stable closed loop system using a range of theoretical considerations. There are two separate problems to consider. In a tuning problem it is assumed that the process to be controlled is constant but with unknown parameters. In an adaptation problem the parameters are changing. The tuning problem is much easier to deal with since the parameter convergence has a final endpoint, while the adaptation problem has not. The estimation algorithms for tuning and adaptation are similar in form but they are applied differently. A common parameter estimation algorithm form is defined by Equation 11.1.

$$\hat{\theta}(t+1) = \hat{\theta}(t) + P(t)\varphi(t)(y(t+1) - \varphi^T(t)\hat{\theta}(t)) \tag{11.1}$$

where:

$\hat{\theta}(t)$ = estimate of the parameter vector at time t.

$\varphi(t)$ = vector of functions of measured signals in the system (regressors).

$y(t+1)$ = measurement signal at time $(t+1)$.

$P(t)$ = gain matrix (also governed by a difference equation).

In the tuning case the gain matrix $P(t)$ goes to zero as t increases, whereas in the adaptation case it is not allowed to converge to zero.

An important theoretical consideration for parameter convergence is to establish the conditions under which the recursive parameter estimation process will work. The conditions must provide a persistent excitation or sufficient richness in the input signal to ensure that the process dynamics can be captured.

Many adaptive algorithms rely on the fact that the parameters change more slowly than the state variables of the system. When investigating the behaviour of the states the parameters are seen as constants. Therefore they are often replaced with their mean values. In Equation 11.1, this is the same as replacing the term $P(t)\varphi(t)(y(t+1) - \varphi^T(t)\hat{\theta}(t))$ with its mean value. The rate of adaptation of the parameters can be controlled by the selection of a gain constant. The averaging method works best when a small adaptation gain is used. In many cases the difference between the true and averaged equations is proportional to the adaptation gain. It is believed that these types of averaging methods may eventually lead to a unification of analysis of adaptive systems.

It is possible to consider a unified theoretical structure for adaptive systems by using nonlinear stochastic theory in which the system and its environment are described by a stochastic model. In this structure the parameters are introduced as state variables and their uncertainty is modelled by stochastic models. According to this model an unknown parameter constant can be modelled by the differential equation $d\theta / dt = 0$ or the equivalent difference equation $\theta(t+1) = \theta(t)$ with an initial probability distribution that models the parameter uncertainty. Parameter drift is described by simply adding random variables to the right sides of these two equations. Next, a rule is developed to minimise the expected value of a loss function, which is made to be a scalar function of the states and controls.

It is difficult to find a control that minimises the expected loss function. If it can be assumed that a solution exists it is possible to formulate the optimal loss function by using dynamic programming. Dynamic programming involves solving a functional equation called the Bellman equation. The optimal regulator formed from stochastic control theory can be represented by Figure 11.1 (Åström and Wittenmark 1995).

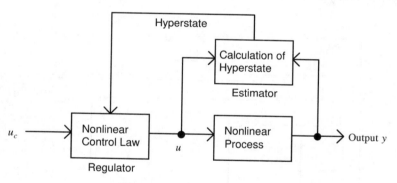

Figure 11.1. Adaptive Regulator

The controller is composed of a nonlinear estimator and a feedback regulator. The conditional probability distribution of the state (the hyperstate) is generated from the measurements by the estimator. The nonlinear feedback regulator maps the hyperstate into the space of control variables. The regulator's nonlinear function can be computed off-line because it changes more slowly, whereas the hyperstate must be up-dated on-line. The hyperstate is usually a high dimensional quantity, which allows a structurally simple control solution. However the disadvantage is that updating of the hyperspace can require the solution of a complex nonlinear filtering problem. The advantage of this approach is that there is no distinction between parameters and the other state variables, which means that the regulator can handle very rapid parameter changes. The control attempts to drive the output to its desired value and it also introduces probing perturbations when the parameters are uncertain. Thus, the optimal control automatically gives a good balance between maintaining adequate control and small control errors. This control property is called dual control, which inherently improves the quality of the estimates and the future controls.

Most work on stability has been done in relation to the MRAS. A typical system may be composed of a linear system with a nonlinear feedback because this is a classical configuration for which stability results are available. Here, it can be said that the closed loop system is stable if the linear part is strictly positive real and the nonlinear part is passive.

11.2 Introduction to Model-reference Adaptive Systems

The MRAS is one of the main adaptive control approaches. When the system specifications are given in terms of a reference model that tells how the process output should ideally respond to command signals it is then possible to use the MRAS. Figure 11.2 shows the block diagram of Whitaker's original MRAS (Whitaker *et al* 1958). The original MRAS was derived for the servo problem in deterministically continuous-time systems. Since then the theory has been extended to discrete-time systems and systems with stochastic disturbances.

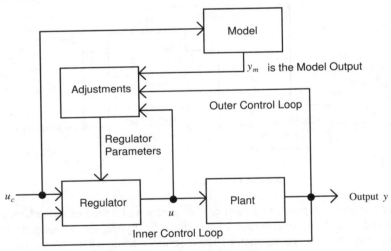

Figure 11.2. Model-Reference Adaptive System

The MRAS regulator has an inner and outer loop. The inner loop is an ordinary feedback loop composed of the process and the regulator. The outer loop is also a regulator loop, it adjusts the parameters in such a way as to make the error e between the process output y and the model output y_m small. It is a nontrivial problem to formulate a stable adjustment mechanism for the MRAS that is guaranteed to reduce the error to zero.

There are three basic approaches to the analysis and design of a MRAS. These are the gradient method, Lyapunov functions and passivity theory. Whitaker's original MRAS was based on the gradient approach and the assumption that the parameters change more slowly than the other variables in the system. Unfortunately, this gradient approach will not necessarily guarantee a stable closed loop system for models based on both poles and zeros. It was because of this that Lyapunov's stability theory and the passivity theory was introduced to modify the adaptation mechanism to ensure stability (Whitaker *et al* 1958).

The general MRAS problem is called the model following problem and it can be stated as follows. Given a system with adjustable parameters as shown in Figure 11.2 the model-reference adaptive method provides a method for the adjustment of the parameters such that the closed loop transfer function will be close to some prescribed model. Perfect model following occurs only when the error is zero for all command signals. In practice the issue becomes one of how small the error can be made.

11.2.1 The Gradient Approach

In the original MRAS idea the so called MIT adjustment rule was used. The MIT rule is so called because it was originally developed at the Massachusetts Institute of Technology (MIT) and it is based on a gradient approach as defined by Equation 11.2.

$$\frac{d\theta(t)}{dt} = -\gamma \frac{\partial \xi(\theta)}{\partial \theta} = -\gamma e \frac{\partial e}{\partial \theta} \tag{11.2}$$

where:

$\theta(t)$ = the parameter vector at time t.

e = model error.

γ = adaptation rate parameter (step size).

$\xi(\theta) = \frac{1}{2}e^2$ = error criterion.

The parameter adjustments are a function of the derivatives of the error with respect to the adjustable parameters θ. The parameters are incrementally adjusted in the opposite direction to the gradient of half the error squared, $\xi(\theta) = \frac{1}{2}e^2$. In this way the error between the process outputs and the reference model slowly approaches zero. If it can be assumed that the parameters θ change more slowly than the other variables in the system, then it can be assumed that θ is constant when computing the derivative $\partial e / \partial \theta$. The MIT rule, Equation 11.2, can be interpreted as a linear filter for computing the sensitivity derivatives of the error with respect to the adjustable parameters. This filter is derived from the process inputs and outputs and includes a multiplier and an integrator as represented in Figure 11.3 (Åström and Wittenmark 1995).

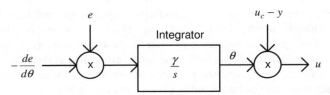

Figure 11.3. MRAS Error Model

A simple example can be used to illustrate how the MRAS attempts to adjust the parameters so that the correlation between the error e and the sensitivity derivative becomes zero. Assume that it is desired to adjust a simple feedforward gain for a system that has a model transfer function $G_m(s) = \theta_1 G(s)$, where θ_1 is a known constant and $G(s)$ is a known transfer function. The MRAS setup for this example is shown in Figure 11.4 (Åström and Wittenmark 1995).

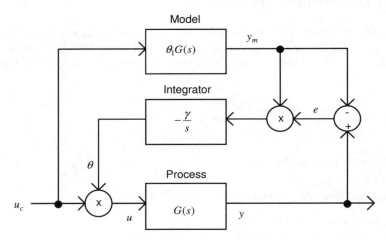

Figure 11.4. MRAS Feedforward Gain Adjustment using the MIT Rule

The error e is defined by Equation 11.3.

$$e = y - y_m = [G(p)\theta - G_m(p)]u_c = G(p)[\theta - \theta_1]u_c \qquad (11.3)$$

where:

u_c = the command signal.

y_m = the model output.

y = the process output.

$p = \dfrac{d}{dt}$, the differential operator.

The sensitivity derivative is defined by Equation 11.4.

$$\frac{\partial e}{\partial \theta} = G(p)u_c = \frac{y_m}{\theta_1}$$

(11.4)

For this example the MIT rule reduces down to Equation 11.5.

$$\frac{d\theta(t)}{dt} = -\frac{\gamma e y_m}{\theta_1} \rightarrow -\gamma e y_m$$

(11.5)

The fixed constant θ_1 in Equation 11.5 has been included in the variable γ. The rate of parameter change must be made proportional to the product of the error and the model output as shown in Figure 11.4. In this very simple example the sensitivity derivatives can be computed exactly. In more complex problems it is often necessary to use approximations for the sensitivity derivatives.

11.2.2 Least Squares Estimation

Equation 11.2 represents a simple estimator in which the parameters are adjusted by following the negative gradient of e^2. The ultimate purpose of this is to reduce the squared error and consequently the error which indicates that the system has converged to a solution. This approach stems from the Least Squares Estimation (LSE) method, which along with its variants forms much of the basis to adaptive learning rules. In the context of the adaptive control problems the LSE method can be illustrated as follows. Assume that a process can be described by the difference equation defined by Equation 11.6.

$$y(t+1) = \theta_1 y(t) + u(t)$$

(11.6)

where:

θ_1 is an unknown parameter.

The predicted output y at time $t+1$ can be estimated based on the estimate θ of θ_1 and the current output $y(t)$ via the model defined by Equation 11.7.

$$\hat{y}(t+1) = \theta y(t) + u(t)$$

(11.7)

The least squares error or loss function is based on data up to and including t and is defined by Equation 11.8.

$$\xi(t) = \frac{1}{2} \sum_{k=0}^{t} e^2[k]$$

(11.8)

where:

$$e(t) = y(t) - \hat{y}(t) = \theta_1 y(t-1) - \hat{\theta}y(t-1) = y(t) - u(t-1) - \hat{\theta}y(t-1)$$

If Equation 11.8 is differentiated with respect to $\hat{\theta}$ and then set to equal zero the least squares estimate of the unknown parameter $\hat{\theta}(t)$ is determined, as defined by Equation 11.9.

$$\hat{\theta}(t) = \frac{\sum\limits_{k=0}^{t} y[k](y[k+1]-u[k])}{\sum\limits_{k=0}^{t-1} y^2[k]} \tag{11.9}$$

Equation 11.9 represents the best estimate in the least squares sense of the unknown parameter given that the process is described by Equation 11.6.

11.2.3 A General Single-input-single-output MRAS

A model reference control law based on the gradient approach can be derived for a general Single-Input-Single-Output (SISO) system. The system model is described by Equation 11.10 and the desired system is characterised by Equation 11.11.

$$Ay(t) = Bu(t)$$

or $\qquad\qquad\qquad\qquad\qquad\qquad\qquad$ (11.10)

$$Ay[n] = Bu[n]$$

where:
   u = the control signal as a function of time.
   y = the output signal as a function of time.
   A and B are polynomials in terms of differential or shift operators.
   The order of A is greater than or equal to the order of B.
   A is monic.

$$A_m y_m = B_m u_c \tag{11.11}$$

where:
   u_c = the command signal as a function of time.
   y_m = the model output as a function of time.

The appropriate linear closed loop control structure for this model is shown in Figure 11.5. Here, the general control law can be described by Equation 11.12.

$$Ru = Tu_c - Sy \tag{11.12}$$

where:
   u_c = the command signal.
   y = the model output.
   R, S and T are polynomials in terms of differential or shift operators.

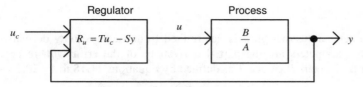

Figure 11.5. Closed Loop with Linear Controller

This control law represents a negative feedback with transfer operator $-S/R$ and a feedforward with the transfer operator T/R. Combining Equations 11.10 and 11.12 gives Equation 11.13 for the closed loop system.

$$(AR + BS)y = BTu_c \tag{11.13}$$

In order to obtain the desired closed loop response, A_m must divide $AR+BS$. The process zeros, given by $B = 0$, will also be closed loop zeros unless they are cancelled by corresponding closed loop poles. Unstable or poorly damped zeros cannot be cancelled, therefore the polynomial B is factored as $B = B^+ B^-$ where B^+ is monic and contains those factors that can be cancelled, and B^- the remaining factors of B. The zeros of B^+ must be stable and well damped. The Diophantine equation is the characteristic polynomial of the closed loop and is defined by Equation 11.14.

$$AR + BS = B^+ A_0 A_m \tag{11.14}$$

It follows that B^+ divides R hence $R = B^+ R_1$. Dividing Equation 11.14 by B^+ gives Equation 11.15.

$$AR_1 + B^- S = A_0 A_m \tag{11.15}$$

The relation, in Equation 11.13, between the command signal u_c and the process output y should be equal to the desired closed loop response given by Equation 11.11. The specifications must also be such that B^- divides B_m hence $B_m = B^- B_m'$ and $T = A_0 B_m'$. In order that there exist solutions to Equation 11.15 that give a proper or causal control law it is necessary that,

$$\text{order } A_0 \geq 2 \text{ order } A - \text{order } A_m - \text{order } B^+ - 1,$$

and,

$$\text{order } A_m - \text{ order } B \geq \text{order } A - \text{order } B$$

The closed loop control system is described by Equations 11.16.

$$y = \frac{BT}{AR + BS} u_c, \text{ and, } u = \frac{AT}{AR + BS} u_c \tag{11.16}$$

The error between the system and the model output is defined by Equation 11.17.

$$e = y - y_m \tag{11.17}$$

The parameter adjustment law is determined from the sensitivity derivatives. The sensitivity parameters are the partial derivatives of the error with respect to the regulator parameters R, S, and T as defined by Equations 11.18 to 11.20.

$$\frac{\partial e}{\partial r_i} = -\frac{BTAp^{k-i}}{(AR + BS)^2} u_c = -\frac{Bp^{k-i}}{AR + BS} u, \quad i = 1,....,k \tag{11.18}$$

$$\frac{\partial e}{\partial s_i} = -\frac{BTBp^{q-i}}{(AR + BS)^2} u_c = -\frac{Bp^{q-i}}{AR + BS} y, \quad i = 1,....,q \tag{11.19}$$

$$\frac{\partial e}{\partial t_i} = \frac{Bp^{m-i}}{AR + BS} u_c, \quad i = 1,....,m \tag{11.20}$$

where:

r_i, s_i, t_i = coefficients of polynomials R, S, and T respectively.

k, q, m = order of polynomials R, S, and T respectively.

$p = \dfrac{d}{dt}$, the differential operator.

Once these sensitivity derivatives are computed for specific polynomials A, B, R, S, and T the parameter adjustment equations are defined by Equations 11.21 to (11.23).

$$\frac{dr_i(t)}{dt} = -\gamma e \frac{\partial e}{r_i} \tag{11.21}$$

$$\frac{ds_i(t)}{dt} = -\gamma e \frac{\partial e}{s_i} \tag{11.22}$$

$$\frac{dt_i(t)}{dt} = -\gamma e \frac{\partial e}{t_i} \tag{11.23}$$

The MIT rule optimises the loss function, $\xi(\theta) = \dfrac{1}{2} e^2$, however it is possible to extend it to optimise a more general loss function. This can be done by firstly specifying a model and a regulator with adjustable parameters. Then a parameter adjustment law is formed by computing the gradient of the loss function with respect to the parameters making the rate of change of parameters in the opposite direction to the gradient. Although this is straightforward in principle, in practice it may be complicated to do. One problem is that to compute the sensitivity derivatives it is necessary to know the model parameters. As this is not realistic it is necessary to make some approximations such as replacing the process parameters with their estimates.

The MIT rule is basically a gradient procedure whose rate of descent is determined by the user chosen parameter γ and the magnitude of the command signal. Modified gradient rules can be formed that do not depend on the magnitude of the command signal. One way to do this is to introduce a normalisation as defined by Equation 11.24.

$$\frac{d\theta(t)}{dt} = -\gamma \frac{e\frac{\partial e}{\partial \theta}}{a + \left(\frac{\partial e}{\partial \theta}\right)^T \left(\frac{\partial e}{\partial \theta}\right)} \tag{11.24}$$

where:
$a > 0$

The parameter a is arbitrarily introduced to avoid a possible division by zero.

In control problems with significant measurement noise it may be desirable to have the parameter adjustment rate depend on the magnitude of the command signal for small levels but not for higher ones. This effect can be achieved by introducing a saturation function $f(.)$ as defined by Equations 11.25 and 11.26.

$$\frac{d\theta(t)}{dt} = -\gamma f \left(\frac{e\frac{\partial e}{\partial \theta}}{a + \left(\frac{\partial e}{\partial \theta}\right)^T \left(\frac{\partial e}{\partial \theta}\right)}, \beta \right) \tag{11.25}$$

$$f(x,\beta) = \begin{cases} -\beta, & x < -\beta \\ x, & |x| \le \beta \\ \beta, & x > \beta \end{cases} \tag{11.26}$$

11.2.4 Lyapunov's Stability Theory

In the gradient methods of parameter adjustment rules the outer loop of the MRAS is first designed and then shown to make the model error go to zero. Another approach to the problem is to develop a rule where the error is guaranteed to go to zero by invoking stability theory, and in particular Lyapunov's stability theory.

Lyapunov stability theory is a direct method of investigating the stability of a solution to a nonlinear differential equation. The key idea is that the equilibrium will be stable if a real function can be found on the state space whose equal level curves enclose the equilibrium such that the derivative of the state variables along the curves always points toward the interior of the curves.

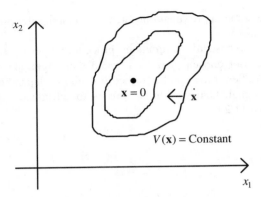

Figure 11.6. Lyapunov's Stability Method

For a state vector $\mathbf{x}$ of dimension n let the differential equation be defined by Equation 11.27.

$$\dot{\mathbf{x}} = f(\mathbf{x},t), \quad f(\mathbf{0},t) = \mathbf{0} \tag{11.27}$$

Lyapunov's stability theorem applied to Equation 11.27 is as follows. Let the function $V: R^{n+1} \to R$ satisfy the conditions,

1. $V(\mathbf{0},t) = 0$ for all $t \in R$.

2. V is differentiable in $\mathbf{x}$ and t.

3. V is positive definite, i.e., $V(\mathbf{x},t) \geq g(\|\mathbf{x}\|) > 0$ where,

    $g: R \to R$ is continuous and increasing with $\lim\limits_{\mathbf{x} \to \infty} g(\mathbf{x}) = \infty$

A sufficient condition for uniform asymptotic stability of the system defined by Equation 11.27 is that the function $\dot{V}(\mathbf{x},t)$ is negative definite, as defined by Equation 11.28.

$$\dot{V}(\mathbf{x},t) = f^T(\mathbf{x},t)\,\text{grad}\,V + \frac{\partial V}{\partial t} < 0 \text{, for } \mathbf{x} \neq \mathbf{0} \tag{11.28}$$

If it is assumed that all state variables of a system are measured, the Lyapunov stability theorem can be used to design adaptive control laws that guarantee the stability of the closed loop system. A simple example to illustrate the basic idea is a first order MRAS where the parameters are known. The model equation is defined by Equation 11.29 and the desired system is characterised by Equation 11.30.

$$\frac{dy}{dt} = -ay + bu \tag{11.29}$$

$$\frac{dy_m}{dt} = -a_m y_m + b_m u_c \tag{11.30}$$

If the parameters are known, perfect model-following can be achieved with the controller defined by Equation 11.31.

$$u(t) = t_0 u_c(t) - s_0 y(t) \tag{11.31}$$

where:

$$t_0 = \frac{b_m}{b}$$

$$s_0 = \frac{a_m - a}{b}$$

The feedback is positive if $a_m < a$, i.e., if the desired model is slower than the process.

A procedure to find the appropriate parameter gains t_0 and s_0 for a model-reference system when the parameters a and b are not known is as follows. The error is as before, $e = y - y_m$. Taking derivatives of the error equation and using Equations 11.29 and 11.30 and some other considerations to eliminate the derivatives of y and y_m results in Equation 11.32 (Åström and Wittenmark 1995).

$$\frac{de}{dt} = -a_m e + (a_m - a - bs_0)y + (bt_0 - b_m)u_c \tag{11.32}$$

To construct the parameter adjustment rules that will drive the parameters t_0 and s_0 to the desired values it is necessary to first identify a suitable Lyapunov equation, which surrounds the equilibrium point, as defined by Equation 11.33.

$$V(e, t_0, s_0) = \frac{1}{2}\left(e^2 + \frac{1}{\gamma b}(bs_0 + a - a_m)^2 + \frac{1}{\gamma b}(bt_0 - b_m)^2 \right) \tag{11.33}$$

The Lyapunov Equation 11.30 is zero when e is zero and the controller parameters are equal to the optimal values. The derivative of Equation 11.33 is defined by Equation 11.34.

$$\frac{dV}{dt} = e\frac{de}{dt} + \frac{1}{\gamma}(bs_0 + a - a_m)\frac{ds_0}{dt} + \frac{1}{\gamma}(bt_0 - b_m)\frac{dt_0}{dt}$$

$$= -a_m e^2 + \frac{1}{\gamma}(bs_0 + a - a_m)\left(\frac{ds_0}{dt} - \gamma y e \right) + \tag{11.34}$$

$$\frac{1}{\gamma}(bt_0 - b_m)\left(\frac{dt_0}{dt} + \gamma u_c e \right)$$

If the parameters are updated as defined by Equations 11.35 to 11.36 then Equation 11.37 follows.

$$\frac{dt_0}{dt} = -\gamma u_c e \qquad (11.35)$$

$$\frac{ds_0}{dt} = -\gamma y e \qquad (11.36)$$

$$\frac{dV}{dt} = -a_m e^2 \qquad (11.37)$$

Therefore, since the function V will decrease so long as the error is not zero it can be concluded that the error will go to zero. This does not mean that parameter t_0 and s_0 will necessarily converge to the equilibrium values. To achieve parameter convergence more conditions need to be imposed. This rule is therefore similar to the MIT rule except that the sensitivity derivatives are replaced with other functions. The adjustment rules defined by Equations 11.35 to 11.36 are very similar to the MIT rule equations and they can both be represented by Equation 11.38.

$$\frac{d\theta}{dt} = \gamma \varphi e \qquad (11.38)$$

For the Lyapunov rule $\varphi = \begin{bmatrix} -u_c & y \end{bmatrix}^T$ and for the MIT rule $\varphi = \frac{1}{p + a_m} \begin{bmatrix} -u_c & y \end{bmatrix}^T$, which is the negative value of the gradient of the loss function .

This example demonstrates that stable parameter adjustment rules can be obtained for systems in which all the state variables are measured.

11.3 Introduction to Self-tuning Regulators

A Self-Tuning Regulator (STR) is a little different to the MRAS in that the regulator parameters are obtained from the solution of a design problem after the process parameters are updated. Figure 11.7 (Åström and Wittenmark 1995) shows a typical block diagram of a STR. The STR is based on the idea of separating the task of estimating unknown parameters from the design of the controller. A recursive estimation method is used to estimate the unknown parameters on-line and then these are used to design the control signal. The estimated parameters are taken as true with no account taken of their uncertainties.

There are two loops in the STR, an inner and an outer loop. The inner loop consists of the process and an ordinary linear feedback regulator. The outer loop consists of a recursive parameter estimator and a design calculation. To produce good estimates it is often necessary to introduce perturbation signals. For the sake of simplicity this mechanism is not shown in Figure 11.7. The system in Figure

11.7 can be seen as an automation of process modelling and design in which the process model and control design are updated during each sampling interval. STRs were originally developed for sampled data systems, but continuous time and hybrid systems have also been developed. The STR is so called because the controller automatically tunes its parameters to obtain the desired properties of the closed loop system.

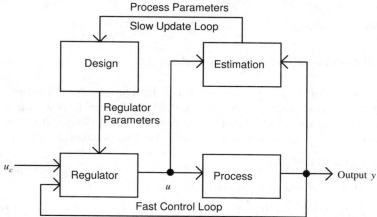

Figure 11.7. Self-tuning Regulator

The block in Figure 11.7 labelled "Design" performs an on-line solution to a design problem for a system with known parameters once they are estimated. When the parameters of the transfer function of the process and the disturbances are estimated first, before the design, this leads to what is called an indirect adaptive algorithm. It is often possible to reparameterize the process so that the model can be expressed in terms of the regulator parameters directly, therefore obviating the need for the "Design" block. This is then called a direct adaptive algorithm.

Many different estimation methods, including stochastic approximation, least squares, extended and generalised least squares, instrumental variable, and maximum likelihood methods, can be used for the parameter estimation. Given the parameters, many different control design methods can be used, including minimum variance, linear quadratic, pole placement, and model-following methods. Different combinations of estimation and design methods will lead to STRs with different properties. The methods chosen really depend on the specifications of the closed loop system.

The design of indirect and direct self-tuning regulators can be illustrated by using a SISO model for a known system. The single-input-single-output process can be described by Equation 11.39.

$$A(z)y(t) = B(z)u(t) + C(z)e(t)$$
$$A^*(z^{-1})y(t) = B^*(z^{-1})u(t) + C^*(z^{-1})e(t) \tag{11.39}$$
$$A^*(z) = z^n A(z^{-1}), \quad n = \text{order of polynomial } A.$$

where:

u = the input signal.

y = the output signal.

$\{e(t)\}$ = sequence of independent, equally distributed Gaussian variables.

A, B, and C are polynomials in terms of forward shift operator z.

Order of A = order of C.

Order of A − order of B = d_0.

11.3.1 Indirect Self-tuning Regulators

The most straightforward way to build a self-tuning regulator for a SISO process is to estimate the parameters of the polynomials A, B, and C and then use the estimates to design the regulator. For the process model described by Equation 11.36 let the desired closed loop response be defined by Equation 11.40.

$$A_m(z)\,y(t) = B_m(z)u_c(t) \tag{11.40}$$

The controller is defined by Equation 11.41.

$$R(z)u(t) = T(z)u_c(t) - S(z)y(t) \tag{11.41}$$

The solution to the Diophantine equation is R_1 and S as defined by Equation 11.42, where the conditions defined by Equations 11.43 to 11.45 hold.

$$AR_1 + B^-S = A_0 A_m \tag{11.42}$$

$$B = B^+B^-, \text{ and, } B_m = B^-B_m' \tag{11.43}$$

$$R = B^+R_1 \tag{11.44}$$

$$T = A_0 B_m' \tag{11.45}$$

Given the specifications in the form of a desired closed loop pulse transfer operator B_m / A_m and a desired observer polynomial A_0 proceed with the following design steps (Åström and Wittenmark 1995).

1. Estimate the coefficients of the polynomials A, B, and C in Equation 11.39 recursively using the least squares method or some other method.

2. Replace A, B, and C with the estimates obtained in Step 1 and solve Equation 11.42 to obtain R_1 and S. Calculate R by Equation 11.44 and T by Equation 11.45.

3) Calculate the control signal from Equation 11.41

This algorithm has some problems, which include,

1. Either the orders or at least upper bounds of the orders of the polynomials in Equation 11.39 must be known.

2. Stability of the closed loop system must be guaranteed.

3. The signals need to be persistently exciting the system to ensure parameter convergence.

11.3.2 Direct Self-tuning Regulators

The design calculations for indirect self-tuning regulators may often be time-consuming and their stability properties may be difficult to analyse. However, if the specifications are made in terms of the desired locations of the poles and the zeros the design step becomes trivial and the model is effectively reparameterized. If the Diophantine Equation 11.42 is multiplied by $y(t)$ and the model of Equation 11.39 is used the result is Equation 11.46.

$$
\begin{aligned}
A_0 A_m y(t) &= R_1 A y(t) + B^- S y(t) \\
&= R_1 B u(t) + B^- S y(t) + R_1 C e(t) \\
&= B^- \left(R u(t) + S y(t) \right) + R_1 C e(t)
\end{aligned}
\tag{11.46}
$$

Equation 11.46 is now a process model that is parameterized in terms of B^-, R, and S. Clearly, the estimation of these parameters gives the solution to the regulator polynomials R and S directly. The control signal is computed from Equation 11.41 together with Equation 11.45. Equation 11.46 is problematic if B^- is not constant because it then becomes nonlinear.

The problem can be parameterized another way by writing Equation 11.46 as defined in Equation 11.47.

$$
A_0 A_m y(t) = \overline{R} u(t) + \overline{S} y(t) + R_1 C e(t)
\tag{11.47}
$$

where:

$\overline{R} = B^- R$, polynomial R is monic but $\overline{R}$ is not.

$\overline{S} = B^- S$.

$\overline{R}$ and $\overline{S}$ have a common factor representing damped zeros, which should be cancelled before calculating the control law.

An algorithm for a direct self-tuning regulator based on Equation 11.47 and the analysis above is as follows (Åström and Wittenmark 1995),

1. Estimate the coefficients of the polynomials $\overline{R}$ and $\overline{S}$ in the Equation 11.47.

2.   Cancel possible common factors in $\overline{R}$ and $\overline{S}$ to obtain R and S.

3.   Calculate the control signal by using Equation 11.41 and the results of Step 2.

4.   Repeat Steps 1, 2, and 3 for each sample interval.

This algorithm based on the rearranged Equation 11.47 avoids the nonlinear estimation problem, but it does involve the estimation of more parameters compared to using Equation 11.46. This means that Step 2 may be difficult since it involves the estimation of the parameters of polynomial B^- twice.

In the special case where B^- is constant the calculations are simpler. If all the zeros can be cancelled and $B^- = b_0$ Equation 11.46 can be written as Equation 11.48 and the desired response as Equation 11.49.

$$A_0 A_m y(t) = b_0 \left(R u(t) + S y(t) \right) + R_1 C e(t) \tag{11.48}$$

$$A_m y_m(t) = b_0 T u_c(t) \tag{11.49}$$

where:
   Order $A = n$.
   A_0 divides T.

The error is then defined by Equation 11.50.

$$\varepsilon(t) = y(t) - y_m(t)$$
$$= \frac{b_0}{A_0 A_m} \left(R u(t) + S y(t) - T u_c(t) \right) + \frac{R_1 C}{A_0 A_m} e(t) \tag{11.50}$$

Suffice it to say that there are a number of different cases based on these new equations that may be considered.

11.4 Relations between MRAS and STR

MRAS theory was spawned from a deterministic servo problem and the STR from a stochastic regulation problem. Although the MRAS and STR originated from different problems they are nevertheless closely related. They both have an inner and out control loop where the inner loop is an ordinary regulator feedback loop in both cases. This process regulator has adjustable parameters set by the outer loop, which bases its operation on feedback from the process inputs and outputs. The differences are in the methods used for the design of the inner loop and the techniques used to adjust the parameters in the outer loop. In the MRAS in Figure 11.2 the regulator parameters are updated directly and in the STR in Figure 11.7 they are updated indirectly via parameter estimation and design calculations. This is not really a fundamental difference because the STR may be modified to make the regulator parameters be updated directly by a reparameterization of the process parameters to the regulator parameters.

11.5 Applications

Since the mid 1950s there have been a number of applications of adaptive feedback control. Initially the implementations were analogue but since the 1970s and the advent of the minicomputer, and subsequently microprocessors, applications have increased significantly. Adaptive control techniques only started to have real impact in industry since the 1980s. The areas of application include military missiles and aircraft, aerospace, process control, ship steering, robotics, and many other industrial control systems. In many applications adaptive control has been found to be useful whereas in others the benefits are minimal. Quite often a constant-gain feedback system can do just as well as an adaptive regulator without the added complexity of the adaptive structure. It is not possible to judge the need for adaptive control from the variations of the open loop dynamics over the operating range. It is necessary to test and compare solutions. This is quite easy to check as it often requires minimal effort to apply a general-purpose adaptive regulator to a problem to see if some advantage may be possible. Nevertheless, it is worthwhile evaluating a constant-gain feedback solution first before spending time trying an adaptive solution.

Adaptive control is not a completely mature technology. Adaptive control systems are not panaceas to difficult control problems but are still used in combination with other more established and proven control methods. Real applications still require many fixes to ensure adequate operation under all possible operating conditions even though the main principles are straightforward. It is necessary to introduce adequate stability mechanisms and often quite complex supervision logic to mitigate any possible disastrous conditions.

PART V. NONCLASSICAL ADAPTIVE SYSTEMS

There are at least three main types of nonclassical processing systems which do not rely on linear modelling techniques. These are Artificial Neural Networks (ANN), so called Fuzzy Logic (FL) and Genetic Algorithms (GA). Some forms of ANNs are similar to the classical adaptive systems in that they have a set of parameters which are optimised based on the minimisation of a scalar error function. Fuzzy logic systems on their own are not strictly adaptive but they can be integrated with ANNs to produce hybrid adaptive systems. Genetic algorithms are a little different in their form and function nevertheless they have various types of adaptive mechanisms, based on analogy with evolutionary processes, designed to search for optimal solutions to learning or optimisation problems.

ANNs are characterised by their massively parallel computing architecture based on brain-like information encoding and processing models. They come as supervised and unsupervised training or learning types. The supervised types have a desired output behaviour that the ANN tries to learn as it is exposed to input training data. The ANN then tries to generalise that behaviour after training, not completely unlike classical adaptive filters do. The network learning is inherent in the weights (analogous to adaptive filter coefficients), which continue to change as training or adaptation proceeds. Although the ANN does learn automatically and directly from the training data there are issues related to network size and structure that need to be considered by a designer to avoid over training and under training due to inadequate interpolation between training samples. However, there are some principles based on Statistical Learning Theory (SLT) and other similar theories that can help estimate suitable network size and structure based on finite training data for some types of ANNs (Cherkassky and Mulier 1998, Vapnik 1998, 2001).

FL effectively mimics human control logic but was initially conceived as a better method for sorting and handling data. Its main application has been for complex control applications since it uses an imprecise (fuzzy) but very descriptive language to deal with input data more like a human operator does and it is very robust and forgiving of operator and data input. FL is based on the idea that people do not require precise, numerical information input, yet they are still capable of highly adaptive control functionality. Therefore, it is reasonable to assume that if feedback controllers could be programmed to accept noisy, imprecise input, they may be much more effective and perhaps even easier to implement. A FL model is empirically-based on a designer's experience rather than on his/her technical understanding of the system and therefore the FL design process fills in the required control structure and refinements to ensure stability and satisfactory operation.

Genetic programming (GP) is a systematic machine learning method for directing computers to automatically solve algorithmic problems. It starts from a high-level statement of what needs to be achieved and from that automatically creates a computer program to solve the problem via a complex search in the solution space. GP is a derivative of Genetic Algorithms (GAs), which represent a learning or adaptation method that is analogous to biological evolution and can be described as a kind of simulated evolution. The Darwinian principle of natural selection is used to breed a population of improving solutions over many generations. Starting with a collection or initial population of say computer programs, the search for an acceptable program solution proceeds from one generation to the next by means of operations inspired by biological processes such as mutation and sexual recombination. GAs are often used to solve complex optimization problems, which are either very difficult or completely impractical to solve, using other methods. As GP requires a minimal input by a designer to produce very good automated solutions it offers a possible way of identifying and investigating what might be the crucial aspects of "intelligence" in design. Although the GP design intelligence is mostly inherent in the human designer's contribution to the problem setup it may also have something to do with the way the GP search proceeds toward the solution.

12. Introduction to Neural Networks

The material in this Chapter has been summarised from the book "Neural Networks for Intelligent Signal Processing" (Zaknich 2003a), which has a general introduction to Artificial Neural Networks (ANN). Most emphasis is placed on the Multi-Layer Perceptron (MLP), as the generic ANN, because it has strong similarities with the other adaptive filters described in this book in relation to general structure, principles of operation and adaptation rules.

12.1 Artificial Neural Networks

The human brain computes in an entirely different way to the highly successful conventional digital computer, yet it can perform very complex tasks relatively quickly and very efficiently. The brain is a highly complex, nonlinear and parallel computer that consists of approximately 10^{10} neurons having over 6 x 10^{13} interconnections. Neural events occur at millisecond speeds whereas events in conventional computers occur in fractions of nanoseconds. The brain however, can make up for this slow speed through its massive number of neurons and interconnectivity between them as compared to the computer's. The computer has fewer elements by about five orders of magnitude and very much less interconnectivity between them. This number is growing quickly with new technology, and it is estimated that by about the year 2030 the numbers of elements might be comparable to that of the brain (Kurzweil 1999). However, it is not yet clear how these elements might be exploited to achieve similar processing power to that of the brain. This is still a very open research question.

The brain, for example, can recognise a familiar face embedded in an unfamiliar scene in approximately 100-200 ms, whereas a conventional computer can take much longer to compute less difficult tasks. This recognition of the brain's power has led to the interest in the development of ANN technology. If it is possible to eventually build ANN based machines with as little as only 0.1% of the performance of the human brain they will be extraordinary information processing and controlling machines. Current ANN machines much less powerful than this can still perform useful engineering tasks which are difficult to achieve with other technologies.

12.1.1 Definitions

ANNs represent an engineering discipline concerned with nonprogrammed adaptive information processing systems that develop associations (transforms or mappings) between objects in response to their environment. That is, they learn from examples. ANNs are a type of massively parallel computing architecture based on brain-like information encoding and processing models and as such they can exhibit brain-like behaviours such as,

1.   Learning.

2.   Association.

3.   Categorisation.

4.   Generalisation.

5.   Feature Extraction.

6.   Optimisation.

Given noisy sensory inputs, they build up their internal computational structures through experience rather than preprogramming according to a known algorithm. A more formal definition of an ANN according to Simon Haykin (Haykin 1999) is,

"A neural network is a massively parallel distributed processor that has a natural propensity for storing experiential knowledge and making it available for use. It resembles the brain in two respects,

1.   Knowledge is acquired by the network through a learning process.
2.   Interneuron connection strengths known as synaptic weights are used to store knowledge."

Usually the neurons or Processing Elements (PEs) that make up the ANN are all similar and may be interconnected in various ways. The ANN achieves its ability to learn and then recall that learning through the weighted interconnections of those PEs. The interconnection architecture can be very different for different networks. Architectures can vary from feedforward, and recurrent structures to lattice and many other more complex and novel structures.

12.1.2 Three Main Types

Broadly speaking there are three main types of ANN based on the learning approach,

1.   Supervised learning type.

2.   Reinforcement learning type.

3.   Self-organising (unsupervised learning) type.

There are also other types, and even within these three main types there are numerous variants. Supervised learning ANNs are trained to perform a task by a "teacher" repeatedly showing them representative examples of the inputs that they will receive, paired with the desired outputs. During each learning or training iteration the magnitude of the error between the desired and the actual network response is computed and used to make adjustments to the internal network parameters or weights according to some learning algorithm. As the learning proceeds the average error is gradually reduced until it achieves a minimum or at least an acceptably small value. This is very similar to the way that some of the linear and nonlinear adaptive filters previously covered in this book work, and they share the same framework as shown in Figure 12.1.

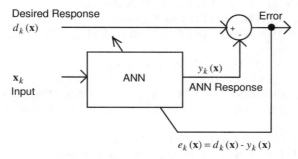

$$e_k(\mathbf{x}) = d_k(\mathbf{x}) - y_k(\mathbf{x})$$

Figure 12.1. A Supervised Learning Scheme

ANNs that learn by reinforcement do not need to compute the exact error between the desired and the actual network response; rather for each training example the network is given a pass/fail signal by the "teacher." If a fail is assigned the network continues to readjust its parameters until it achieves a pass or continues for a predetermined number of tries, whichever comes first. A new training example is then presented to the ANN and so on until a satisfactory general learning has been achieved. Reinforcement learning is sometimes thought of as a special type of supervised learning (Sutton and Barto 1999) and it has some loose similarities with Genetic Algorithms (GAs).

Self-organising ANNs take examples of the inputs and form automatic inter-groupings or clusterings of the input data based on some measure of closeness or similarity. It is then sometimes possible to assign some meaning to those clusters in proper context with the nature of the data and problem involved. The input data is usually represented in vector form so that the measures of closeness can be computed using a vector norm such as the Euclidean norm, defined previously by Equation 2.35. A good overview of up to date advances in unsupervised learning can be found in (Hinton and Sejnowski 1999).

12.1.3 Specific Artificial Neural Network Paradigms

An ANN's learning recall mechanism can vary based on the design; it can have either a feedforward recall or feedback recall mechanism. Examples of these supervised and unsupervised ANNs categorised according to their learning recall mechanism are listed in Table 12.1. Some of the more historically prominent ANN paradigms with the names of their inventors and dates of development are as follows,

1. PERCEPTRON, (1957-Rosenblatt).
2. MADALINE, (1960-62-Widrow).
3. AVALANCHE, (1967-Grossberg).
4. CEREBELLATION, (1969-Marr, Albus & Pellionez).
5. BACKPROPAGATION (BPN), (1974-85-Werbos, Parker, Rumelhart), more commonly referred to as MULTI-LAYER PERCEPTRON (MLP).
6. BRAIN STATE IN A BOX, (1977-Anderson).
7. NEOCOGNITRON, (1978-84-Fukushima).
8. ADAPTIVE RESONANCE THEORY (ART), (1976-86-Carpenter, Grossberg).
9. SELF-ORGANISING MAP, (1982-Kohonen).
10. HOPFIELD, (1982-Hopfield).
11. BI-DIRECTIONAL ASSOCIATIVE MEMORY, (1985-Kosko).
12. BOLTZMANN/CAUCHY MACHINE, (1985-86-Hinton, Sejnowsky, Szu).
13. COUNTERPROPAGATION, (1986-Hecht-Nielsen).
14. RADIAL BASIS FUNCTION, (1988-Broomhead, Lowe).
15. PROBABILISTIC (PNN), (1988-Specht).
16. GENERAL REGRESSION NEURAL NETWORK (GRNN), (1991-Specht).

Table 12.1

	Feedback Recall	**Feedforward Recall**
Supervised Learning	Brain-state-in-a-box Fuzzy Cognitive Map	Perceptron Multi-layer Perceptron ADALINE, MADALINE Boltzman Machine CMAC Radial Basis Function Network Probabilistic Neural Network General Regression Neural Network
Unsupervised Learning	Adaptive Resonance Theory Hopfield Neural Network Bidirectional Associative Memory	Kohonen Couterpropagation Fuzzy Associative Memory

There are many more ANNs than these. The most common and popular ANN in use today is MLP neural network (previously known as the Backpropagation neural network) because it is simple to understand and it generally works well. It can be used as a classifier or for nonlinear continuous multivariate function mapping or filtering. However, its disadvantages are that it can only be used with supervised training, it needs abundant examples to train and the training can be slow. Nevertheless, researchers have discovered ways of improving the training speed by new learning laws and by putting constraints on some of the weights during learning.

Radial Basis Function Neural Networks are now becoming popular. They can be trained faster than the MLP neural network and have many other desirable features for engineering applications.

12.1.4 Artificial Neural Networks as Black Boxes

From an engineering perspective many ANNs can often be thought of as "black box" devices for information processing that accept inputs and produce outputs. Figure 12.2 shows the ANN as a black box, which accepts a set of N input vectors paired with a corresponding set of N output vectors. The input vector dimension is p and the output vector dimension is K where $p, K \geq 1$. The output vector set may represent the actual network outputs given the corresponding input vector set or it may represent the desired outputs.

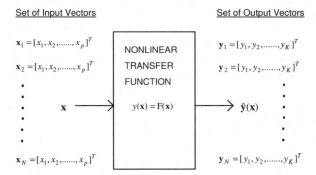

Figure 12.2. An Artificial Neural Network as a Black Box

This "black box" model is not the only model for ANNs. ANNs exist that are fully interconnected where the outputs are connected back to the inputs and also have various dynamic mechanisms which add and subtract neurons and interconnections as they operate. There are many possible model forms but these are usually dependent on the requirements of the problem being solved.

Artificial neural networks in their many forms can be used for a number of different purposes including,

1.   Probability function estimation.

2.   Classification or pattern recognition.

3.   Association or pattern matching.

4.   Multivariate function mapping.

5.   Time series analysis.

6.   Nonlinear filtering or signal processing.

7.   Nonlinear system modelling or inverse modelling.

8.   Optimisation.

9.   Intelligent or nonlinear control.

12.1.5 Implementation of Artificial Neural Networks

ANNs can be implemented either as software simulations on conventional computers or in hardware, known as neurocomputers. The two types of neurocomputer are the fully implemented type and the virtual type. The fully implemented type has a dedicated processor for each neuron (PE) and has the advantage of high speed but the disadvantages of being expensive and inflexible. The virtual type uses a single controlling micro-computer surrounded by fictitious or virtual neurons, which are implemented as a series of look-up tables. The tables maintain lists of interconnections, weights and variables for the neural network's differential equations. Software simulations on conventional computers are most useful after the more time consuming task of network training has been performed on a neurocomputer. Usually the time required to run many useful ANN developments on a conventional computer is acceptably fast, especially now as processing speed is increasing with technological advances.

ANN technology is still too young to be able to identify the best general hardware realisations to perform network calculations for any given application. VLSI technology has advanced to the point where it is possible to make chips with millions of silicon-based connections. However, there is still no clear consensus on how best to exploit VLSI capabilities for massively parallel ANN applications. The basic operations in neural networks are quite different to the more common multiply-accumulate used in classical digital signal processing algorithms. Also, neural network learning algorithms continue to be developed as there are no really ideal ones for large networks.

In the past most ANN implementations used analogue or continuous hardware, in deference to the nervous system design. Analogue circuits are widely used for high speed processing in television systems. They are frequently more efficient than digital implementations when the required computational operation can be performed by natural physical processes. There is also the question of noise in high frequency circuits to consider but the question of analogue versus digital cannot be fully determined until the algorithms and their required accuracies are known. Extremely high throughput can be achieved with either analogue or digital technologies if the functionality of the design can be sufficiently restricted.

Optoelectronic implementations of ANNs were first introduced in 1985 and they remain a very promising approach. Optical systems offer the massive interconnectivity and parallelism required by neural networks that even VLSI methods cannot equal. Ultimately it is desirable that entire ANNs be implemented with optics but fully optical decision devices are not yet available. Instead, it is necessary to convert signals from optical to electronic form for the decision making stages. Previous research in this area has been limited to relatively small hybrid optoelectronic systems. To achieve the full potential of optoelectronics for larger and faster networks integrated optoelectronic technology needs to be developed beyond that for current high speed optical communication.

12.1.6 When to Use an Artificial Neural Network

There are three general types of problems in which an ANN may be used (Eberhart and Dobbins 1990), where,

1.   An ANN provides the only practical solution.

2.   Other solutions exist but an ANN gives an easier or better solution.

3.   An ANN solution is equal to others.

ANNs should only be used for type 1 and 2 problems. There is no sensible reason to use an ANN where there already exists a well established and efficient problem solution method. ANNs can provide suitable solutions for problems which generally are characterised by,

1.   Nonlinearities.

2.   High dimensionality.

3.   Noisy, complex, imprecise, imperfect and/or error prone sensor data.

4.   A lack of a clearly stated mathematical solution or algorithm.

12.1.7 How to Use an Artificial Neural Network

The specification and design of an ANN application should aim to produce the best system and performance overall. This means that conventional methods should be used if and where possible and ANNs used to supplement them or only if they can add some benefit. The heart of a neural network design involves at least five main tasks,

1.   Data collection.

2.   Raw data preprocessing.

3.   Feature extraction and input vector formation from preprocessed data.

4.   Selection of an ANN type and topology (architecture).

5.   ANN training, testing and validation.

After suitable data is collected and preprocessed the features are chosen by the designer based on his/her knowledge and experience with the problem. Features should be chosen because they are believed to have some correlation with the desired output. It can be useful to eliminate or prune redundant or ineffective features. It is also possible to determine which sets of features are the most significant by comparative analysis. An ANN design should incorporate a minimum of three sets of independent input-output feature vector pairs that are representative of the process. There should be a training, testing and validation vector set. The training set is used to do the network training. Either during training and/or after training the testing set is used to check, or cross-validate, that the trained network is able to adequately generalise its learning to new data. Finally, when the training and testing has been done the validation set is used as a check on the ANN's generalisation and accuracy.

12.1.8 Artificial Neural Network General Applications

ANNs have application in wide ranging areas of human interests including finance, engineering and medicine. They are no longer just scientific curiosities as they have already been applied in many and various real products. They have been used to,

1.   Perform optical character recognition for reading cheques.

2.   Score bank loan applications.

3.   Forecast stock market performance.

4.   Detect credit card fraud.

5.   Plan optimal routes for intelligent vehicles.

6.   Recognise speech and fingerprints in security systems.

7.   Control robotic motion and manipulators.

8.   Stabilise stealth aircraft (stealth technology is aerodynamically unsound).

9.   Predict fatigue life and failure of mechanical components.

10. Filter, equalise, echo cancel communication channels.

11. Control traffic flow and switching in communication channels.

12. Classify radar and sonar signals.

13. Classify blood cell reactions and blood analysis.

14. Detect cancer; diagnose heart attacks.

15. Perform brain modelling.

ANNs are still in their infancy as far as their full design and application potential is concerned. It is very likely that they will eventually have at least some application in almost all real systems. This is because real systems are actually nonlinear although many have been adequately modelled and solved previously by using well established classical linear theory.

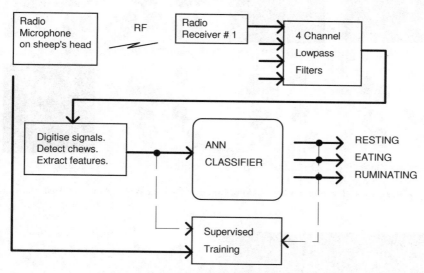

Figure 12.3. Sheep Eating Phase Monitor

12.1.9 Simple Application Examples

Some simple ANN examples of classification, function mapping, nonlinear filtering and control are presented here to provide some introductory exposure of how they can be used.

12.1.9.1 Sheep Eating Phase Identification from Jaw Sounds

When sheep are in the field they are likely to be doing one of three things, resting, eating or ruminating. These activities can be determined quite accurately by monitoring their jaw sounds with a radio microphone attached to their skulls (Zaknich and Baker 1998). An ANN can be trained to distinguish these three phases by extracting suitable frequency spectral features from the jaw sound signals to feed into the ANN classifier as shown in Figure 12.3.

12.1.9.2 Hydrate Particle Isolation in SEM Images

Automatically isolating alumina hydrate particles in Scanning Electron Microscope images is very difficult because the region between touching particles can often look like a particle surface feature. It is possible to use an ANN classifier to help discriminate between boundaries of touching particles and other features to effect a suitable particle separation (Zaknich 1997). Figure 12.4 shows an image sequence in the process of particles isolation.

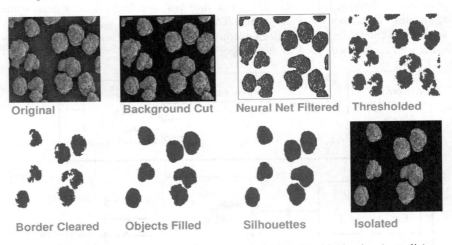

Figure 12.4. Particle Isolation (Compliments of Alcoa World Alumina Australia)

Figure 12.5. Oxalate Detection (Compliments of Alcoa World Alumina Australia)

12.1.9.3 Oxalate Needle Detection in Microscope Images

In the Alumina production process it is very important to identify and count relatively small numbers of oxalate needles in microscope images dominated by hydrate particles. Once again an ANN classifier can be trained to distinguish between the needle shaped oxalate and blob shaped hydrate by using suitable shape features (Zaknich and Attikiouzel 1995a). The original image with oxalate needles and hydrate blobs is shown in Figure 12.5 along with the resulting needle detection image.

12.1.9.4 Water Level Determination from Resonant Sound Analysis

It is possible to use an ANN to determine the water level in a glass by creating a multivariate functional map from features extracted from resonant sounds to water level. Although the resonant sounds of a water filled glass are complex and dependent on how they are excited there is a definite functional relationship between them and the associated water level that the ANN can capture from training data. Figure 12.6 shows the system block diagram.

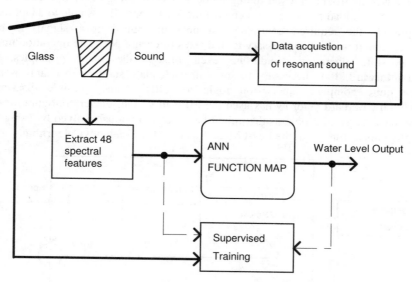

Figure 12.6. Water Level Measurement

12.1.9.5 Nonlinear Signal Filtering

Short wave radio signals are broadcasted over long distances at relatively low power outputs. Consequently they are very susceptible to various nonlinear and linear effects such as fading and various types of noises such as white noise from receiving equipment and impulse noise from local electricity power systems. For radio transmission of simple sinusoidal tone burst signals as used in international time signals, Morse code and telemetry signals, a nonlinear ANN filter can be effectively used to solve this problem and recover the original tone bursts (Zaknich and Attikiouzel 1995b). Figure 12.7 compares the results achieved with a linear filter against an ANN filter.

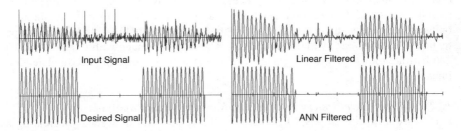

Figure 12.7. Short Wave Signal Filtering

12.1.9.6 A Motor Control Example

In the case where a motor is driving a variable nonlinear load it is possible to use a MLP ANN controller in a self-tuning mode as shown in Figure 12.8 to control the system. As the load characteristics change the MLP controller is able to adapt itself to provide the required control given only the speed error signal. In real applications it would be necessary to add fixes to ensure adequate operation under all possible operating conditions even though the main principles are straightforward. It is necessary to introduce adequate stability mechanisms and often quite complex supervision logic to mitigate any possible disastrous conditions. Bernard Widrow has used the MLP in various clever configurations to solve these types of problems that might have been previously solved by using the adaptive linear filter with the Least Mean Squares (LMS) adaptive algorithm.

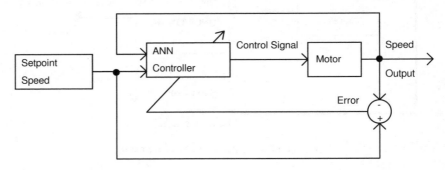

Figure 12.8. Self-tuning ANN Motor Controller

12.2 A Three-layer Multi-layer Perceptron Model

The three-layer feedforward Multi-layer Perceptron has a parallel input, one parallel hidden layer and a parallel output layer. The input layer is only a "fan-out" layer, where the input vector is distributed to all the hidden layer PEs. There is no real processing done in this layer. The hidden layer is the key to the operation of the MLP. Each of the hidden nodes is a single PE, which implements its own decision surface. The output layer is a set of decision surfaces in which each of its PEs has decided what part of the decision space the input vector lays. The role of the output layer is essentially to combine all of the "votes" of the hidden layer PEs and decide upon the overall classification of the vector. The nonlinearity provided is by the nonlinear activation functions of the hidden and output PEs and this allows this network to solve complex classification problems that are not linearly separable. This is done by forming complex decision surfaces by a nonlinear combination of the hidden layer's decision surfaces.

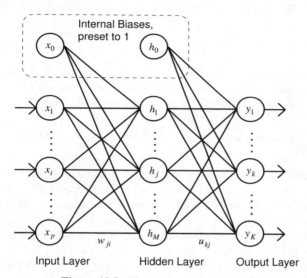

Figure 12.9. Three-layer Feedforward MLP

Figure 12.9 represents a three-layer feedfoward MLP model. After training the feedforward equations relating the inputs to the outputs are described by the general matrix Equation 12.1.

$$\mathbf{y} = \mathbf{f}(\ \mathbf{U}\ \mathbf{f}(\ (\mathbf{W}\ \mathbf{x})\ )\ )$$
(12.1)

where:

$\mathbf{x}$ = $[1, x_1, x_2,..., x_i,..., x_p]^T$ input vector $((p+1) \times 1)$.
$\mathbf{W}$ = matrix of weights w_{ji} between input-hidden nodes $((M+1) \times (p+1))$.
$\mathbf{U}$ = matrix of weights u_{kj} between hidden-output nodes $(K \times (M+1))$.
$\mathbf{y}$ = output vector $(M \times 1)$.
$\mathbf{f}(.)$ = multivariate activation function.
p = number of real input nodes.
M = number of real hidden nodes.
K = number of output nodes.

Equation 12.1 can be expressed in detail by Equations 12.2, 12.3 and 12.4.

$$h_j = f_j(\sum_{i=0}^{p} w_{ji}x_i), \qquad \text{for } j = 1,2,...,M \tag{12.2}$$

$$y_k = f_k(\sum_{j=0}^{M} u_{kj}h_j), \qquad \text{for } k = 1,2,...,K \tag{12.3}$$

The so called activation function $f(z)$ that is often used in Equations 12.2 and 12.3 is defined by Equation 12.4. It is this type of nonlinear activation function applied to hidden and output nodes that introduces the nonlinearity into the ANN.

$$f(z) = f_j(z) = f_k(z) = \frac{1}{1+e^{-z}} \tag{12.4}$$

The derivative of this particular nonlinear activation function, based on the natural exponential function, is a function of itself as defined by Equation 12.5. This useful property will be used later in the MLP learning algorithm development.

$$f' = \frac{df(z)}{dz} = \frac{e^{-z}}{(1+e^{-z})^2} = \frac{1}{1+e^{-z}}(1-\frac{1}{1+e^{-z}}) = f(1-f) \tag{12.5}$$

The outputs y_k are a function of hidden outputs h_j and the weights u_{kj} between the hidden layer and the output. The outputs h_j are a function of the inputs x_i and the weights between the inputs and the hidden layer. Note that the weights between the inputs and the internal bias node h_0 are zero and that the bias inputs x_0 and h_0 are set to equal 1. For classification problems the desired outputs are chosen to have values of about 0.9 to signify a class membership and about 0.1 for non-class membership. The number of output nodes are made to be equal to the number of classes, therefore the desired output vector for say the first class of three classes would be $[0.9, 0.1, 0.1]^T$. Input values may be any positive or negative real number but often they are normalised to range between -1 and $+1$ for convenience. Of course after training the output vector for a given class input vector will not be exactly like the desired output vector, it may for example be something like $[0.83, 0.23, 0.008]^T$. Here, this example vector would be recognised as a class one vector because the value in the first class position is obviously larger than the other two. Various strategies are used to select the class, either simply take the highest value or the highest value only if it is greater than any of the others by a certain margin, else it can just be specified as unclassified.

12.2.1 MLP Backpropagation-of-error Learning

The MLP stores its knowledge in the weights. The problem is to adjust these weights in such a way that will produce the required knowledge and solutions to required classification problems. Because the type of classification problems of general interest are too complex to solve *a priori* by analytic techniques it is necessary to develop an adaptive training algorithm that is driven by example data. The hope is that if there are adequate features, number of PEs and sufficient representative training data samples the weights will slowly adjust correctly through training. They should adjust in such a way as to end up with a set of network weights that will give a satisfactory classification performance for other inputs that the network has not seen during training. This optimisation can be achieved most effectively by adjusting the weights to minimise the Mean Square Error (MSE) of the network outputs compared with desired responses. This can be very time consuming if it is necessary to compute the MSE of all the training pairs before the weights can be incrementally adjusted once. Alternatively, it is possible to use Backpropagation-of-error learning which is based on the gradient descent optimisation technique. The main idea behind Backpropagation-of-error learning is to adjust the weights a little each time as a new random training input-output vector pair is presented to the network. This is done repeatedly until a satisfactory convergence occurs. The local gradient of the error function, Equation 12.6, for each given input vector **x** is computed for K output nodes and used to adjust the weights in the opposite direction to the gradient. This Backpropagation-of-error learning based on the local gradient is basically the same idea as the LMS algorithm except in this case it is propagated back to hidden layers, whereas in the LMS algorithm it is only applied to the output layer of the linear combiner.

$$E_{\mathbf{x}} = \frac{1}{2} \sum_{k=1}^{K} (d_k - y_k)^2$$
(12.6)

Moving in the opposite direction to the gradient is the direction that on the whole makes the overall network error smaller. The main problem with gradient descent optimisation is that it can be prone to converging to a local minimum instead of the global minimum. There are a number of techniques including so called "simulated annealing" that have been developed to try to solve this problem. For more comprehensive details on the Backpropagation-of-error algorithm refer to the works of the three co-inventors, (Parker 1985), (Werbos 1974, Werbos 1990) and (Rumelhart 1986), and also (Haykin 1999).

12.2.2 Derivation of Backpropagation-of-error Learning

After training it is required that the MSE be minimal for the whole training set of input-output vector pairs. To achieve this it is necessary to adjust the two sets of network weights, the output layer weights u_{kj} and the hidden layer weights w_{ji} in concert. The gradient of the error needs to be calculated in the whole weight space. This can be done using partial derivatives and the chain rule to calculate the contribution that each of the weights makes on the total error as developed in the following Sections.

12.2.2.1 Change in Error due to Output Layer Weights
The partial derivative of the error Equation 12.6 with respect to the output layer weights u_{kj} is defined by Equation 12.7.

$$\frac{\partial E_\mathbf{x}}{\partial u_{kj}} = \frac{\partial E_\mathbf{x}}{\partial y_k} \cdot \frac{\partial y_k}{\partial u_{kj}}$$
(12.7)

Equation 12.7 is made up from the partial derivative of the error function $E_\mathbf{x}$ multiplied by the derivative of the output generating function y_k. If The error function Equation 12.6 is substituted into Equation 12.7 then Equations 12.8 to 12.11 result.

$$\frac{\partial E_\mathbf{x}}{\partial u_{kj}} = \frac{\partial}{\partial y_k}[\frac{1}{2}\sum_{a=1}^{K}(d_a - y_a)^2].\frac{\partial}{\partial u_{kj}}[f_k(\sum_{b=0}^{M} u_{kb}.h_b)]$$
(12.8)

$$\frac{\partial E_\mathbf{x}}{\partial u_{kj}} = (y_k - d_k).f_k{}'(\sum_{b=0}^{M} u_{kb}.h_b).h_j$$
(12.9)
$$= (y_k - d_k).y_k(1 - y_k).h_j$$

$$\frac{\partial E_\mathbf{x}}{\partial u_{kj}} = \delta y_k.h_j$$
(12.10)

$$\delta y_k = (y_k - d_k).y_k(1 - y_k)$$
(12.11)

Equation 12.11 represents the backpropagating error related to the hidden layer output.

12.2.2.2 Change in Error due to Hidden Layer Weights

The calculation of the change in error as a function of the hidden layer weights w_{ji} is more difficult because there is no way of getting "desired y_k outputs" for the hidden layer PEs. It is only known what the network outputs should be. The partial derivative is similar to before but just a little more complex. The required equations are the Equations 12.12 to 12.17.

$$\frac{\partial E_\mathbf{x}}{\partial w_{ji}} = \frac{\partial}{\partial w_{ji}}[\frac{1}{2}\sum_{a=1}^{K}(d_a - y_a)^2]$$
(12.12)

$$\frac{\partial E_\mathbf{x}}{\partial w_{ji}} = \sum_{a=1}^{K}\frac{\partial}{\partial w_{ji}}[\frac{1}{2}(d_a - y_a)^2]$$
(12.13)

$$\frac{\partial E_\mathbf{x}}{\partial w_{ji}} = \sum_{a=1}^{K}[\frac{\partial}{\partial y_a}(\frac{1}{2}(d_a - y_a)^2).\frac{\partial y_a}{\partial h_j}.\frac{\partial h_j}{\partial w_{ji}}]$$
(12.14)

$$\frac{\partial E_\mathbf{x}}{\partial w_{ji}} = \sum_{a=1}^{K}[(y_a - d_a).f_a{}'(\sum_{b=0}^{M} u_{ab}.h_b).u_{aj}].f_j{}'(\sum_{b=0}^{P} w_{jb}.x_b).x_i$$
(12.15)

$$\frac{\partial E_x}{\partial w_{ji}} = \delta h_j . x_i$$

(12.16)

$$\delta h_j = \sum_{a=1}^{K} [(y_a - d_a).f_a'(\sum_{b=0}^{M} u_{ab}.h_b).u_{aj}].f_j'(\sum_{b=0}^{P} w_{jb}.x_b)$$

$$= \sum_{a=1}^{K} [(y_a - d_a).y_a(1 - y_a).u_{aj}].h_j(1 - h_j)$$

(12.17)

Equation 12.17 represents the backpropagation of the error from the output layer to the hidden layer.

12.2.2.3 The Weight Adjustments

In order to minimise the error it is necessary to adjust all the weights in the opposite direction to the error gradient each time a training input-output vector pair is presented to the network as defined by Equations 12.18 to 12.21.

$$\Delta u_{kj} = -\eta.\frac{\partial E_x}{du_{kj}} = -\eta.\delta y_k.h_j$$

(12.18)

$$u_{kj}^{new} = u_{kj}^{old} + \Delta u_{kj}$$

(12.19)

$$\Delta w_{ji} = -\mu.\frac{\partial E_x}{dw_{ji}} = -\mu.\delta h_j.x_i$$

(12.20)

$$w_{ji}^{new} = w_{ji}^{old} + \Delta w_{ji}$$

(12.21)

where:

μ and η are positive valued scalar gains or learning rate constants.

The learning rate is controlled by the scalar constants μ and η. These should be made relatively small, i.e., μ and $\eta < 1$. If they are too small the rate of convergence is slow, but if they are too large it may be difficult to converge once in the vicinity of a minimum since the estimate of the gradient is only valid locally. If the weight change is too great then it may be in a direction not reflected by the gradient. The ideal learning strategy may be to use relatively high values of learning rates to start with and then reduce them as the training progresses. When there is only a finite training vector set it is advisable to continually select the individual training vector input-output pairs at random from the set rather than sequence through the set repeatedly. The training may require many 100,000s or even 1,000,000s of these iterations, especially for very complex problems.

For these equations to work an activation function is required that is differentiable and if possible one whose derivative is easy to compute. The sigmoid function of Equation 12.5 is a suitable function because not only is it continuously differentiable its derivative is a simple function of itself, as seen in Equation 12.6. Therefore, the weight adjustment Equations 12.18 and 12.20 can be rewritten simply as Equations 12.22 and 12.24 respectively.

$$\Delta u_{kj} = -\eta . \delta y_k . h_j$$
$$= -\eta . [(y_k - d_k).y_k.(1-y_k)].h_j, \quad \text{for } k = 1,..,K, \text{and } j = 0,..,M \tag{12.22}$$

where:

$$y_k = f_k = f_k(\sum_{b=0}^{M} u_{kb}.h_b) = \frac{1}{1 + e^{-(\sum_{b=0}^{M} u_{kb}.h_b)}} \tag{12.23}$$

$$\Delta w_{ji} = -\mu . \delta h_j . x_i$$
$$= -\mu . \sum_{a=1}^{K} [(y_a - d_a).y_a(1-y_a).u_{aj}].h_j(1-h_j).x_i \tag{12.24}$$
$$= -\mu . \sum_{a=1}^{K} [\delta y_a.u_{aj}].h_j.(1-h_j).x_i, \quad \text{for } j = 1,..,M, \text{and } i = 0,.,p$$

where:

$$h_j = f_j = f_j(\sum_{b=0}^{p} w_{jb}.x_b) = \frac{1}{1 + e^{-(\sum_{b=0}^{p} w_{jb}.x_b)}} \tag{12.25}$$

The weight adjustment Equations 12.22 and 12.24 are now in terms of the actual input, output and desired values and can therefore be computed by simple arithmetic.

A summary of the three-layer MLP Backpropagation-of-error learning is,

Parameters: p = number of real input nodes, plus one hidden.
K = number of output nodes, plus one hidden.
M = number of real hidden nodes.
η, μ = step sizes.
W = $((M+1) \times (p+1))$ weight matrix between input and hidden nodes.
U = $(K \times (M+1))$ weight matrix between hidden and output nodes.

Initialisation: $x_0 = 1$.
$h_0 = 1$.
W = a set of very small random values.
U = a set of very small random values.

Weight updates: For each new training vector pair compute,

$$u_{kj}^{\ new} = u_{kj}^{\ old} - \eta.(y_k - d_k).y_k.(1 - y_k).h_j,$$
$$\text{for } k = 1,..,K, \text{ and } j = 0,..,M$$

$$w_{ji}^{\ new} = w_{ji}^{\ old} - \mu. \sum_{a=1}^{K} [(y_a - d_a).y_a.(1 - y_a).u_{aj}].h_j.(1 - h_j).x_i,$$
$$\text{for } j = 1,..,M, \text{ and } i = 0,..,p$$

where: $$y_k = \frac{1}{1 + e^{-(\sum\limits_{b=0}^{M} u_{kb}.h_b)}}, \text{ and,}$$

$$h_j = \frac{1}{1 + e^{-(\sum\limits_{b=0}^{p} w_{jb}.x_b)}}$$

12.2.2.4 Additional Momentum Factor

When the network weights approach a minimum solution the gradient becomes small and the step size diminishes too, giving very slow convergence. If a so called momentum factor is added to the weight update equations the weights can be updated with some component of past updates. This can reduce the decay in learning updates and cause the learning to proceed through the weight space in a fairly constant direction. The benefits of this, apart from faster convergence toward the minimum, is that it may even be possible to sometimes escape a local minimum if there is enough momentum to travel through it and over the following hill in the error function.

Adding a momentum factor to the gradient descent learning equations results in Equations 12.26 and 12.27 respectively.

$$\mathbf{W}(k+1) = \mathbf{W}(k) - \mu \, \partial E_x/\partial \mathbf{W} + \alpha(\mathbf{W}(k)-\mathbf{W}(k-1)) \qquad (12.26)$$

$$\mathbf{U}(k+1) = \mathbf{U}(k) - \eta \, \partial E_x/\partial \mathbf{U} + \beta (\mathbf{U}(k)-\mathbf{U}(k-1)) \qquad (12.27)$$

where:
μ, η, α and β are positive valued scalar gain or learning rate constants, all less than 1.

When the gradient has the same algebraic sign on consecutive iterations the weight change grows in magnitude. Therefore, momentum tends to accelerate descent in steady downhill directions. When the gradient has alternating algebraic signs on consecutive iterations the weight changes become smaller, thus stabilising the learning by preventing oscillations.

12.2.3 Notes on Classification and Function Mapping

The three-layer feedforward MLP developed above is specifically crafted for pattern classification. It has the same nonlinear activation functions on both the hidden and output PEs. More layers can be added by simply adding more parallel sets of PEs and the associated weights coming from the previous layer. More layers provide a more complex ANN model but in practice it is rarely needed, or advisable, to go beyond about four or five layers. The design equations for four or five layer networks, although a little more complex, can be derived in a similar manner to that of the three-layer model developed above.

The activation functions associated with the hidden layers are sigmoids whereas the activation functions associated with the outputs are usually sigmoidal for classification applications, and linear (or no activation function) for function mapping applications. When developing a network for function mapping it may then be necessary to add at least another hidden layer with activation functions to increase the nonlinear model complexity. For classification problems it is common to assign one output node for each separate class, with a training output value of 0.9 representing class selection and 0.1 class rejection. Normally values of 1.0 and 0.0 would be used, but for MLP systems using sigmoidal activation functions this may introduce unnecessary training time due to saturation effects. This is because the sigmoidal function defined by Equation 12.4 ranges between saturation levels of 0 for a $-\infty$ input and 1 for a $+\infty$ input to it.

For function mapping the sigmoidal function defined by Equation 12.28 is often used in preference to Equation 12.4 because it ranges between values of -1 and $+1$, and therefore can more conveniently represent positive and negative excursions.

$$f(z) = \frac{1 - e^{-z}}{1 + e^{-z}} \tag{12.28}$$

12.2.4 MLP Application and Training Issues

The key issues involved with the practical application of the MLP include the following,

1.   Design of training, testing and validation data sets.

2.   Determination of the network structure.

3.   Selection of the learning rate.

4.   Problems with under-training and over-training.

The typical MLP design needs to have a minimum of three sets of independent vectors taken at random from the process. There should be a training, a testing and a validation vector set of approximately equal proportions. The training set is used to train the network weights. At regular intervals during training the mean square error between the desired and network response for the entire training and testing sets is calculated and plotted on a graph as shown in Figure 12.10. The MSE should on average get progressively smaller as the network learns. If the network is complex enough, eventually there will be a point where the testing set error will begin to increase while the training set error will continue to get smaller or remain constant. The training should then be stopped and the network weights fixed. At that point the validation data set should be run through the network and the MSE checked to see that it is acceptably low and that the network accuracy is acceptably high. If so, then the network design has been completed and the network can be put into service. If not, it may be necessary to alter the number of hidden nodes and repeat the process. If it is still not acceptable it may be necessary to improve the pre-processing and feature extraction from the raw data or increase the training data set size.

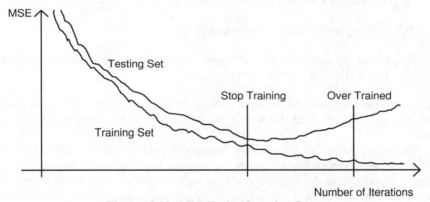

Figure 12.10. MLP Typical Learning Curves

If the MSE of the training set begins to increase while the MSE of the testing set continues to reduce (refer to Figure 12.10) this can also signify that the complexity or size of the network is probably too large for the problem. Another way to deal with this is to reduce the network a bit and retrain it until the training and testing MSEs reduce down to approximately the same values as the number of training iterations increase.

MLP training is facilitated by making sure that there are approximately equal numbers of training vectors for each class irrespective of the actual *a priori* probability of occurrence for the classes when in operation. This does not necessarily apply to all neural networks. Once the training vector set has been selected, the training involves taking vectors one at a time, on a completely random basis from the whole set. Feeding the training set into the network sequentially over and over will probably result in dismal failure.

The MLP divides up the vector space of each layer with hyperplanes. Unfortunately, there are no general rules for arbitrary problems to help decide exactly how many hyperplanes, hidden layers and hidden nodes are needed. This is a function of the data and the problem. HNC incorporated, a neurocomputing company, has implemented many commercial MLP solutions. As a result of this HNC has developed some rules of thumb about this (Hecht-Nielson 1990),

1. Start with no hidden layers and input nodes connected to outputs.

2. Next, try one hidden layer, having one fourth the number of nodes in it as the number of the input dimension plus the number of output categories. Again, connect the inputs to the outputs.

3. Try decreasing the number of hidden nodes.

4. Try increasing the number of hidden nodes

5. Try adding a second hidden layer with one or two nodes.

6. Try no input-to-output connections.

7. Generally only very unusual applications find success with three hidden layers. Try this only as a last resort.

After implementing any of the above Steps, if good results are achieved during training, try the next Step. Stick with those changes which seem to improve the results.

The initial values of the network weights should be fairly small and set to random values. Depending on the exact learning law used, the learning rate and other factors such as the momentum factor need to be chosen by the designer. The larger these values are set to the larger the weight adjustments per iteration. Initially these could be large enough to allow the network to quickly get close to the right general solution. They could then be reduced to allow slower more uniform convergence toward the optimum solution.

12.3 Exercises

The following Exercises identify some of the basic ideas presented in this Chapter.

12.3.1 Problems

12.1.   What are two most important aspects of artificial neural network technology?

12.2.   Name some general characteristics of ANNs.

12.3.   What are the main types of ANN?

12.4.   Can ANNs be seen as magic boxes which can be applied to virtually any problem?

12.5.   Specify some possible applications for ANNs.

12.6.

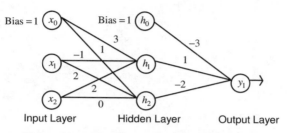

Figure 12.11. Three-layer Feedforward Perceptron

The three-layer Perceptron network shown in Figure 12.11, when properly trained, should respond with a desired output $d = 0.9$ at y_1 to the augmented input vector $\mathbf{x} = [1,x_1,x_2]^T = [1,1,3]^T$. The network weights have been initialised as shown in Figure 12.11. Assume sigmoidal activation functions at the outputs of the hidden and output nodes and learning gains of $\eta = \mu = 0.1$ and no momentum term. Analyse a single feedforward and backpropagation step for the initialised network by doing the following,

a.   Give the weight matrices $\mathbf{W}$ (input-hidden) and $\mathbf{U}$ (hidden-output).
b.   Calculate the output of hidden layer, $\mathbf{h} = [1, h_1, h_2]^T$ and output y_1.
c.   Compute the error signals δy_1, δh_1 and δh_2.
d.   Compute all the nine weight updates, Δw_{ij} and Δu_{ij}.

Use the activation function $f(z) = \dfrac{1}{1 + e^{-z}}$.

13. Introduction to Fuzzy Logic Systems

Fuzzy logic systems on their own are not strictly adaptive but they can be integrated with ANNs to produce hybrid adaptive systems. The resulting adaptive hybrid system is called a Fuzzy Artificial Neural Network (FANN). The principles of artificial neural networks have been covered in the previous Chapter. This Chapter introduces the principles of basic fuzzy logic and fuzzy logic control followed by a short description of FANNs and their basic applications.

Basic fuzzy logic can be classified into time-independent and time-dependent categories. Time-independent fuzzy logic can be described as a generalised combinatorial or sequential logic, where the passage of time is not directly relevant. Time-dependent fuzzy logic is called Temporal Fuzzy Logic (TFL) and it is an extension of basic time-independent fuzzy logic with the inclusion of time-dependent membership functions (Kartalpoulos 1996). Only time-independent fuzzy logic is considered in this Chapter.

Fuzzy logic can be used for control applications, in which case it is referred to as Fuzzy Logic Control (FLC). FLC is used in applications involving complex systems where mathematical models are not available but where the system is controllable by human experts using situation-action control rules and policies. Approximately half the applications are used to enhance conventional control systems and the remainder to enable a non-conventional control approach. Advantages of FLC are that it implements expert knowledge, it can provide robust nonlinear control, and the system development and maintenance time is reduced. One disadvantage has previously been the lack of FLC stability theory. However, this problem has now been formally addressed post 1992. Other disadvantages are that problem domain knowledge must exist and this knowledge needs to be explicitly expressed in FLC terms.

A suitable introductory book on FL for beginners is (Mukaidono 2001).

13.1 Basic Fuzzy Logic

Fuzzy logic is a logic that is used to infer a crisp outcome from so called fuzzy input values. An example of crisp logic is binary logic where variables have only one of two possible exact values "1" or "0" (true or false). Binary logic can be extended to multivalue logic, where the variables can have one of many possible crisp values. In contrast to crisp logic there is propositional logic where the variables are defined by meaningful but uncertain terms such as "fast," "moderate," "slow," "very hot," "hot," "cool," "cold" etc.. Propositional logic can be represented by a fuzzy logic, which can be seen as a generalised logic that includes crisp variable values as well as all the possible values in between them. For example, the crisp binary variables "1" and "0" may each be represented by a number in the range $[0, \alpha]$, indicating the degree to which the variable is said to have the attribute represented by the crisp value. Thus a relationship is defined to express the distribution or degree of truth of a variable. For example, a propositional variable "slow" may be defined as a distribution around some value whereby any value within the distribution may be interpreted as "slow" but with different degrees of slowness.

13.1.1 Fuzzy Logic Membership Functions

A fuzzy set F of a universe of discourse $X = \{x\}$ is defined as the mapping defined by Relation 13.1, i.e., the membership function $\mu_F(x) \in [0, \alpha]$.

$$\mu_F(x): X \to [0, \alpha] \tag{13.1}$$

Each x is assigned a number in the range $[0, \alpha]$ indicating the extent to which x has the attribute F. When $\alpha = 1$ the membership function is normalised and the fuzzy logic is called normal. Only normal logic will be used from here on in. In the special case when the distribution is of zero width the membership function is reduced to singularities or simply to crisp logic. Crisp binary logic has two possible singularities and n-variable crisp logic has n possible singularities.

Given that X is a time-invariant set of objects x then a fuzzy set $\tilde{F}$ in X may be expressed as a set of ordered pairs $(x, \mu_{\tilde{F}}(x))$ defined by Equation 13.2.

$$\tilde{F} = \left[(x, \mu_{\tilde{F}}(x)) \mid x \in X \right] \tag{13.2}$$

where:

$\mu_{\tilde{F}}$ is the membership function that maps X to a membership space $M = [0,1]$.

$\mu_{\tilde{F}}(x) \in [0,1]$ is the grade of membership of x in $\tilde{F}$.

For example, take a set of temperatures $X = \{10,20,30,40,50,60,70,80,90,100\}$, where x is the temperature. Temperatures around 10 are considered cool $(\tilde{A})$, around 40 hot $(\tilde{B})$ and around 70 very hot $(\tilde{C})$. The attributes cool, hot and very hot are not crisply defined. Normalised fuzzy sets can be defined for cool, hot and very hot in pairs $(x, \mu_{\tilde{F}}(x))$ as follows,

$$\tilde{A} = \{(0,1), (10,1), (30,0)\}$$

$$\tilde{B} = \{(20,0),(30,1),(50,1),(70,0)\}$$
$$\tilde{C} = \{(50,0),(70,1),(100,1)\}$$

The normalised membership functions for cool $(\mu_{\tilde{A}}(x))$, hot $(\mu_{\tilde{B}}(x))$, and very hot $(\mu_{\tilde{C}}(x))$, can now be represented as exemplified in Figure 13.1.

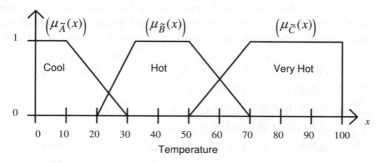

Figure 13.1. Normalised Membership Functions for Cool, Hot and Very Hot

Figure 13.1 represents the fuzzification of the three variables, which spreads the variables into relevant distribution profiles that include all temperatures between 0 and 100. A temperature of 60 belongs to the variable hot with a confidence of 0.5 and to variable very hot with a confidence of 0.5. This type of overlap of variables is typically used in fuzzy logic problems.

In a given problem it may be difficult to choose suitable membership function structures. Questions related to function type, shape, number and aggregations must be answered. Often these are arrived at by some measure of trial and error guided by the adequacy of the final design performance. In any event, it is strongly recommended that the crossover points between adjoining functions should be at greater than or equal to the maximum possible function output (sum of overlapping variables).

13.1.2 Fuzzy Logic Operations

In fuzzy logic operations, unlike Boolean logic operations, the results are not crisp. The output of fuzzy logic operations exhibit a distribution described by the membership function. The operations are analogous to Boolean union (OR) and intersection (AND) and are described by min-max logic. The fuzzy output of the union of a set of fuzzy variables is the maximum membership function value of any of the variables. The fuzzy output of the intersection of a set of fuzzy variables is the minimum membership function value of any of the variables. The complement operation (equivalent to NOT) is the complement of 1, i.e., 1 minus the membership function value. For example given fuzzy inputs, $A = 0.2$, $B = 0.7$ and $C = 0.5$, then,

$$A \text{ OR } B \text{ OR } C = \max(A, B, C) = 0.7$$
$$A \text{ AND } B \text{ AND } C = \min(A, B, C) = 0.2$$
$$\text{NOT } A = (1 - A) = 0.8$$

13.1.3 Fuzzy Logic Rules

Fuzzy logic rules are developed based on *a priori* knowledge and the designer's past experience. In a given problem all the possible input-output relations must be known. These can then be expressed in terms of a complete set of IF-THEN rules, e.g.,

IF A_1 *AND / OR* B_1, THEN H_{11}, else ......

IF A_2 *AND / OR* B_2, THEN H_{22}, else ......

IF A_1 *AND / OR* B_2, THEN H_{12}, else ......

IF A_2 *AND / OR* B_1, THEN H_{21}, else ......

if A_2 *AND / OR* B_2, then H_{22}, else ......

where:

A_i and B_j are fuzzy inputs and H_{ij} are the actions for each rule.

The set of rules for two variables can be economically tabulated as shown in Table 13.1.

Table 13.1. Tabulation of Fuzzy Rules having Two Variables

A_1	H_{11}	H_{12}
A_2	H_{21}	H_{22}
	B_1	B_2

Rules can often have more than two variables, in which case statement decomposition can be used to simplify the tabulation. For example, Table 13.2 shows the tabulation for three variables based on the following rule decomposition,

rule, IF A_i *AND / OR* B_j *AND / OR* C_k, THEN H_{ijk}, is decomposed to,

IF A_i *AND / OR* B_j, THEN H_{ij}

IF H_{ij} *AND / OR* C_k, THEN H_{ijk}

Table 13.2. Tabulation of Fuzzy Rules having Three Variables

A_1	H_{11}	H_{12}	C_1	H_{111}	H_{121}	H_{211}	H_{221}
A_2	H_{21}	H_{22}	C_2	H_{112}	H_{122}	H_{212}	H_{222}
	B_1	B_2		H_{11}	H_{12}	H_{21}	H_{22}

A fuzzy system having N inputs, one output and M membership functions at each input can have a total of M^N rules. This can be a very large number of possible rules, some of which may not contribute significantly to the problem solution. Therefore good judgement must be used to eliminate the unnecessary rules.

13.1.4 Fuzzy Logic Defuzzification

Defuzzification is the process where the relevant membership functions are sampled to determine the grade of memberships. There may be more than one output variable membership function chosen for a given set of inputs based on the results of the fuzzy logic rules. It is often the case that two adjoining membership functions for the output variable are chosen. The grade of memberships is used in the fuzzy logic equations and an outcome region is defined from which a crisp output is computed. There are at least three main techniques that can be used to produce the crisp output. These include taking the maximum of the chosen membership functions, a weighted average of possible outputs (chosen membership function centres of mass), or computing the centre of mass under a portion of the chosen membership function region. However, there are many other possible methods as well.

For example, let the membership functions in Figure 13.1 represent the fuzzy input variable and let the membership functions in Figure 13.2 represent the output variable for a very simple air blower controller. Figure 13.2 shows two possible output membership functions for the blower speed, "low fan speed" and "high fan speed."

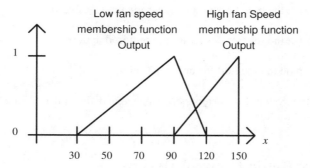

Figure 13.2. Membership Functions for Output to Airblower

If the input temperature was measured to be 58 degrees Celsius and the fuzzy logic rules were such that the selected input membership function was "hot" and the corresponding output membership function was selected to be "high fan speed" the defuzzified output is computed through the use of Figure 13.3. In this example there are two possible outputs, the output projected down from the "high fan speed" curve at the level projected across from the "hot" input membership function for an input temperature of 58 degrees, and the output projected down from the centre of mass of the portion of the "high fan speed" function below the projected line.

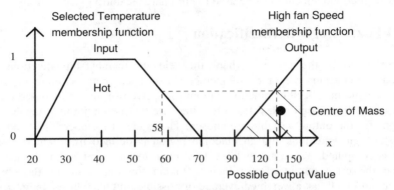

Figure 13.3. Defuzzification Example

In general a fuzzy logic problem is solved by applying the following sequence of Steps (Kartalpoulos 1996),

1. Define the details of the problem to be solved.

2. Determine all the relevant input and output variables and their ranges.

3. Establish the membership profiles for each variable range

4. Establish the fuzzy rules including required actions.

5. Determine the method for defuzzification.

6. Test the system for correct performance and finish if it is satisfactory, else go back to Step 3.

13.2 Fuzzy Logic Control Design

The FL control rules or fuzzy IF-THEN rules are often based on the knowledge and experience of human operators that have been involved in the operation of the system to be controlled. Given the control rules the control strategy is realised using fuzzy reasoning, which becomes the very structure of the FL controller. Each different possible method of fuzzy reasoning produces its own fuzzy controller. Fuzzy reasoning methods can be classified as direct or indirect but the most popular methods are the direct ones because they employ simple structures often based on min- and max- operations as described above. Indirect methods conduct reasoning by truth-value space but they are fairly complex. The most popular of the direct methods is Mamdani's direct method (Tanaka 1997).

Direct methods uses inference rules such as,

$$\text{IF } \quad x \text{ is } \tilde{A} \quad AND \quad y \text{ is } \tilde{B} \quad \text{THEN} \quad z \text{ is } \tilde{C}$$

where:

$\tilde{A}$, $\tilde{B}$ and $\tilde{C}$ are fuzzy sets.

This says that given the premise (x is $\tilde{A}$ AND y is $\tilde{B}$) then the consequence is (z is $\tilde{C}$), where x and y are premise variables and z is a consequence variable. A practical example of this might be IF the temperature x is a "little high" AND the humidity y is "quite high" THEN increase airconditioner setting z to "high." The fuzzy sets $\tilde{A}$ ="little high," $\tilde{B}$ ="quite high" and $\tilde{C}$ ="high" can be replaced with more explicitly fuzzy numbers such as, $\tilde{A}$ ="about 20 degrees," $\tilde{B}$ ="about 80% humidity" and $\tilde{C}$ ="about setting 9."

13.2.1 Fuzzy Logic Controllers

For Single-Input-Single-Output (SISO) fuzzy logic control it is necessary to construct the control rules by first writing down the operator's actions in the IF-THEN format and then it is also possible to add more from the response characteristics of the system. The procedure may go as follows (Tanaka 1997),

1. Construction of initial control rules and fuzzy sets.

2. Parameter tuning to determine the fuzzy sets.

3. Validation of the system.

4. If not satisfactory revise control rules and go to Step 2, else stop.

The choice of control rules also includes the identification of the parameters for the fuzzy sets, which sometimes may not be optimal. Parameter tuning involves tuning the parameters for the fuzzy sets used in the control rules to try to improve performance. If this process is unable to achieve satisfactory performance then it may require control rule revision.

13.2.1.1 Control Rule Construction

To construct the fuzzy control rules it is first necessary to select the input variables associated with required operating targets, which in turn determine the outputs. Conventional linear Proportional-Integral-Derivative (PID) controllers use deviation from a setpoint to construct input variables, i.e., variation (derivative) over a derivative time T_d and accumulation (integral) over an integral time T_i of this deviation. It is possible to construct a PD, PI and P, as well as the full PID controller depending on the design equation. For example a PD controller can be defined by Equation 13.3.

$$u = K_p(e + T_d\dot{e})$$ (13.3)

where:
  u is the position output.
  e is the deviation from the setpoint.
  $\dot{e}$ is the time derivative of e.
  K_p is the proportional gain.
  T_d is the differential time.

A PI controller can be defined by Equation 13.4.

$$\dot{u} = K_p(\dot{e} + \frac{1}{T_i}e)$$ (13.4)

where:
  $\dot{u}$ is the speed output.
  e is the deviation from the setpoint.
  $\dot{e}$ is the time derivative of e.
  K_p is the proportional gain.
  T_i is the integral time.

The general fuzzy rule for a fuzzy PI controller is a follows,

IF e is A AND $\dot{e}$ is B THEN $\dot{u}$ is C

For such a fuzzy PI controller the fuzzy input variables e and $\dot{e}$ are typically made to have about three positive sets, three negative sets and a zero as indicated in Figure 13.4

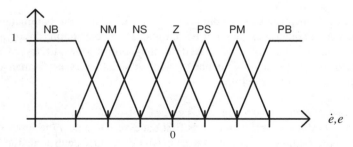

Figure 13.4. Membership Functions for Inputs e and $\dot{e}$ of Fuzzy PI Controller

The fuzzy PI controller may have a set of simple rules as follows,

IF	e is NB	AND	$\dot{e}$ is Z	THEN	$\dot{u}$ is NB
IF	e is NM	AND	$\dot{e}$ is Z	THEN	$\dot{u}$ is NM
IF	e is NS	AND	$\dot{e}$ is Z	THEN	$\dot{u}$ is NS
IF	e is Z	AND	$\dot{e}$ is Z	THEN	$\dot{u}$ is Z
IF	e is PB	AND	$\dot{e}$ is Z	THEN	$\dot{u}$ is PB
IF	e is PM	AND	$\dot{e}$ is Z	THEN	$\dot{u}$ is PM
IF	e is PS	AND	$\dot{e}$ is Z	THEN	$\dot{u}$ is PS
IF	e is NB	AND	$\dot{e}$ is NB	THEN	$\dot{u}$ is NB
IF	e is NM	AND	$\dot{e}$ is NM	THEN	$\dot{u}$ is NM
IF	e is NS	AND	$\dot{e}$ is NS	THEN	$\dot{u}$ is NS
IF	e is PB	AND	$\dot{e}$ is PB	THEN	$\dot{u}$ is PB
IF	e is PM	AND	$\dot{e}$ is PM	THEN	$\dot{u}$ is PM
IF	e is PS	AND	$\dot{e}$ is PS	THEN	$\dot{u}$ is PS

However, to make a better controller it may be useful to add more rules. Although, more rules may provide a better chance of achieving the control target, it is not recommended to have anymore rules than necessary. This is because fewer rules result in simpler parameter tuning. The extra following rules,

IF	e is PB	AND	$\dot{e}$ is NS	THEN	$\dot{u}$ is PM
IF	e is NB	AND	$\dot{e}$ is PS	THEN	$\dot{u}$ is NM

will improve the initial response. On the other hand, the extra following rules,

IF	e is PS	AND	$\dot{e}$ is NB	THEN	$\dot{u}$ is NM
IF	e is NS	AND	$\dot{e}$ is PB	THEN	$\dot{u}$ is PM

will dampen the overshoot.

13.2.1.2 Parameter Tuning

Once the control rules are determined then it is necessary to tune them. Parameter tuning affects the shape of the fuzzy sets, which are typically triangular, as shown in Figure 13.4, or exponential or trapezoidal. Triangular sets are much easier to work with and are therefore more popular. There are three progressively simpler ways that tuning may be performed as follows (Tanaka 1997),

1. Tuning of three parameters of each fuzzy set.

2. Tuning of only one parameter of each fuzzy set.

3. Tuning of scale factor for the total set.

Triangular sets are each described by three parameters, the two base points and the peak, which may each be moved left or right. Only the peak of the zero set stays fixed at zero but its base points may be moved. Type 1 tuning in Figure 13.4 would involve adjusting 18 of these points. Notice that the end sets *NB* and PB only have two parameters each since they are not really triangles. In the type 2 tuning there are only six parameters, the peak points for all except the zero set. For type 3 tuning there is only a single scale factor to adjust and is quite effective when there are a large number of variables to be tuned.

The performance of the example fuzzy PD controller can be assessed by applying a step response. The ideal response requires a fast rise time to the new setpoint with minimal overshoot. For such a controller, type 3 tuning, or adjusting the fuzzy set scale factor, can regulate its gain to some extent.

13.2.1.3 Control Rule Revision
If the fuzzy control rules are wrong or inappropriate then parameter tuning will have virtually no effect on performance. This may be remedied by adding necessary rules or deleting deleterious ones. However, if more rules are added than necessary there will be no further performance gain. Rules that are affecting the performance can be identified on-line by observing the control effect when they are called into play.

13.3 Fuzzy Artificial Neural Networks

Fuzzy systems and artificial neural networks, although very different in their basic principles of operation, are related in that they can both work with imprecise and noisy data spaces. Fuzzy systems are capable of taking fuzzy input data and providing crisp outputs according to a set of *a priori* fuzzy rules. This approach does not involve any learning as such. Artificial neural networks can also take fuzzy inputs and produce crisp outputs by training with known input-output pairs and there is no need to know the underlying rules. Both approaches have their benefits and shortcomings. Designing with fuzzy systems requires a thorough understanding of the fuzzy variables and membership functions of the desired input-output relationships as well as experience and knowledge of how to select the most significant fuzzy rules. On the other hand designing with artificial neural networks requires the development of a good sense for the problem through a large degree of experimentation and practice with the network complexity, learning algorithms, acceptable specifications and data collection and preprocessing. These shortcomings of both approaches can to some degree be reduced or even overcome by merging the two approaches. This can be done by either incorporating the learning and classification of neural networks into fuzzy systems or the logic operations of fuzzy systems into neural networks. The resulting hybrid system may be called a Fuzzy Artificial Neural Network (FANN).

In the FANN the neural network part is mainly used to automatically generate the fuzzy logic rules during the training period and for subsequent rule adaptation as the process changes. The fuzzy logic part is used to infer and provide the crisp or defuzzified output. Figure 13.5 shows a possible general structure for a FANN. The forward network is fuzzy and the feedback is a neural network. The neural network accepts both inputs and outputs from which it creates new classifications and input-output associations and thereby generates new rules. The new rules are transferred to the forward network at the appropriate times. There are many other possible FANN structures that may be used based on the designer's creativity.

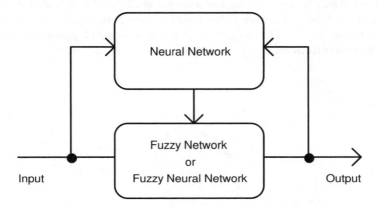

Figure 13.5. A Closed-loop FANN Control System

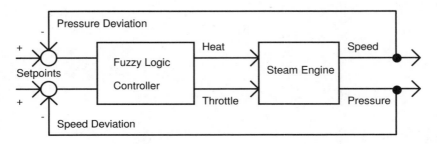

Figure 13.6. Fuzzy Controller for a Steam Engine

13.4 Fuzzy Applications

Although the application areas for fuzzy logic are still growing it has already been successfully applied in many commercial products. These include control systems related to trains, washing machines, cameras, televisions, vacuum cleaners, automobiles and communications systems. They have also been used for pattern recognition related to financial and commercial transactions, speech recognition, optical character recognition, person recognition via fingerprints, voice and face recognition, robotics, and many others. Fuzzy logic systems generally have application in the same type of areas as neural networks. Of course, the FANN is a useful hybrid of fuzzy logic and neural networks for application in the same areas.

Fuzzy logic control was first used to control a steam engine (Mamdani 1974). Mamdani regulated the outlet pressure and engine speed to constant values using the control setup shown in Figure 13.6. The input to the steam engine was the supplied heat to the boiler and the throttle of the engine. A linear PID controller was also developed to control the engine but it was less effective than the fuzzy controller because a steam engine has a nonlinear characteristic. Another problem with steam engines is that their characteristics change over time requiring control parameter tuning. The FLC system required less frequent parameter tuning to maintain good performance.

14. Introduction to Genetic Algorithms

Genetic Algorithms (GAs) represent a learning or adaptation method that is analogous to biological evolution according to Darwin's theory of evolution (Darwin 1859) and can be described as a kind of simulated evolution. Evolution based computational approaches have been investigated ever since the early days of computing in the 1950s. GAs can be seen as general optimisation methods. Although they are not guaranteed to find "optimum" solutions in the classical sense they often succeed in finding solutions meeting a high measure of fitness or acceptability. GAs have been applied successfully, not only to machine learning problems including function approximation and learning network topologies, but also to problems such as printed circuit board layout, work scheduling and many others.

GAs are often used to solve complex optimisation problems that are either very difficult or completely impractical to solve using other methods. In their most common form GAs work with hypotheses described by bit strings whose interpretation relates to the application. However, hypotheses may also be described by symbolic expressions, computer programs, specific model parameters, collections of rules etc. When the hypotheses are computer programs the evolutionary computing process is called Genetic Programming (GP).

GP is a method for automatically creating computer programs. It starts from a high-level statement of what needs to be done and uses the Darwinian principle of natural selection to breed a population of improving programs over many generations (Koza *et al* 2003). Given a collection or population of initial hypotheses the search for an acceptable hypothesis proceeds from one generation to the next by means of operations inspired by processes in biological evolution such as random mutation and crossover. A measure of "fitness" is required to evaluate the relative worth of the hypotheses in each generation. For each generation the most "fit" hypotheses are selected probabilistically as seeds for producing the next generation by mutating and then recombining their components. Thanks to John Holland's seminal work in 1975 (Holland 1995) the very powerful ideas of having populations and sexual recombination were introduced at that time.

The interest in GAs lies in the fact that evolution is known to be a successful and robust method for biological adaptation. However, evolutionary process tends to be very slow, due to the random mutations required, so there needs to be more justification for its use than just the fact that it works. However, as computer processing capacity increases this becomes much less of a concern since many generations can be progressed much more quickly. GAs using evolutionary process have been found to be very useful in applications where hypotheses contain complex interacting parts, where the impact of each part on overall hypothesis fitness may be difficult to understand or model. GAs can also take advantage of parallel computer hardware since they lend themselves to computational subdivision into parallel subparts.

A suitable introductory book on GAs is (Coley 1999).

14.1 A General Genetic Algorithm

Given a predefined qualitative fitness measure and a space of candidate hypotheses GAs are ultimately used to discover the best hypothesis according to that fitness measure. This process of discovery is actually an iterative process of choosing a set of the best current population of hypotheses, using this set to create another population of hypotheses to choose from, and so on for generation after generation until a sufficiently suitable hypothesis is arrived at. In an artificial neural network learning problem a set of hypotheses may be a set of artificial neural network models and the fitness measure may be the overall accuracy of a model in relating the input and output values of a given training data set. The problem may be one of choosing a suitable hypothesis that can play tic-tac-to. In this case the fitness factor may be the number of games won by an individual hypothesis when it plays many games against the others.

Implementations of GAs are of necessity different in detail depending on the problem, however, they do typically follow a similar overall structure. A population of hypotheses are iteratively updated generation after generation by probabilistically selecting the most fit individuals. Some of these most fit individuals are carried over into the next generation intact while the others are used as a basis for creating new offspring by applying genetic operations such as crossover and mutation. This structure is demonstrated in the following prototypical GA (Mitchell 1997).

Given the following variables,

1.   Fitness: the fitness measure.

2.   Acceptable_Fitness: the fitness termination threshold.

3. n: the number of hypotheses in the population.

4. r: the fraction of hypotheses to be replaced by crossover at each Step.

5. m: the mutation rate as a percentage.

Then,

1. Initialise a population P of hypotheses h_i, for $i = 1,...., n$, by random selection.

2. Evaluate the fitness function Fitness(h_i) for each of the n hypotheses h_i.

3. While the maximum Fitness(h_i) < Acceptable_Fitness create new generations P_s by repeating the Steps i. to v. below,

   i. Select: Probabilistically select $(1 - r)n$ members of P to add to P_s. The probability of selecting hypothesis h_i from P is Pr(h_i),

   $$\Pr(h_i) = \frac{\text{Fitness}(h_i)}{\sum\limits_{j=1}^{p} \text{Fitness}(h_j)}$$

   The probability of selection of a hypothesis is directly proportional to its fitness.

   ii. Crossover: Probabilistically select $(r.n)/2$ pairs of hypotheses from P, according to Pr(h_i). For each pair <h_1, h_2>, produce two offspring by applying the crossover operator. Add all the offspring to P_s.

   iii. Mutate: Choose m percent of the members of P_s with uniform probability. For each, invert one randomly selected bit in its representation.

   iv. Update: Update the population P to be P_s.

   v. Evaluate: Compute Fitness(h) for every h in P.

4) Return from P the hypothesis with the highest fitness.

14.2 The Common Hypothesis Representation

The hypotheses in GAs can be complex and of many forms. They may be sets of IF-THEN statements or they may be symbolic descriptions representing specific models, model parameters or even computer programs. These hypotheses are commonly reduced down to suitable bit string representations so that they can be conveniently manipulated by the genetic operators, which include the mutation and crossover operators.

IF-THEN rules can be represented by strings of bits by simply designing a logical rule encoding structure that allocates substrings for each rule precondition and postcondition. An IF-THEN rule is structured as follows,

IF (*precondition*) THEN (*postcondition*)

For example, if there is a rule structure related to controlling the temperature in a room temperature preconditions can be selected such as temperature hot or cold. The postconditions are, turn the air conditioning on or turn the air conditioning off. Therefore, the following IF-THEN rules apply,

IF (t*emp=hot*) THEN (*air on*)
IF (t*emp=cold*) THEN (*air off*)

These two IF-THEN rules can be represented with bits, respectively, as follows,

preconditions	*postcondition*
10	1
01	0

A logic bit sequence 10 represents the precondition *temperature = hot* and logic sequence 01 represents the precondition *temperature = cold*. A logic bit 1 represents the postcondition *turn air conditioner on* and logic bit 0 represents the postcondition *turn air conditioner off*. A precondition bit sequence of 00 would indicate that the temperature was neither hot nor cold and a precondition sequence of 11 would indicate that it did not matter whether it was hot or cold.

To make a more effective temperature control system it is necessary to introduce a heating element control and an extra precondition related to how hot or cold the temperature is before control is applied. To achieve this the IF-THEN rules can be modified as follows,

IF(*temp=hot*) ∧ (|(*temp*)-(desired temp)|≥*threshold*) THEN (*air on*) ∧ (*heat off*)
IF(*temp=hot*) ∧ (|(*temp*)-(desired temp)|<*threshold*) THEN (air *off*) ∧ (*heat off*)
IF(*temp=cold*) ∧ (|(*temp*)-(desired temp)|<*threshold*) THEN (*air off*) ∧ (*heat off*)
IF(*temp=cold*) ∧ (|(*temp*)-(desired temp)|≥*threshold*) THEN (*air off*) ∧ (*heat on*)

If a logic bit 1 represents the precondition |(*temperature*)-(*desired temperature*)| ≥ *threshold* and a logic bit 0 represents the precondition |(*temperature*)-(*point*)| < *threshold* these new rules can be represented by concatenating the extra precondition bit to the previous ones. An extra postconditioning bit related to the control of the heater element can also be concatenated. The four new If-THEN rules can again be represented with bits, respectively, as follows,

preconditions		*postconditions*	
10	1	1	0
10	0	0	0

```
01  0              0   0
01  1              0   1
```

If symbols are used, to represent hypotheses instead of bits, it is possible to have more generic encoding of hypotheses that are separate computer algorithms or even models etc..

14.3 Genetic Algorithm Operators

Given the hypotheses bit or symbol sequence encodings they are manipulated via genetic operators to produce successor generations. The genetic operators are idealised versions of biological genetic operations, including crossover and mutation.

The crossover operator produces two new offspring from two parent sequences by copying selected bits or symbols from each parent. The selected bits or symbols for crossover are identified by the use of a binary mask string. Some crossover operators include the single-point crossover, the two-point crossover, and the uniform crossover as illustrated in Figure 14.1.

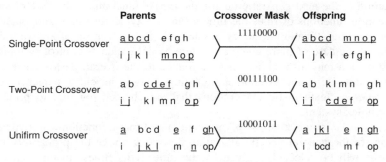

Figure 14.1. Cross-over Operators

The single-point crossover takes the first n symbols or bits from the first parent and the remaining symbols or bits from the second parent, according to the mask bits, to form the first offspring. The second offspring is formed using the same crossover mask but with a switch of parental roles. This results in the second offspring being formed from the remaining symbols or bits not used by the first offspring. In Figure 14.1 a single-point crossover is illustrated for $n = 4$. Every time the single-point crossover is used n is chosen at random, the mask is created and then applied as before.

In the two-point crossover, offspring are formed by substituting intermediate segments of one parent into the middle of the second parent. The crossover mask in this case is made up of n_0 zeros, followed by n_1 contiguous ones, followed by enough zeros to complete the mask string. The values for n_0 and n_1 are chosen randomly each time the operation is performed. In the example in Figure 14.1 $n_0 = 2$ and $n_1 = 4$.

Uniform crossover combines symbols or bits sampled uniformly from the two parents. The crossover mask in this case is generated by independent and random selection of each bit in the mask string.

The mutation operator, with point mutation, produces an offspring from a single parent. For example, in Figure 14.2, a single bit in the parent is chosen at random and negated to produce the offspring, or in the case of a symbol, changed to another randomly selected symbol.

	Parents	Crossover Mask	Offspring
	101 <u>1</u> 0100	_____	101 <u>0</u> 0100
Point Mutations			
	abc <u>d</u> efgh	_____	abc <u>m</u> efgh

Figure 14.2. Mutation Operator

14.4 Fitness Functions

Potential hypotheses in each generation are ranked and probabilistically chosen for propagation to the next generation using the fitness function. In the case of simple classification functions or rules the fitness function typically has a component that evaluates the classification accuracy of each rule for a given set of known training samples. In the case of more complex system models, such as control systems, the fitness function may also include criteria related to the overall performance of the system model. Here, the symbol or bit strings may represent a sequence of IF-THEN rules or mathematical operations that are chained together to produce a more complex system.

There are a number of ways of using the fitness function for hypothesis selection. In the general genetic algorithm described above the probability that a hypothesis will be selected is given by the ratio of its fitness to the sum of the fitness of other hypotheses. This is fitness proportionate selection, which is sometimes called roulette wheel selection. A more diverse population selection is achieved with what is called tournament selection. Here, two hypotheses are first selected at random from the population. The more fit of the two is selected according to some predefined probability p and the less fit according to probability $(1-p)$. Rank selection is another method. Here, the hypotheses are first sorted by fitness and then the probability of selection is based on the rank in the sorted list rather than on the fitness ratio.

14.5 Hypothesis Searching

The hypothesis search method used in GAs is a randomised beam search method to seek out a maximally fit hypothesis. It is a randomised, parallel, hill-climbing search for hypotheses that optimise a predefined fitness function. The method is not a smooth steady movement toward an optimum solution, but rather, one where offspring from generation to generation can be quite abrupt and different from the parents. Although this method is not prone to falling into something like local minima, characteristic of the gradient descent optimisation method, it does have other potential difficulties. One of these is called crowding.

Crowding is a problem that may occur when an individual is found that is much more highly fit than the others in the population. If fitness proportionate selection is used, then this individual may quickly reproduce such that copies of the individual and similar individuals take over a large fraction of the population. The effect of this is that population diversity is reduced, thus possibly slowing GA progress. This problem can be reduced by replacing fitness proportionate selection by tournament selection or rank selection. Another way to reduce crowding is to reduce the fitness of an individual by the presence of other similar individuals. A third way is to restrict the kinds of individuals allowed to recombine to form offspring. Allowing only the most similar individuals to recombine will encourage the formation of multiple clusters of similar individuals. If individuals are spatially distributed and only nearby individuals are allowed to recombine this will also produce multiple clusters.

14.6 Genetic Programming

Genetic programming is used to find optimal computer programs or engineering solutions that can be expressed in algorithmic form. This is a most significant form of GAs that has some relation to the adaptive and other computational intelligent systems. It is possible to use GP to optimise these types of systems and find engineering solutions that employ them. There are two main parts to GP, as for the other adaptive and computational intelligent systems, the human produced preparatory part that is specific to the problem to be solved and the problem independent executional steps. The real intelligent part of any of our current Artificial Intelligent (AI) solution processes is actually in the preparatory part. The special feature of GP is that it typically requires no more of a designer's intelligence, preparatory steps, than other methods, but it can produce remarkable and often surprising results. It can produce good solutions to problems on a routine basis that are sometimes a revelation to the designer and that entitles it to be called a systematic problem-solving method..

According to (Koza et al 2003) the five preparatory steps required of a human designer in GP are to establish the following things.

1. The set of independent variables of the problem, zero-argument functions and random constants for each branch of the required program.

2. The set of primitive functions for each branch of the required program.

3. The fitness measure needed to evaluate each program candidate.

4.   Certain parameters for controlling the GP run.

5.   A termination criterion and final program selection.

Doing the first two Steps essentially establishes the search space of the problem by specifying the primitive components that are used to create the programs and the ways they can be combined. Often this requires very little high level knowledge, only what components may be relevant to the required solution. For example to produce a solution for an electronic circuit it is only necessary to specify what range of elemental electronic components are relevant to the required solution and all the ways that they can be interconnected. The third Step embodies the high-level problem statement, what is to be achieved, by providing a criterion to guide the solution search in the desired direction and a way to rank potential candidates. Candidates are chosen, using the fitness measure, to participate in the various genetic operations such as, crossover, reproduction, mutation, and the architecture altering operations. The fitness measure may be multiobjective, in which case it is necessary to prioritise the different tradeoffs that may be relevant due to conflicting requirements. It is often convenient to blend the components of the fitness measure into a weighted single numerical value, like the Mean Square Error (MSE) that is used for tuning adaptive filters and ANNs. The major run control parameters of the fourth Step are population size and the number of generations to be run. Although theses can be analytically determined they are usually selected on a basis of how much computer time can be spent on the problem. Minor control parameters are usually selected based on experience with other similar problems. The final fifth Step simply requires a selection of what is an acceptable solution or failing that when to stop the process.

Although the preparatory steps change from one problem to another the main GP execution steps remain the same. Furthermore, it is often not a major transition from problem to problem in the same domain or even from domain to domain. The GP problem-solving approach is not based on a logically sound procedure yet it works so well, as well as being a general problem-solving method. It is interesting to note that neither logic nor determinism govern either the human inventive process or natural evolution yet they both produce logically consistent results. Of course, GP does require logical processes in the preparatory stages as does human inventiveness, but often the breakthrough happens when a seemingly illogical step is taken at some point in the process. The active maintenance of inconsistent and contradictory alternatives is a key to the success of GP as well as to natural evolution on which it is based. Furthermore, neither process is aware of established wisdom, which allows it to freely investigate possibilities that a human following convention may never consider.

14.7 Applications of Genetic Programming

According to (Koza *et al* 2003) genetic programming is now capable of routine human-competitive machine intelligence. In their book they show how GP is applied to a variety of problems. Examples are given where GP have created results that either infringe or duplicate the functionality of previously patented inventions such as the reinvention of negative feedback. The book describes fifteen instances where GP developments either infringe or duplicate the functionality of a previously patented 20th-century invention, six instances where it has done the same with respect to post-2000 patented inventions, two instances where GP has created a patentable new invention, and thirteen other human-competitive results.

In their book (Koza *et al* 2003) demonstrate the power and generality of genetic programming by solving problems from various fields including,

1.   Control.

2.   Analogue electric circuits.

3.   Placement and routing of circuits.

4.   Antennas.

5.   Genetic networks.

6.   Metabolic pathways.

7.   Synthesis of networks of chemical reactions.

In previous publications by Koza and other researchers GP is shown to solve many other problems in diverse and numerous areas including robotics. It is accurate to describe GP as a generic systematic problem solving method. In (Bräunl 2003) GP is applied to the solution of a walking gait for a legged robot, a problem for which there is no known deterministic algorithm.

14.7.1 Filter Circuit Design Application of GAs and GP

Koza *et al* (Koza *et al* 1996) developed a GA approach to designing electronic filter circuits. They developed a system which transform a simple fixed seed circuit into a final circuit design. The primitive functions used by the GP to construct its programs were functions that edit the seed circuit by inserting or deleting circuit components and wiring connections. The fitness of each offspring circuit was tested at 101 different input frequencies to determine how closely it conformed to the desired filter specification. The fitness measure was the sum of the 101 magnitudes of errors between the desired and actual filter outputs. A population size of 640,000 was maintained at each generation where offspring were produced in the proportion of 10% by selection, 89% by crossover, and 1% by mutation. The initial circuit selections were so unreasonable that it was not even possible to simulate the behaviour of 98% of the circuits produced. The percentage of unreasonable circuits dropped to 84.9% in the second generation, 75% in the third generation and to an average of 9.6% in the remaining generations. A very good final circuit, producing an output very similar to the desired response, was produced after only 137 generations.

14.7.2 Tic-tac-to Game Playing Application of GAs

Another application involves optimising a Multi-Layer Perceptron (MLP) artificial neural network to play tic-tac-to (Burns 1996). A MLP structure having nine inputs and one output was found that evaluates the best next move from any board position by presenting to it all the next possible moves from that position. The input vector $\mathbf{x}$ to the MLP had a dimension of nine and the elements were ether "-1," "+1," or "0" signifying a cross, a nought or an empty space respectively. The input vector elements were defined according to the vector $\mathbf{x} = [x_1, x_2, x_3, x_4, x_5, x_6, x_7, x_8, x_9]^T$ and the scheme depicted in Figure 14.3.

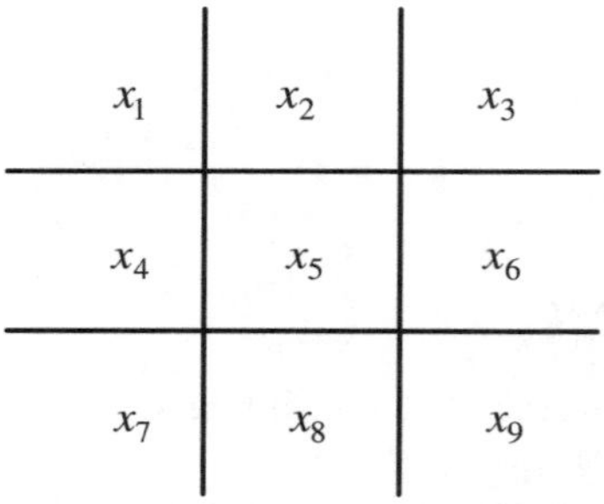

x_1	x_2	x_3
x_4	x_5	x_6
x_7	x_8	x_9

Figure 14.3. Input Vector for Tic-Tac-To

For example, the board position specified by Figure 14.4 is represented by the vector $\mathbf{x} = [-1, +1, -1, 0, +1, 0, 0, +1, -1]^T$.

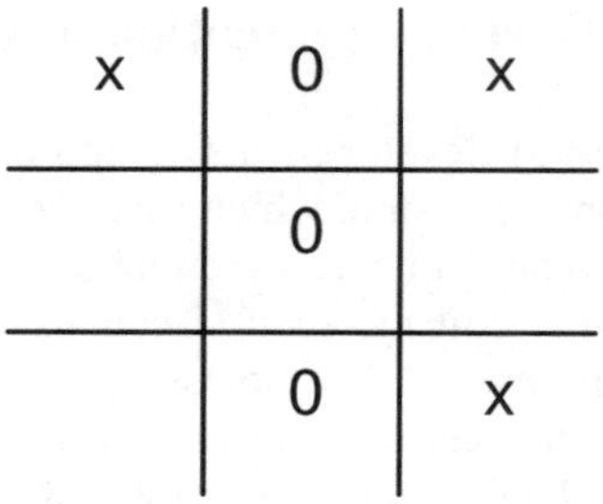

X	0	X
	0	
	0	X

Figure 14.4. Example Tic-Tac-To Board Position

There are three next possible moves for crosses from the position depicted in Figure 14.4 and these may be represented by the three vectors as follows,

$$\mathbf{x}_1 = [-1, +1, -1, -1, +1, 0, 0, +1, -1]^T$$

$$\mathbf{x}_2 = \left[-1,+1,-1,0,+1,-1,0,+1,-1\right]^T$$
$$\mathbf{x}_3 = \left[-1,+1,-1,0,+1,0,-1,+1,-1\right]^T$$

If these three input vectors are presented, one at a time, to a MLP having nine input nodes and one output node the next move is signified by the vector that gives the highest network output value.

The GA used in this application used a basic evolutionary strategy process as follows,

1.   Create an initial population.

2.   Repeat until satisfied Steps i, ii, and iii below,

    i.   Evaluate population performance.

    ii.   Select the best individuals to form a new population.

    iii.   Mutate to form the new population.

Mutation is normally seen as a method of compensating for the decrease in genetic material caused by selection and crossover. However, in this application it has been used as the primary search technique. Each generation is made similar to the previous one with only a small perturbation on average. A Gaussian mutation operator with a zero mean was used to ensure that large perturbations occurred less often than small ones. The mutation was performed by adding a random zero mean vector to the MLP weight vector (a vector which included all the network's weight values). The standard deviation of the Gaussian mutation operator was stored from generation to generation and itself mutated by a global Gaussian mutation operator. The global mutation operator standard deviation was set to 0.1.

The fitness function was based on the number of wins and losses of each MLP network at each generation playing against a random player. Network A was considered a better player than network B if A had more wins than B and A had fewer losses than B, accounting for draws. The GA system sorted the networks into three categories according to the degree of fitness: two copies, one copy, or no copy for the next generation. Within this restriction the numbers of individuals chosen for each new generation were kept approximately constant at twenty. From a population of twenty networks the top five had two offspring, ten had one offspring, and the remaining five had no offspring. Networks played a total of ten games against a random player per generation for a fixed total of 800 generations.

The result was that the GA improved MLP networks, from winning on average only 56.4% (±4.98%) of tic-tac-to games to an average of 93.6% (±1.57%) games. This application demonstrates that the GA approach is able to train feedforward networks without having training data available so long as the overall system performance can be measured in some way. A GA can be applied to optimise any set of model parameters as long as the fitness function of a given solution can be measured and it is "relatively" smooth.

PART VI. ADAPTIVE FILTER APPLICATION

This Part IV is involved with issues to do with the practical application of adaptive filters. The theory of adaptive filters is based on ideal notions, which in practice are often difficult to comply with. The main one being that of having access to the desired filter response signal. Consequently it is essential to show how adaptive filters can be applied in various ways that demonstrate how to either avoid strict theoretical necessities or how to achieve them in specific application contexts. A range of common adaptive filter applications are investigated to this end followed by descriptions of two generic adaptive filter structures that utilise multiple individual adaptive filters in their make up.

There are two generic adaptive filter structures, the Sub-Band Adaptive Filter (SBAF) and the Sub-Space Adaptive Filter (SSAF) that have considerable utility for solving practical problems by virtue of how they subdivide the data space. In the SBAF the data space is subdivided in the frequency domain and the SSAF subdivides the data vector space itself. The SBAF splits wide-band input signal spaces into independent equal bandwidth frequency sub-bands via a set of parallel band-pass filters. Each sub-band can then be down sampled without any information loss and processed separately by an adaptive FIR filter. Then it is transformed back up to its normal frequency range before reconstructing the total signal output as the sum of all the processed sub-bands. The main advantage of doing this is that the overall effective adaptive FIR filter length can be reduced with a consequential gain in speed of convergence.

The SSAF model (Zaknich 2003b) is derived from the Modified Probabilistic Neural Network (MPNN) (Zaknich 1998) and is similar to the MPNN extension called the Tuneable Approximate Piecewise Linear Regression (TAPLR) model (Zaknich and Attikiouzel 2000). The TAPLR model can be adjusted by a single smoothing parameter continuously from the best piecewise linear model in each vector sub-space to the best approximately piecewise linear model over the whole data space. A suitable value in between ensures that all neighbouring piecewise linear models merge together smoothly at their boundaries. The SSAF model was developed by altering the form of the MPNN, a radial basis function network initially developed for general nonlinear regression. The MPNN's special structure allows it to be used to model a process by appropriately weighting piecewise linear models associated with each of the network's radial basis functions as is done in the TAPLR. The SSAF extends this idea by allowing each piecewise linear model section to be adapted separately (separate parallel adaptive filters) as new data flows through it, thereby reducing a single complex nonlinear adaptation problem down to set of simpler ones in parallel with each other. The SSAF model represents a learning/filtering method for nonlinear processes that provides one solution to the stability-plasticity dilemma associated with nonlinear adaptive learning systems and standard adaptive filters.

15. Applications of Adaptive Signal Processing

The standard adaptive filter needs three fundamental signals, an input $x[k]$, an output signal $y[k]$ and the desired signal $d[k]$. In this standard context, the desired signal is used like a training signal to drive the filter to convergence, much like is done for feedforward neural network training. However, if the desired response is known, why then is the adaptive filter needed at all? In practice the desired signal is not usually known explicitly but it is often possible to derive or find a suitable signal to use which is strongly correlated to the desired signal. Furthermore, this correlated signal can usually be supplied to the adaptive filter in real-time along with the input signal, allowing it to function in a practically useful way. Since there is no general solution to this problem of determining a suitable desired signal it is necessary to study specific examples to gain insight into adaptive filtering practice. Seven common adaptive filtering applications worthy of consideration are,

1. Adaptive prediction.

2. Adaptive modelling and inverse modelling.

3. Adaptive echo cancelling.

4. Adaptive equalisation of communication channels.

5. Adaptive self-tuning filters.

6. Adaptive noise cancelling.

7. Adaptive array processing.

These seven applications are described in the following Sections, in turn, to show some of the main ways of configuring and solving important classical adaptive filtering problems.

15.1 Adaptive Prediction

Adaptive prediction is the process of estimating future signal samples $x[k+\delta]$ based on having a set of most recent samples $\{x[k], x[k-1],...., x[k-N]\}$. Wiener developed optimum linear least squares filtering techniques for signal linear prediction. When the signal's autocorrelation function is known, Wiener's theory yield's the impulse response of the optimum filter. The autocorrelation function can be determined using a correlator. Otherwise, the optimum prediction filter can be determined directly by adaptive filtering.

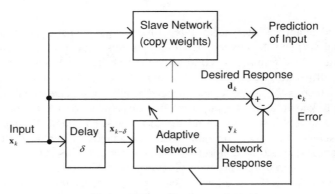

Figure 15.1. Adaptive Prediction

Figure 15.1 shows a typical adaptive prediction system arrangement. The input signal is delayed by δ time units and fed to an adaptive filter, where the undelayed input serves as the desired response. The filter weights continually adapt and maintain convergence to produce a best least squares estimate of the present input signal, given an input that is this very signal delayed by δ. These optimum weights are then periodically copied into a duplicate "slave filter" whose input is undelayed and whose output is therefore a best least squares prediction of the input δ time units into the future. Some important areas of application of adaptive prediction are in,

1.   Speech encoding.

2.   Data and image compression.

3.   Spectral estimation.

4.   Event detection.

5.   Line enhancement.

6.   Data transmission.

Often the purpose for prediction is not driven by an interest in knowing the next sample but rather it is way to achieve data compression. For example, short segments of human speech can be modelled as an autoregressive process. The vocal tract is like a concatenation of concentric uniform lossless tubes of varying width. This vocal tract model fits very well into a lattice filter structure. The reflection coefficients of the lattice filter and the pitch of the signal can be estimated from voiced speech segments by using a method based on linear prediction. Having a small collection of parameters including the reflection coefficients, pitch period and gain parameters the speech can be stored or transmitted very efficiently. The speech can then be reconstructed by exciting the appropriate all-pole lattice filter with white noise or a periodic impulse train, depending on whether the speech was voiced or unvoiced. This technique of speech processing is called Linear Predictive Coding (LPC)

Adaptive prediction is also used in real-time spectral estimation, based on fitting an autoregressive model to the data sequence. The optimum linear predictor coefficients are used to estimate the autoregressive model parameters by using an adaptive predictor. The adaptive predictor has the ability to track the time-varying statistics of nonstationary signals and thus produce better real-time spectral estimates.

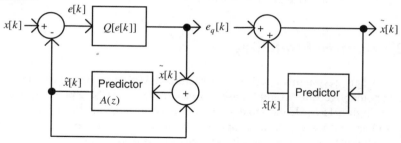

Figure 15.2. DPCM Encoder and Decoder

Differential Pulse Code Modulation (DPCM) is used to efficiently compress and transmit Pulse Code Modulated (PCM) speech signals. Because there is a strong correlation between successive samples it is possible to reduce the signal size by the DPCM system shown in Figure 15.2 (Zelniker and Taylor 1994). The encoder consists of a quantizer Q and a predictor, which is a feedback loop around the quantizer. The prediction signal $\hat{x}[k]$ is generated as a linear combination of the N previous samples of the signal $\tilde{x}[k]$, and the prediction error $e[k]$ is defined by Equation 15.1.

$$e[k] = x[k] - \hat{x}[k] = x[k] - \sum_{m=1}^{N} a_m \tilde{x}[k-m] \tag{15.1}$$

The signal $\tilde{x}[k]$ is defined by $\tilde{x}[k] = \hat{x}[k] + e_q[k]$, where $e_q[k]$ is the quantization prediction error. This error differs from the true prediction error by the quantization error, i.e., $e_q[k] = Q[e[k]] = e[k] - n_q[k]$, where $n_q[k]$ is the quantization error.

The quantization prediction error $e_q[k]$ is then the signal that is transmitted. The predictor is in a loop around the quantizer to avoid accumulation of quantization errors in the decoder, i.e., $n_q[k] = e_q[k] - e[k]$, where $e[k] = x[k] - \hat{x}[k]$, therefore $n_q[k] = e_q[k] + \hat{x}[k] - x[k]$. From the decoder system it is evident that $e_q[k] + \hat{x}[k] = \tilde{x}[k]$, thus it follows that $\hat{x}[k] - x[k] = n_q[k]$. The difference between the predictor input and the input sample is due only to the instantaneous quantization error and does not accumulate. At the receiver end the signal $\tilde{x}[k]$ is reconstructed by the decoder, which has a predictor that is identical to the predictor in the encoder. If the encoder and decoder predictors are started from the same initial conditions the only error in the reconstructed signal will be the unavoidable quantization error.

The data compression of DPCM results from the fact that if the speech signal is highly correlated, the prediction error will have a smaller dynamic range than the signal itself. Consequently, the prediction error can be coded with fewer bits per sample. Adaptive Differential Pulse Code Modulation (ADPCM) replaces the fixed predictor in the DPCM system with an adaptive predictor that is capable of adapting to time-varying input signal statistics.

15.2 Adaptive Modelling

In cases when a system of unknown structure has observable input and output signals an adaptive filter could be used to model the system's impulse response as shown in Figure 15.3. If the input signal is robust in frequency content and if the internal plant noise is small, the adaptive filter will adapt to become a good model of the unknown system. This is known as forward modelling, or system identification, and has applications in,

    1.   Biological, social, economic sciences.

    2.   Adaptive control systems.

    3.   Digital filter design.

    4.   Coherence estimation.

    5.   Geophysics.

Inverse modelling involves developing a filter that is the inverse of the unknown system, as shown in Figure 15.4. The delay block is usually included to account for the propagation delay through the plant and the adaptive processor, assuming that both are casual systems. Inverse modelling is used in,

1.   Adaptive control.

2.   Speech analysis.

3.   Channel equalisation.

4.   Deconvolution.

5.   Digital filter design.

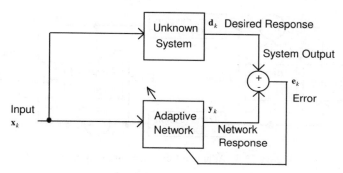

Figure 15.3. System Modelling

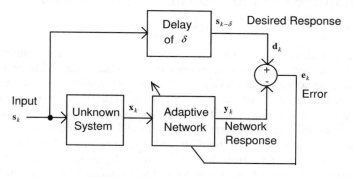

Figure 15.4. Inverse System Modelling

15.3 Adaptive Telephone Echo Cancelling

In long distance telephone circuits echo is natural because of amplification in both directions and series coupling of telephone transmitters and receivers at both ends. Previously, echo suppressors were used to prevent this by giving one-way communication to the party speaking first. To avoid the resulting switching effects and to permit simultaneous two-way transmission of voice and data, adaptive echo cancellers are now used instead of suppressors. Separate circuits are used in each transmission direction. Often hybrid transformers are used to prevent incoming signals from coupling through the telephone set and passing as outgoing signals. However, the hybrid transformers are routinely balanced for the average local circuit so they cannot do their job perfectly for any specific circuit with its own unique path length and electrical characteristics. An adaptive filter is therefore used to cancel any incoming signal that might leak through the hybrid transformer, causing echo. Figure 15.5 shows a typical arrangement. The adaptive filters at each end adapt to the outgoing signal so that any incoming echo signal that is similar to the original outgoing signal is subtracted from the transmitted signal.

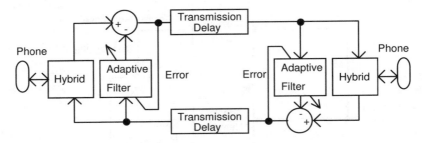

Figure 15.5. Adaptive Echo Cancelling System

15.4 Adaptive Equalisation of Communication Channels

Communication channels can have nonflat frequency responses and nonlinear phase responses in the signal passband. Consequently, a communication channel is often modelled as a linear and time-varying filter whose characteristics are not explicitly known. Sending digital data at high speed through physical communication channels can result in Intersymbol Interference (ISI) caused by channel noise and signal pulse smearing in the dispersive channel medium. An equaliser is a system that essentially reverses the effects of the ISI and therefore aids in the detection process at the receiver. For channels that are stationary or slowly varying in time, the equaliser is implemented as a linear filter. The equaliser is often an integral part of a modem system, which includes a linear adaptive filter. A modem's adaptive filter can adapt itself to become the channel's inverse by using decision-directed learning as shown in Figure 15.6 (Widrow and Winter 1988). Since without equalisation telephone channels can still provide Bit Error Rates (BERs) of 10^{-1} or less, the quantized binary output can therefore be used as the effective desired response for training. When the adaptive channel equalisation filter eventually converges the bit error rate will typically reduce to 10^{-6} or less.

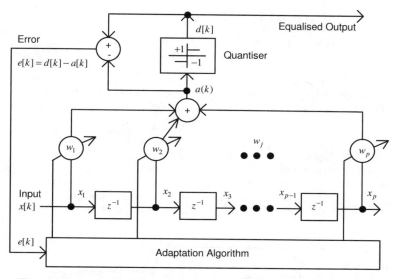

Figure 15.6. Adaptive Channel Equaliser with Decision-directed Learning

Another equaliser arrangement for noisy communication channels is described by Figure 15.7. The equaliser $E(z)$ must reverse the effects of the channel distortion represented by the channel transfer function $H(z)$ and the additive noise $n[k]$. The purpose of the equaliser is to process the receiver signal $y[k]$ and produce as good an estimate of the transmitted signal $x[k]$ as possible. To adapt the equaliser $E(z)$ it is fed with regular sequences of a known (training) input signal for a short period of time every so often.

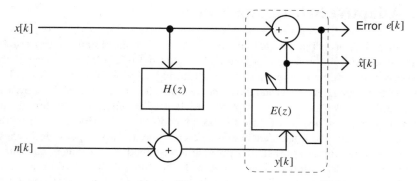

Figure 15.7. Adaptive Equaliser

When the noise is white with a variance of σ_n^2 the Wiener solution of the equalisation problem is defined by Equation 15.2 (Zelniker and Taylor 1994).

$$E(z) = \frac{\sigma_x^2 H(z^{-1})}{H(z)H(z^{-1})\sigma_x^2 + \sigma_n^2} \qquad (15.2)$$

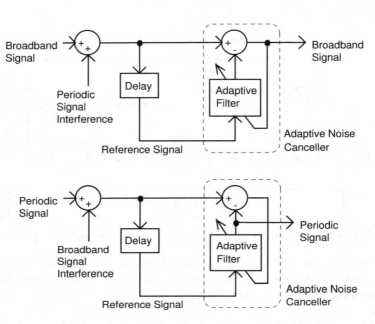

Figure 15.8. Self-tuning Adaptive Filter Arrangements

15.5 Adaptive Self-tuning Filters

Where there is a broadband signal corrupted by periodic interference or a periodic signal corrupted by a broadband signal and no external reference input free of the signal it is still possible to apply an adaptive filtering solution (Widrow *et al* 1975). It can be done by applying an adaptive noise canceller in arrangements shown in Figure 15.8 to solve these types of problems. If it is assumed that the broadband noise is random it is possible to use the broadband plus the periodic signal as the reference signal for the noise canceller's reference input. This is possible because only the periodic signal content will provide a sustained signal that is correlated with the noise (the periodic signal) that is to be filtered out. The removal of narrowband noise from a broadband signal is sometimes called line enhancement.

15.6 Adaptive Noise Cancelling

A major problem in ECG measurement is the appearance of mains hum at a frequency 50 or 60Hz, i.e., power-line interference due to magnetic induction, displacement currents in leads or in the body of the patient and equipment interconnections and imperfections. This problem can be minimised through use of proper equipment grounding and use of twisted pair leads. Another method is to use adaptive noise cancellation as shown in Figure 15.9 (Widrow *et al* 1975).

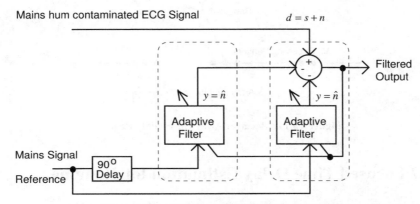

Figure 15.9. Adaptive Mains Cancelling System

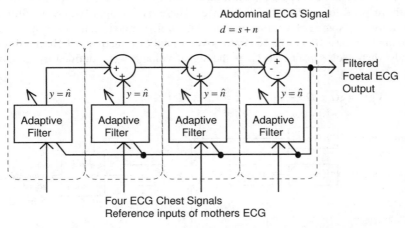

Figure 15.10. Adaptive Maternal Heartbeat Cancelling System

Another ECG measurement problem involves the cancellation of the maternal heartbeat in the measurement of the foetal ECG. This is a difficult filtering problem because the mother's heartbeat is correlated with the foetal heartbeat. The second harmonic frequency of the maternal ECG is close to the fundamental of the foetal ECG. However, the interference of the mother's heartbeat can still be cancelled from the foetal heartbeat through the use of adaptive filtering as shown in Figure 15.10. The abdominal signal has the foetal plus the mother's ECG signals, where the mother's signal is considerably stronger. Consequently, a number of ECG signals are taken from different locations on the chest to act as reference signals for the adaptive filters.

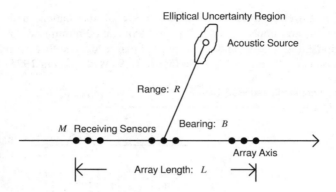

Figure 15.11. Receiver Array Uncertainty Region

15.7 Focused Time Delay Estimation for Ranging

An underwater acoustic point source radiating energy to several collinear receivers (Figure 15.11) can be located in two-dimensional space by a range R and a bearing B with respect to a frame of reference. There is an approximately elliptical uncertainty in source location due to uncoupled range and bearing errors. Assuming there are M sensors separated over a total of L metres and observed for T seconds, then each sensor i receives the signal voltage $V_i(t)$ defined by Equation 15.3.

$$V_i(t) = s(t + D_i) + n_i(t), \quad i = 1,...., M, \ 0 \le t \le T \tag{15.3}$$

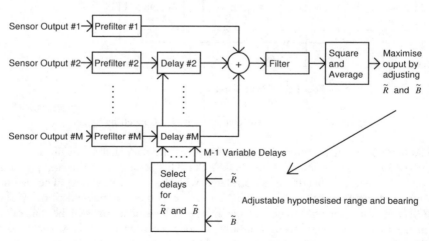

Figure 15.12. Beam Array Focusing System

The signal and noises are uncorrelated and the noises are mutually uncorrelated. For a spatially stationary nonmoving source the signal can be viewed as an attenuated and delayed source signal. Figure 15.12 shows how, by focusing all the time-delay elements at many (hypothesised) range and bearing pairs and watching for the peak output of the Maximum Likelihood (ML) time-delay vector system, the ML position estimate is observed. This is a focused beamformer, which maximises a quantity by adjusting a number of delay parameters such that all delays must intersect in a single hypothesised position. Each receiver input is prefiltered to accentuate a high SNR then delayed and summed. The summed signal is fed to a filter, squared and averaged over the observation time. After an initial input of range and bearing estimate the system can then track the source automatically.

The variances of the range and bearing estimates, $\sigma_A^2(\hat{R})$ and $\sigma_A^2(\hat{B})$, for high SNR are defined by Equations 15.4 and 15.5 respectively.

$$\sigma_A^2(\hat{R}) \cong \frac{K_R R^4}{TMVL_e^4} \quad \text{rads}^2 \tag{15.4}$$

$$\sigma_A^2(\hat{B}) \cong \frac{K_B}{TMVL_e^2} \quad \text{rads}^2 \tag{15.5}$$

where:
  $L_e = L\sin(B)$ the effective array length.
  K_R, K_B are array type constants.
  M is the minimum number of sensors.

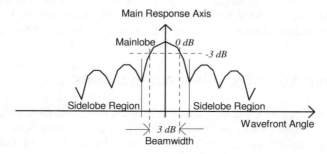

Figure 15.13. Array Beampattern

Array length is a more important factor in bearing estimation than either integration time or the number of sensors. Variance of the range is limited by the range relative to the baseline. The general beamformer response pattern is shown in Figure 15.13, where there is a main response lobe flanked by diminishing side lobes as a function of the wavefront angle.

15.7.1 Adaptive Array Processing

The delay-and-sum beamformer shown in Figure 15.12 is really a primitive type of spatial filtering that is unable to deal with sources of interference (Stergiopoulos

2001). This can be remedied by making the system adaptive in such a way that it places nulls in the directions of the interfering sources. By doing this the system's output Signal to Noise Ratio (SNR) is increased, thereby improving the directional response of the beamformer.

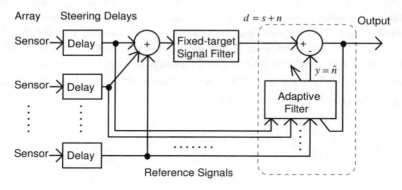

Figure 15.14. Adaptive Array Filter

The adaptive array processor shown in Figure 15.14 is now a type of multiple input interference canceller. The steering delays are used to form a beam and produce a peak array gain in a desired look direction. The noisy target signal $s_k + n_k$ is obtained through a fixed filter. An estimate of the noise $\hat{n}_k$ is obtained through the multiple-input adaptive processor and is used to cancel n_k. Before the first summing junction the noise signals from each transducer are likely to be high in relation to the desired signal and therefore can be used as noise reference signals.

15.8 Other Adaptive Filter Applications

There are a number of good books that have been published of late that cover a range of typical applications of adaptive filtering. These include (Garas 2000) on "Adaptive 3D sound systems," (Brandstein and Ward 2001) on "Microphone arrays," (Benesty *et al* 2001) on "Advances in network and acoustic echo cancellation" and (Benesty and Huang 2003) on "Adaptive signal processing: Applications to real-world problems." Following are short overviews of these four books with emphasis on special features.

15.8.1 Adaptive 3-D Sound Systems

The book (Garas 2000) is an initial investigation into the application of adaptive filters in creating robust virtual sound images through loudspeakers in real-time. A virtual three-dimensional sound image is one that is made to apparently appear to be at a certain point in three-dimensional space where no loudspeaker exists. This is achieved by processing a monophonic sound signal by a matrix of digital filters containing directional cues, which the ears interpret in such a way as to give the impression that the signal is emanating from a specific location. Adaptive filters are most suited for this type of application because they have two significant properties. Firstly, they have a tracking capability that can be used to track moving listeners by the appropriate adjustment of their coefficients. The second property is related to *in-situ* design of filters, which allows the possibility of including the listeners' own Head-Related Transfer Functions (HRTF) in the filter design.

Whereas stereo and conventional 3-D sound systems are designed to be optimal for a specific area of the listening space, the multichannel systems developed in the book make no assumptions regarding the number of listeners nor their positions in standard reverberant listening spaces. This makes the work generally applicable to many applications related to the listening of multiple listeners to multiple audio signals through multiple speakers. Instead of using a fixed matrix filter design, as typical of modern 3-D systems, to invert the matrix of acoustic transfer functions between the loudspeakers and listener's ears and produce the directional information, the fixed filters are replace by adaptive filters. Peoples HRTFs vary significantly among individuals so the average HRTF designed for general use is not optimal for all listeners. Furthermore the HRTF is only valid for a single listening position anyway. However, if microphones are placed near the entrance of the listener's ear canals adaptive filters can deal with their HRTFs directly in real-time independent of head movement and location.

The real-time 3-D systems described in the book are implemented in the frequency domain to reduce the inevitably huge computational burden. The standard LMS adaptation algorithm is replaced with the more efficient adjoint LMS algorithm. Still, for practical application, measures need to be taken to improve the adaptive filter convergence speeds to keep up with listener head movements. Because of the complexity of this application most of the analysis results are gathered from simulation experiments, but this does not detract from value of the work as a good introduction and identification of the important issues toward practical implementations.

A fairy good up-to-date introduction to speech and audio processing is (Gold and Morgan 2000). It covers many aspects of processing and perception of both speech and music and as such offers significant information to the designer of real-time adaptive 3-D audio systems.

15.8.2 Microphone arrays

Although the study and implementation of microphone arrays began over 20 years ago the book (Brandstein and Ward 2001) is one of the first to attempt to provide a single complete reference on this now relatively mature field. It is broken up into four parts; theory of speech enhancement, theory of source localisation, array-based technology applications, and discussion of open questions and future issues.

If speech is acquired by an array of microphones the speaker can freely roam within a room environment and it is still possible to maintain the speech quality against background noise, interference sources and reverberation effects. This arrangement is superficially similar to narrow band conventional array processing as used for radar and sonar, since in both cases the sources and noises are spatially separated. However, in the speech application the signal has an extremely wide bandwidth relative to its centre frequency and there is very significant multipath interference due to room resonance. The most important difference is that in the speech case the speech source and noise signals are located much closer to the array, invalidating conventional far-field assumptions. These significant differences have required the development of new array techniques for such microphone array applications. Some of these new methods incorporate adaptive techniques.

The localisation and tracking of a speech source is the fundamental requirement of microphone array systems. An important application for this is for camera aiming in video-conferencing systems as well as for speech source enhancement against a background of random and coherent noise and other interfering talkers. A number of practical approaches are reviewed and developed on ways to solve this problem, including combining audio and video information to track the motion of a talker.

Some specific microphone array systems available today include the two-dimensional harmonic array installed in the main auditorium of Bell Laboratories, Murry Hill and the 512-element Huge Microphone Array (HMA) developed at Brown University. Large arrays are very effective but arrays consisting of only two to eight microphones over a space of centimetres are much more common and affordable. These smaller systems are used more for close-talking under low to moderate noise conditions for dictating at a workstation or using hands-free telephones. Array techniques for background noise and interference cancelling are also used very effectively to aid the hearing impaired as well as for sound capture in automobiles. A new application for microphone arrays that is fundamentally different to spatial filtering approaches, is the separation of blind mixtures of acoustic signals recorded at a microphone array.

The book has summaries of currently open problems in the field and personal expert views on future trends, offered from both academic research as well as industry perspectives. Specific issues are related to hands-free communication, automotive, desktop, hearing aids teleconferencing, very large arrays and signal sub-space approaches.

15.8.3 Network and Acoustic Echo Cancellation

The hybrid devices used, for many decades, to connect two-wire local and four-wire long distance telephone lines have contributed to the echo problems of the past. This problem was first addressed in the 1960s by using an adaptive filter for echo cancellation. In more recent times of hands-free teleconferencing other echoes appeared due to the coupling between loudspeaker and microphone, which were solved using adaptive echo cancelling. This has also led to multichannel echo cancellation in more recent times.

The book (Benesty et al 2001) covers the subject of adaptive echo cancelling with an emphasis on new ideas to what might be considered an old problem. It is aimed at researchers and developers as well as students. A history of echo cancellation is provided to give the relevant background to the problem followed by more recent developments. These include the class Normalised Least Mean Squares (NLMS) adaptive algorithms, a robust fast recursive least-squares adaptive algorithm, efficient implementation of echo cancellers for a large number of simultaneous channels, telecommunication applications, a Fast Normalised Cross-Correlation (FNCC) method for double-talk detection, a practical stereo conferencing system, a new frequency domain adaptive filtering theory, a frequency domain system for double-talk and echo cancellation, and a theory for the development of a generalised least mean squares algorithm and generalised affine projection algorithm.

15.8.4 Real-world Adaptive Filtering Applications

The book (Benesty and Huang 2003) offers a reference to the latest real-world applications where adaptive filtering techniques play an important role. The subject matter covers applications in acoustics, speech, wireless, and the currently open area of networking.

The specific topics covered are, new directions in adaptive filtering for sparse impulse responses, approaches to feedback cancellation, introduction to single channel acoustic echo cancellation, a new general class of algorithms for multichannel adaptive filtering, noise filtering in speech communications, adaptive beamforming for speech and audio signal acquisition, blind source separation of convolved mixtures of acoustic signals, multichannel time delay estimation, classic adaptive equaliser techniques, adaptive space-time processing for wireless receivers in CDMA networks, an IEEE 802.11 wireless local area network system with multiple receive antennas, and a least square estimate of the difference between the sender and receiver clock frequencies and the fixed delay in the Internet network.

16. Generic Adaptive Filter Structures

This Chapter describes two generic adaptive filter structures, the Sub-Band Adaptive Filter (SBAF) and the Sub-Space Adaptive Filter (SSAF) (Zaknich 2003b). The SBAF is a method of breaking a wide-band input signal into equal bandwidth frequency sub-bands via a set of parallel band-pass filters. By doing this a reduction of the effective FIR filter length can be achieved with a consequential gain in speed of convergence. Breaking the input signal into separate frequency bands allows the complex problem to be broken down into parallel simpler sub-problems that together still solve the problem but with less overall computational burden.

The same general idea can be applied to the input signal space, for nonlinear problems, by breaking the problem into a set of simpler parallel linear problems with consequential processing gains. The SSAF does this by applying a set of decoupled smoothly merged parallel linear adaptive filters that each cover separate regions of the input space. At any given time only one or a small set of neighbouring adaptive filters are active in a part of the overall data space. The region sizes are each made to be small enough for adequate coverage by a single linear adaptive filter. In this way it obviates the need to cover the whole input data space with a single more complex nonlinear adaptive filter.

16.1 Sub-band Adaptive Filters

There are two main problems with LMS type adaptive FIR acoustic echo cancellation systems that work with wide bandwidth signals. Firstly, whether within rooms or underwater acoustic environments the echo delays are relatively long requiring a very large number of FIR adaptive filter coefficients. The long impulse response of the resulting FIR filter and the large eigenvalues of signals (in particular speech signals) result in a fairly slow and uneven rate of adaptive convergence. These problems can be minimised through the use of a sub-band echo cancellation system (Vaseghi 1996).

A sub-band system splits the wide-band input signal into say N equal bandwidth frequency sub-bands via a set of parallel band-pass filters. Each sub-band can then be down sampled without any information loss. For example, if the current sampling rate is F_s samples per second and there are N equal sub-bands it is possible to down sample each sub-band to a sampling rate of D samples per second, where, $\frac{F_s}{N} < D < F_s$. This is done by frequency shifting each sub-band down to the base band by a factor of $R = \frac{F_s}{D}$ and resampling at rate D. Each sub-band is then processed separately (or in parallel) in the baseband at a sampling rate of D and finally transformed back to its normal frequency range and back to the original sampling rate F_s before reconstructing the total signal output as the sum of all N processed sub-bands. The main advantage of doing this is that the FIR filter length is reduced, along with a consequential gain in speed of convergence.

An outline of the sub-band acoustic echo cancellation system is shown in Figure 16.1. The input signal $x[k]$ and the echo path output signal $y[k]$ are passed through the same analysis filter designs to produce N separate sub-band signals, each down sampled by a factor or R. The wide lines in Figure 16.1 show the signal paths for the N multiple sub-band signals. The sub-band filter outputs can be contained in the N-dimensional vector $\hat{\mathbf{y}}[k]$ which forms an estimate of the sub-band echo signals vector $\mathbf{y}[k]$ producing the resulting sub-band error signals vector $\mathbf{e}[k]$. These vector signals are used by the adaptation algorithm to adjust each set of sub-band filter coefficients. The echo cancellation progresses in the usual way as the system attempts to continuously drive the sub-band error signals $\mathbf{e}[k]$ to zero.

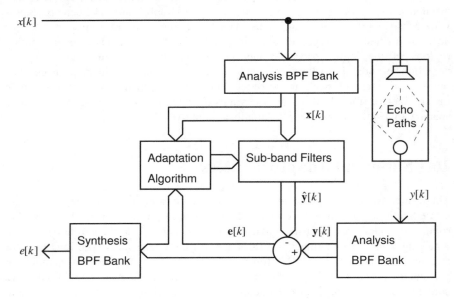

Figure 16.1. Sub-band Adaptive Filter

The impulse response of each sub-band FIR filter can have the same time duration as the original full band filter but the tap length of each sub-band filter is $\frac{1}{R}$ of the full band filter. Obviously, each LMS sub-band FIR adaptive filter will require much less computation because it will have a much smaller product of the filter length and sampling rate. The computational complexity of the sub-band filters will be $\frac{1}{R^2}$ of the full band filter. Hence, the overall computational complexity of the total sub-band system will be $\frac{N}{R^2}$ of the full band system. Convergence speed is inversely proportional to both filter length and the eigenvalue spread of the autocorrelation of the input signal. However, the eigenvalue spread is the more significant factor so it is desirable to have a more flat frequency spectrum (small eigenvalue spread) for fast convergence. The signal within each sub-band is likely to have a flatter spectrum than the full band signal and so this would in theory tend to help convergence. However, in practice the eigenvalue spread is actually larger for sub-band signals than for full band signals. This is because the slopes of the band edges of the analysis bandpass filters cause spectra notches there, thus creating some very small eigenvalues. Nevertheless, the improvement provided by the sub-band structure as a whole is due to the fact that the errors at the band edges get little weight due to the attenuating filter characteristics.

The benefit of the sub-band approach is that a more complex problem is broken down into a set of smaller much less complex problems. This gives an overall computational and speed of convergence gain with only a relatively small administrative overhead of splitting and recombining parts through analysis and synthesis parallel bandpass filter banks respectively. These extra filtering operations do introduce an extra delay in the signal path that can sometimes be problematic (Benesty and Huang 2003). However one way of eliminating the delay is to compute the adaptive weights in the sub-bands and then transform them to an equivalent full band FIR filter (Morgan and Thi 1995). The reduction in computational complexity can be exploited in a number of ways (Benesty et al 2001). Either the overall system bandwidth or the duration of the impulse response to be modelled can be increased. Computational hardware can be reduced to save costs or alternatively a more complex adaptive algorithm can be employed. The sub-band structure also allows for a more efficient parallel processing.

To ensure that sub-band processing will work correctly it is necessary to use bandpass filters for the analysis and synthesis parts with very good pass and stop band characteristics (Bremaud 2002). Obviously since it is not possible to design bandpass filters with perfect zero stopband characteristics, misalignment errors are inevitable. There is misalignment due to the residual aliasing within the sub-band signals and misalignment due to the inevitable truncation of the sub-band filter's impulse responses by the required sharp bandpass filter characteristics. Another requirement for satisfactory operation is that the down sampling factor R must be chosen such that the passband and transition region of the modulated versions of the analysis bandpass filters do not overlap in the frequency domain after down sampling.

The echo canceller in each sub-band is essentially independent of the others so it is possible to use any of the usual adaptive algorithms in each. The most common algorithm used is the Normalised LMS algorithm because it has been determined that its convergence is faster in the sub-band canceller than in the full band implementation. Each canceller is not only independent in its operation but it also acts on separate frequency regions.

A generic extension of this basic structure could be to define sub-spaces rather than frequency sub-bands and have separate adaptive algorithms operating in each. This can be an especially useful thing to do for a nonlinear process, where the process can be broken up into independent approximately linear sub-space regions. This new structure could then weld the sub-spaces together seamlessly as does the sub-band structure. Using a set of linear adaptive algorithms to cover a processing space may provide a processing advantage over trying to deal with the whole process with a single less efficient nonlinear adaptive filter like a high order Volterra filter. This structure might be called a sub-space adaptive filter. Furthermore, if necessary, it would also be possible to create a separate sub-band structure within each sub-space to gain even further processing advantage for such a nonlinear process problem.

16.2 Sub-space Adaptive Filters

Single standard linear or nonlinear adaptive filters have no specific mechanism built into them that prevents old learning to be progressively lost as new data flows through them. If fact, these standard adaptive filters are deliberately designed to readapt to new operating states as they occur. However this can pose a problem if on occasions large state changes are occurring faster than the system can adequately readapt to them. For these special circumstances, it would be very desirable to have an efficient adaptive learning system that retains all the features and benefits of the standard more flexible adaptive system but that is also able to retain a useful degree of old learning. In a learning context this is known as the stability-plasticity dilemma. As the process switches to a new distant state very quickly the system can immediately start adapting from the closest learned state available from previous experiences and continue tracking as quickly as possible. For stationary processes the system would eventually build a complete model of the whole process as it experiences the full range of operating states. Continued adaptation would then only occur as a result of noise fluctuations. For nonstationary processes the system would eventually build up a set of learned starting points available for each major adaptation state change, which may be better than an initial zero or random start. Any learned states that are not revisited for a long time may be removed routinely to save memory or simply retained if memory is not a problem.

A system as described above can be achieved by making a suitable extension to the Tuneable Approximate Piecewise Linear Regression (TAPLR) model (Zaknich and Attikiouzel 2000), which is based on the Modified Probabilistic Neural Network (MPNN) structure (Zaknich 1998) that is typically used for nonlinear regression. The MPNN structure has a set of amplitude weighted Radial Basis Functions (RBFs), each having a common bandwidth, that cover the input data space. It is possible to associate a separate adaptive linear filter model with each RBF and to only adapt the respective filter within the sphere of influence of that RBF at any given point in time. In the standard MPNN, used for regression, the centres of each RBF in the input space are typically computed as the mean of all the input training vectors within a local vicinity of the RBF. Each RBF in the network has a scalar amplitude weight Z_i associated with it that typically represents the density of training points (points used to compute its centre) within each RBFs sphere of influence. Each RBF centre vector is then associated with a suitable desired output scalar and the whole structure builds up a generalisation between the vector input space and scalar outputs. In the proposed multiple Sub-Space Adaptive Filter (SSAF) model vector points used to develop the RBFs are not necessarily directly related to the filter's input vectors and their associated desired filter outputs. The adaptation mechanism can be controlled by any measurable state vector meaningful to the filtering process. Also, the weighting factor Z_i can be used for a number of different possible purposes depending on the needs of the design. It can represent training density as before or represent *a priori* probability of occurrence of data in its RBFs region for filter operation. In the present adaptive filtering context, the most useful thing to do is to set Z_i to be proportional to the distance of separation between neighbouring RBFs, in order to provide an appropriate weighting between neighbouring linear adaptive filter models in the data space.

Other multiple model approaches similar to the proposed method have been developed in recent years. One of these is based on a combination of piecewise polynomials (Heredia and Arce 2000). This method also uses an additive combination of multiple kernels, which constitute localised models to cover the data space. The main difference is related to kernel shape parameter selection. In the proposed SSAF model the kernels are all RBFs having a common bandwidth. Whereas, Heredia and Arce's method finds appropriate parameters for its more complex kernels via linear optimisation. Another somewhat similar approach is the off-line multiple model approach used by (Simani, Fantuzzi, Rovatti and Beghelli 2000). Although they don't use local kernels, they do combine multiple affine models to cover the data space and force continuity constraints among local affine models by solving an optimisation problem. Their parameter estimation algorithm is based on the well established Frisch scheme (Frisch 1934).

An on-line approach that arranges linear filters and thresholds in a tree structure was previously proposed by (Gelfand and Ravishankar 1993). They developed a stochastic gradient based training algorithm to adapt both filter coefficients and thresholds at the nodes of the tree as well as to prune the tree. Their method performs a sequential, hierarchical partitioning of the input vector space into polygonal domains and a pruning procedure selects a suitable tree size. Although this method does offer robust estimation and fast adaptation of linear filters it suffers from the problem that training at parent nodes is not completed prior to training at offspring nodes. Therefore data at non-root node filters have a complicated nonstationary and dependent character, and consequently stabilisation with penalty methods is required. Nevertheless it is a conceptually simple and computationally efficient method (Gelfand, Krogmeier and Balasubramanian 1995).

The SSAF model has many applications but its application to audio and underwater acoustic signal modelling, equalisation and filtering systems is of particular interest. This is because SSAF model can be used to provide a convenient and practical model of system responses, a sound propagation medium model or a model of signal propagation through either a changing homogeneous or nonhomogeneous three-dimensional medium based on measured data. In these types of applications it is likely that only the properties of arbitrary locales of the medium change at any time, in which case adaptation or training processes and filters in other locales need not be affected. Consequently, these types of environments can be very difficult to analytically model throughout, especially when their properties are changing over space as well as in time in quite complex ways due to varying thermal and other environmental conditions.

The main offering of the SSAF is a practical and conceptually simple method for smoothly coupling and selecting multiple linear (or affine) or nonlinear adaptive filter models using common bandwidth RBFs to cover the operating data space. Although the method is strictly not optimal in any sense it does have a convenient way of fine tuning through adjustment of a single smoothing parameter, being the common bandwidth parameter of the RBFs, based on representative training data.

In the following Sections reviews of the MPNN and APLR models are provided as a background to the subsequent development of the SSAF model. Also, some hints for suitable applications are provided with some discussion and description of some representative problem developments. Suitable training strategies as well as possible extensions of the idea are discussed in the later Sections of the Chapter.

16.2.1 MPNN Model

The Modified Probabilistic Neural Network was originally developed for general regression and application to nonlinear signal processing problems (Zaknich 1998). It is effectively a generalisation of Specht's General Regression Neural Network (GRNN) (Specht 1991) and is related to his Probabilistic Neural Network (PNN) classifier (Specht 1990). Both the MPNN and GRNN have fundamental similarities with the method of (Moody & Darken 1989); the method of radial basis functions (Powell 1985); and a number of other nonparametric kernel based regression techniques inspired by the work of (Nadaraya 1964) and (Watson 1964).

If it can be assumed that for each local region in the input vector space, represented by the centre vector c_i, there is a corresponding scalar output y_i that it maps into, then a convenient general model to use for all forms of the general MPNN and its subset the GRNN is Equation 16.1.

$$\hat{y}(\mathbf{x}) = \frac{\sum\limits_{i=1}^{M} Z_i y_i f_i (\|\mathbf{x} - \mathbf{c}_i\|, \sigma)}{\sum\limits_{i=1}^{M} Z_i f_i (\|\mathbf{x} - \mathbf{c}_i\|, \sigma)} \tag{16.1}$$

where:

$f_i (\|\mathbf{x} - \mathbf{c}_i\|, \sigma)$ is a common bandwidth RBF.

$\mathbf{c}_i$ is the network trained centre vector i in the input space.

σ is the single learning parameter chosen during training.

y_i is the scalar training output i related to $\mathbf{c}_i$.

M is the number of RBF centre vectors $\mathbf{c}_i$.

Z_i is the number of training vectors $\mathbf{x}_j$ associated with each $\mathbf{c}_i$.

L is the total number of training vectors $\mathbf{x}_j$, $L = \sum\limits_{i=1}^{M} Z_i$.

Equation 16.1 represents the GRNN if all the $Z_i = 1$, the y_i are real valued, the centre vectors $\mathbf{c}_i$ are replaced with individual training vectors $\mathbf{x}_i$ and $M = L$. A Gaussian RBF, with a bandwidth parameter of σ (typically modelling the signal noise standard deviation), is often used for $f_i(x)$ as defined by Equation 16.2. There are many other RBFs that can be used but the Gaussian RBF is often adequate, although not very computationally efficient.

$$f_i \left(\|\mathbf{x} - \mathbf{c}_i\|, \sigma \right) = \exp \frac{-(\mathbf{x} - \mathbf{c}_i)^T (\mathbf{x} - \mathbf{c}_i)}{2\sigma^2} \tag{16.2}$$

The RBF Equation 16.2 can be represented more economically by Equation 16.3.

$$f_i(d_i, \sigma) = \exp \left(\frac{-d_i^2}{2\sigma^2} \right) \tag{16.3}$$

where:

$$d_i = \|\mathbf{x} - \mathbf{c}_i\| = \sqrt{(\mathbf{x} - \mathbf{c}_i)^T (\mathbf{x} - \mathbf{c}_i)}.$$

One simple way that the MPNN set of network vector pairs $\{(\mathbf{c}_i, y_i) \mid i=1, ..., M\}$ can be formed is through a form of vector quantization, where the input space is firstly partitioned into uniform hypercubes (Zaknich 2003a). Then, $\mathbf{c}_i$ is made to be the mean of all training input vectors $\mathbf{x}_j$ in each hypercube that map to each y_i. The value y_i is usually computed as the mean of the outputs y_j associated with their corresponding inputs $\mathbf{x}_j$. These corresponding outputs y_j must be sufficiently close to each other to be adequately represented by their mean. In this way a local group of vectors in the input space can be replaced with a single centre vector that maps to a single mean scalar output value. The value Z_i is simply the number of associated input vectors $\mathbf{x}_j$ that are averaged to make centre vector $\mathbf{c}_i$. If training samples are taken randomly from the process, Z_i could be interpreted as being proportional to their *a priori* probability of occurrences. A number of different hypercube sizes can be systematically tested to choose the best one, else a reasonable guess often suffices. There are various strategies for constructing the network based on Equation 16.1 depending on the application and its requirements (Zaknich 1998) and these must be considered in the context of the problem.

Training then simply involves finding the single optimal learning parameter σ giving the minimum Mean Square Error (MSE) of the network output minus the desired output for a representative testing set of known sample vector pairs $\{(\mathbf{x}_k, y_k) \mid k=1, ..., \text{NUM}\}$. In typical applications there is often a unique σ value that produces the minimum MSE between the network output and the desired output for the testing set (independent of the training set) and it can be found quite easily by trial and error. Alternatively, since the relationship between σ and MSE is usually smooth with a broad minimal MSE vs. σ section σ can often be found very quickly by a convergent optimisation algorithm based on recurrent parabolic curve fitting (Zaknich and Attikiouzel 1993). Also, because the relation between σ and MSE is usually smooth with a broad minimal MSE section the value of σ is often not overly critical for adequate performance anyway.

The MPNN model can best be described as a semiparametric model, which produces more efficient networks than the nonparametric GRNN model, but it can still be quite large and inefficient compared to the Multi-Layer Perceptron (MLP). A way to make the MPNN smaller without compromising accuracy is to convert it to the Approximate Piecewise Linear Regression (APLR) model as described in the next Section.

16.2.2 Approximately Piecewise Linear Regression Model

The APLR model defined by Equation 16.4 can be formed by first partitioning the input space into uniform hypercubes as is done for the MPNN. When the centre vectors $\mathbf{c}_i$ are found the Z_i number of corresponding input vectors $\mathbf{x}_j$ are also used to create a set of best fit least squares linear regression models $l_i(\mathbf{x})$ associated with each of the centres. The outputs y_i in Equation (16.1) are then replaced with the outputs of the subsequent linear models $l_i(\mathbf{x})$ for all input vectors during operation. When this is done the linear model outputs provide more accurate mappings from the input to the output space within each hypercube than to the fixed averaged means y_i of the MPNN. Consequently, the input space hypercubes can be made much larger and fewer, resulting in a much smaller network size (smaller M) for comparable regression accuracy.

In the APLR model the adjustment of σ during training controls the degree of weighting of each linear model associated with each centre or RBF. Input vectors closest to a centre will activate the associated linear model more than for those further away. For very small σ the linear model associated with the centre closest to the current input point will dominate, resulting in a linear response in the local space of that centre. For very large σ the network output will approach an unweighted biased average of all the linear models. Somewhere in between an optimal model will result that provides approximately linear operation close to each centre, and possibly deviating significantly from linearity close to boundaries regions between centres. With an appropriate choice of σ a sufficiently smooth merging of neighbouring linear models occurs at the boundaries.

$$\hat{y}(\mathbf{x}) = \frac{\sum_{i=1}^{M} Z_i l_i(\mathbf{x}) f_i(\|\mathbf{x} - \mathbf{c}_i\|, \sigma)}{\sum_{i=1}^{M} Z_i f_i(\|\mathbf{x} - \mathbf{c}_i\|, \sigma)} \tag{16.4}$$

where:

$f_i(\|\mathbf{x} - \mathbf{c}_i\|, \sigma)$ is a common bandwidth RBF.
$\mathbf{X}$ is an arbitrary input space vector.
$\mathbf{c}_i$ is the trained centre vector i in the input space.
σ is the single learning parameter chosen during training.
$l_i(\mathbf{x})$ is the trained linear output model related to centre $\mathbf{c}_i$.
M is the number of RBF centre vectors $\mathbf{c}_i$.
Z_i is the number of training vectors $\mathbf{x}_j$ associated with each $\mathbf{c}_i$.

L is the total number of original training vectors $\mathbf{x}_j$, $L = \sum\limits_{i=1}^{M} Z_i$.

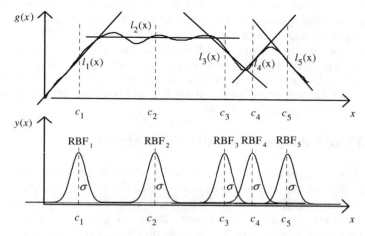

Figure 16.2. Illustrative 1-D Example of the APLR Model

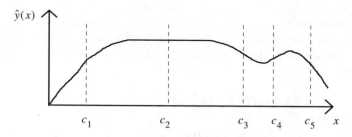

Figure 16.3. APLR Model Regression Result

A simple one-dimensional illustration of an APLR model, typical of what might approximate a loudspeaker amplitude frequency response, is illustrated by Figure 16.2 and Figure 16.3. Figure 16.2 shows an arbitrary continuous and differentiable scalar function $g(x)$, which is modelled by five ($M = 5$) scalar linear (actually affine) sections ($i = 1, \ldots, 5$) each associated with its own RBF_i and scalar centre c_i in the scalar input space x. Figure 16.3 shows what the resulting APLR model might look like after selection of suitable σ and Z_i values. In this example the Z_i

values would be proportional to the distances of separation between RBFs. For example, Z_1 would be larger than say, Z_4 because RBF$_1$ needs to span more space that RBF$_2$ in order to provide appropriate weighting between respective linear models $l_1(x)$ and $l_4(x)$ and their respective neighbouring models. For such a model to work well the relative spans between RBFs should not be too great, say the smallest to the largest span should not exceed a ratio of about 2 to 10 (Zaknich 2004).

Linear model $l_2(x)$ in Figure 16.2 provides more smoothing than the others and consequently contributes more overall regression error to the total result. This may still be acceptable for a given design specification, but if not it is just a matter of adding more linear (or affine) models within region two to improve regression accuracy. Another quite acceptable solution may be to make model $l_2(x)$ a suitable order of polynomial or even a MLP. There is nothing in the APLR structure that prevents the mixing of linear and nonlinear models in appropriate sections. However, retaining a set of linear models throughout may provide a simpler design at the expense of model efficiency. It can be easily appreciated that each of the models $l_i(x)$ may either be computed off-line and fixed and/or simply adapted as new training data becomes available.

16.2.3 The Sub-space Adaptive Filter Model

A slight modification to the APLR model (Zaknich and Attikiouzel 2000) in conjunction with an adaptive mechanism forms the basis to the proposed SSAF model (Zaknich 2003b). The linear models $l_i(\mathbf{x})$ in the SSAF Equation 16.5 can now be defined as adaptive filter models. Refer to Figure 16.4 for a diagrammatic representation of Equation 16.5. The main difference between Equation 16.4 and Equation 16.5 is that in Equation 16.5 the RBFs are formed in a process state space $\mathbf{z}$, which is not necessarily the same as the input space $\mathbf{x}$, but in practice often is. The process state space $\mathbf{z}$ represents some aspect of the process that can be measured, monitored and used to determine when a particular filter model $l_i(\mathbf{x})$ should take operational effect. Z_i becomes a filter model weight that can be arbitrarily adjusted or set independently of training vector numbers as required by the problem.

$$\hat{y}(\mathbf{x}, \mathbf{z}) = \frac{\sum_{i=1}^{M} Z_i l_i(\mathbf{x}) f_i(\|\mathbf{z} - \mathbf{cz}_i\|, \sigma)}{\sum_{i=1}^{M} Z_i f_i(\|\mathbf{z} - \mathbf{cz}_i\|, \sigma)} \tag{16.5}$$

where:

$f_i(\|\mathbf{x} - \mathbf{cz}_i\|, \sigma)$ is a common bandwidth RBF.

$\mathbf{x}$ is an arbitrary filter input space vector.

$\mathbf{z}$ is an arbitrary process state vector associated with $\mathbf{z}$. Often $\mathbf{z} \equiv \mathbf{x}$.

$\mathbf{cz}_i$ is the trained RBF centre vector i in the process state space.

σ is the single learning parameter chosen during tuning.

$l_i(\mathbf{x})$ is the adaptive linear filter model output related to centre $\mathbf{cz}_i$.

M is the number of RBF centre vectors $\mathbf{cz}_i$.

Z_i is the weight associated with adaptive linear model $l_i(\mathbf{x})$ and centre $\mathbf{cz}_i$.

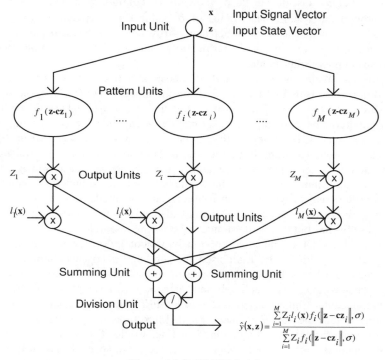

Figure 16.4. SSAF Model

The complete SSAF model is established from Equation 16.5 by adding a suitable adaptation mechanism similar to that of the MPNN (Zaknich and deSilva 1997). In MPNN the network parameters associated with a particular RBF$_i$ were adapted, for a given input-output pair, only if,

$$Z_i f_i\left(\left\|\mathbf{x}-\mathbf{c}_i\right\|,\sigma\right) \geq Z_k f_k\left(\left\|\mathbf{x}-\mathbf{c}_k\right\|,\sigma\right), \ \forall \ k \neq i.$$

For the SSAF model the adaptation condition, for a given input-output pair and process state vector $\mathbf{z}$, is now defined by Equation 16.6. (This equation was incompletely written in (Zaknich 2003b), where the Z_i and Z_k factors were incorrectly omitted.) The SSAF model is basically the adaptive version of the Integrated Sensory Intelligent System (ISIS) model described in (Zaknich 2003c). In (Zaknich 2003c) the ISIS model is nonadaptive and used to implement a fixed nonlinear detector to solve an underwater Doppler shifted chirp signal detection problem.

$$Z_i f_i \left(\left\| \mathbf{z} - \mathbf{cz}_i \right\|, \sigma \right) \geq Z_i f_k \left(\left\| \mathbf{z} - \mathbf{cz}_k \right\|, \sigma \right), \ \forall \ k \neq i \tag{16.6}$$

In most designs it would be expected that σ is determined and fixed before adaptation begins, else it can be periodically or continually adjusted as the data flows through the system. The range of σ value for acceptable performance is often quite broad and as such is not necessarily an overly critical parameter. Its main purpose is to ensure a sufficiently smooth interconnection between neighbouring linear filter models.

There are many possible standard linear adaptive filters and associated adaptation mechanisms that may be used on the models $l_i(\mathbf{x})$ (Stergiopoulos 2001, Garas 2000, Haykin 1999, Diniz 1997, Haykin 1996, Hayes 1996), depending on design requirements. Most applications may be adequately served by using Finite Impulse Response (FIR) filters with a suitable variant of the Least Mean Squares (LMS) adaptation algorithm. For applications that may require more efficient filter models and a faster adaptation capability the Recursive Least Squares (RLS) adaptation algorithm, or one of its variants, used in conjunction with linear recursive filters may provide a better solution. The use of adaptive versions of time-frequency filters (Hlawatsch 1998) would also add a very useful dimension to design possibilities but that is the subject of ongoing research.

The SSAF model, as indicated previously, is actually somewhat similar in form to an adaptive multiple FIR filter used for sub-band echo cancellation (Vaseghi 1996). Such a filter splits the input signal into a parallel frequency sub-bands having equal bandwidths. Each sub-band can be down sampled without loss of information, and assigned a separate smaller FIR filter. The main advantage of doing this is that the set of reduced length filters operating in parallel can now converge much faster than a single long FIR filter would. The SSAF is really a kind of more general form of this, where the chosen sub-space is not restricted to frequency bands but can be any definable and appropriate sub-division of the operation space that can be somehow related to the filter's operating domain.

16.2.4 Example Applications of the SSAF Model

The SSAF model was originally developed with its application to three-dimensional audio and underwater acoustic signal modelling, equalisation and filtering firmly in mind. These processes are very difficult to model analytically, especially when the properties of the acoustic medium can and do vary nonlinearly with position and time. Consequently, problems such as these are characterised by the need to develop and adapt practical models from data measurements in real-time. The two simple example problems of loudspeaker modelling and adaptive equalisation, and acoustic signal propagation through the ocean can serve to illustrate SSAF applicability and utility. In both these applications the state space vector $\mathbf{z}$ represents the three spatial dimensions and is used to segment and identify three-dimensional spatial sub-spaces of operation. Also, any model adjustments are restricted to local RBF sub-spaces based on updating local measurements while still maintaining good interpolation between the other sub-models.

16.2.4.1 Loudspeaker 3-D Frequency Response Model

A loudspeaker three-dimensional frequency response model can be built up by initially measuring the frequency responses at a number of suitable points in space in front of the loudspeaker. The SSAF model can then be used to both smoothly interpolate the response between measured responses over both the frequency and physical space of interest and then be adapted with respect to the whole model according to any dynamic equalisation requirements. Although loudspeaker amplitude vs. frequency contours can vary at a single spot in three-dimensional space as well as over space in front of the speaker they change relatively smoothly over frequency and space. This means that relatively few spots in front of the speaker need to be initially measured to build an effective and smooth three-dimensional loudspeaker response model.

The shape of Figure 16.3 is similar to a typical loudspeaker amplitude response at one spot in front. Equation 16.5 can be used to very easily model and smooth this type of shape using standard one-twelfth octave frequency amplitude measurements taken at one meter distances from the loudspeaker around it. An interpolation model of the frequency response at a single spot has been developed that is based on centres placed at one-twelfth octave frequencies (Zaknich 2004). In this model each linear model $l_i(\mathbf{x})$ is placed at each one-twelfth octave frequency is taken to be a constant equal to the respective scalar measurement. This is a little simpler than that implied in Figure 16.2 where the $l_i(\mathbf{x})$ models are arbitrary lines with slopes other than zero (constant line). Weighting factors Z_i were set to values defined by Equation 16.7, to span the exponentially increasing gaps between centres. The smoothing factor σ was then be used for final tuning and smoothing of the overall frequency response curve.

$$Z_i = \left[(2)^{\frac{1}{12}} \right]^{i-1} , \ i = 1,..,M \tag{16.7}$$

It was noted in (Zaknich 2004) that due to the exponentially increasing span between one-twelfth octave frequencies a single SSAF model, using constant valued $l_i(\mathbf{x})$, designed to cover the whole frequency range of 20 Hz to 20 KHz was not adequate. The reason for the problem is due to using Gaussian RBFs, whose tails approach zero very quickly. Therefore, the interpolation between RBFs implied in SSAF Equation 16.5 is inadequate over the large span frequency regions when σ is reduced to be optimal for smaller span regions. Of course, if σ is made larger to accommodate the larger spans then there is far too much smoothing in the model over the lower span frequency regions of the model. This problem can easily be seen in Figure 16.5, where the single SSAF model appears staircased (inadequate interpolation/smoothing) at the high frequency end and very smoothed at the lower frequency end. This was solved by making four separate SSAF models to cover the whole frequency range such that the smallest to largest frequency span in any one model was less than about 10.

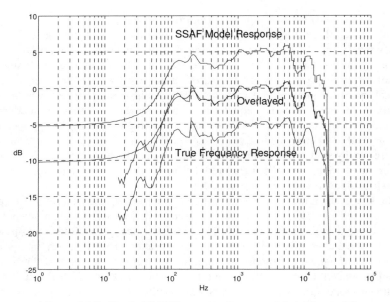

Figure 16.5. Single SSAF Frequency Response Model for $\sigma=20$

Another way to solve this problem is to modify the SSAF model Equation 16.5 to a new more flexible Equation 16.8 by introducing another weight S_i to adjust the σ weighting.

$$\hat{y}(\mathbf{x},\mathbf{z}) = \frac{\sum\limits_{i=1}^{M} Z_i l_i(\mathbf{x}) f_i(\|\mathbf{z}-\mathbf{cz}_i\|, S_i \sigma)}{\sum\limits_{i=1}^{M} Z_i f_i(\|\mathbf{z}-\mathbf{cz}_i\|, S_i \sigma)} \tag{16.8}$$

where:

$f_i(\|\mathbf{x}-\mathbf{cz}_i\|, \sigma)$ is a common bandwidth RBF.

$\mathbf{x}$ is an arbitrary filter input space vector.

$\mathbf{z}$ is an arbitrary process state vector associated with $\mathbf{z}$. Often $\mathbf{z} \equiv \mathbf{x}$.

$\mathbf{cz}_i$ is the trained RBF centre vector i in the process state space.

σ is the single learning parameter chosen during tuning.

$l_i(\mathbf{x})$ is the adaptive linear filter model output related to centre $\mathbf{cz}_i$.

M is the number of RBF centre vectors $\mathbf{cz}_i$.

Z_i is a weight associated with adaptive linear model $l_i(\mathbf{x})$ and centre $\mathbf{cz}_i$.

S_i is a weight associated with adaptive linear model $l_i(\mathbf{x})$ and centre $\mathbf{cz}_i$.

Now, instead of weighting the spans according to Equation 16.7 it is done according to Equation 16.9, while setting all $Z_i = 1$. By doing this it can be seen that a much more adequate model will result because the required RBF bandwidths will be in exact proportion to the frequency spans, allowing a better σ tuning control.

$$S_i = \left[(2)^{\frac{1}{12}} \right]^{i-1} \quad , \quad i = 1,..,M \tag{16.9}$$

Equation 16.8 has more weighting parameters to set, which can help it accommodate to the needs of specific applications. Consequently, Equation 16.8 can be seen as a more generic form of the SSAF model. However more care must be exercised in its application to ensure that proper benefits result without introducing unnecessary complications with final tuning. Nevertheless, one of the virtues of the model is that once it is set the final fine tuning can be done by simply adjusting the single parameter σ.

This loudspeaker frequency response model can form the basis of an adaptive room equalisation system that can make separate filter adaptations within each one-twelfth octave frequency sub-band. Loudspeakers are typically minimum phase systems and consequently their phase characteristics can be computed directly from the magnitude of the amplitude response (Poularikas 1996) obviating the need for extra data memory. However, a similar companion interpolation model can also be defined for the phase values if desired. In like manner it is possible to design a separate SSAF interpolation model to do the spatial interpolation or it is also possible to build the spatial interpolation into the single model. This is the subject of ongoing research and will be reported on in future publications.

16.2.4.2 Velocity of Sound in Water 3-D Model
The timely maintenance of an accurate velocity of sound model to cover a three-dimensional region of sonar operation in the ocean is important in order to calculate good subsequent signal propagation models. The main variation of sound velocity occurs with depth due to thermal variations occurring during the course of the day (Urick 1983). In this case the SSAF model can be used to maintain a good three-dimensional velocity of sound model and keep it adapted during the course of system operation. Often it is sufficient to have a sound velocity model only as a function of depth to achieve good first order results. Measurements of sound velocity may be made at various depths and locations by various means and used to adapt the main model in the sections delineated by the RBFs. These measurements can be done periodically using depth probes or preferably by a sub-ocean vehicle and acoustically transmitted back to the system.

Figure 16.6 Shows a typical ocean sound velocity vs. depth profile (Urick 1983). If depth is divided up into an appropriate number of layers an affine adaptive model can be associated with each layer using Equations 16.5 and 16.6. Weighting for each depth span can be controlled by setting factors Z_i proportional to the respective depth spans between model/layer centres. As new velocity values are measured at any depth the appropriate model is adapted to maintain accuracy while an appropriate value of σ keeps joins at the model boundaries sufficiently smooth. Furthermore, there would be no special problem with even changing the number of layers/models by simply adjusting the number of centres to the number required to keep the overall model accurate if the thermal conditions change dramatically, especially near the surface. At depth the sound velocity is dominated by water pressure and is less affected by any temperature variations.

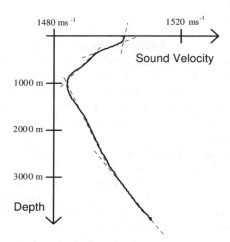

Figure 16.6. Ocean Sound Velocity vs. Depth Profile

In Figure 16.6 are drawn a number of dotted lines that may form the basis of the relatively few linear models $l_i(\mathbf{x})$ required to form a suitable SSAF model. Any sound velocity measurements made within any layer is sure to improve the adaptive local interpolation model without affecting the other layer models until they can be adapted by appropriate remeasurements, if warranted.

16.3 Discussion and Overview of the SSAF

This Chapter has introduced the SSAF model, based on the MPNN and APLR regression models. The SSAF provides some significant benefits for practical design of complex adaptive filtering systems. It can be seen as a development that provides a simple and practical means of solving some important complex nonlinear adaptive filtering problems. However, in order to apply it successfully it still requires intelligent design decisions made by a human designer. The designer must decide on the role and set the values for the Z_i, and also S_i in Equation 16.8, decide on how many RBFs to use and where to place them to adequately cover a chosen process state space.

These decisions can be automated in a similar way as is typically done for the MPNN (Zaknich 2003a). However, a more efficient mechanism needs yet to be developed that may begin to approach some optimal solution, within specified design constraints. For example, such a mechanism must be able to automatically reduce the SSAF down to a single RBF and linear adaptive filter model when solving a linear problem. For nonlinear problems it needs to be able to reduce down to a minimal set of RBFs and adaptive filter models $l_i(\mathbf{x})$ that adequately cover the whole operating space for a given design specification and acceptable error margin. The methods of (Verselinovic and Leenaerts 1996) and (Mattavelli, Amaldi and Gruter 1996) can offer some useful insights as they both deal with the automatic generation of piecewise linear models. The method of (Verselinovic and Leenaerts 1996) is intended for the modelling of nonlinear multivariable scalar functions and uses a specified relative error to guide model generation in a way that trades off between model size and accuracy. A simple combinatorial optimisation approach is used in (Mattavelli, Amaldi and Gruter 1996) that provides natural partitions of time series state-spaces for a given error tolerance. The partitioning of the data space and allocation of centres can also be done using methods related to Vector Quantization and unsupervised artificial neural networks (Kohonen 1990). An unsupervised on-line clustering method (Young, Zaknich and Attikiouzel 2001) previously applied to the MPNN may also provide useful solutions.

In the SSAF model linear filter models are preferred but there is no reason why quadratic or higher order Volterra filters can not also be used in any required combination. The only drawback may be that many more piecewise model parameters must be stored for each RBF. This may not really be much of a drawback if it allows a process to be better approximated with fewer RBF regions, i.e., smaller M. Anyway, memory is relatively cheap and processing power is increasing significantly with advancing technologies to make this less of an issue with time.

A further very useful way to exploit this SSAF structure is to use it to smoothly piece together and merge a set of adaptive MLP filters or models throughout a nonlinear data space. This provides a method of decoupling MLP models such that, as data statistics change in a local region only the MLP related to that region needs to be adapted. Not only does this allow the total model to adapt much faster but it also preserves the training of each of the now much less complex and unaffected local MLPs, which helps with the stability-plasticity dilemma that MLPs are limited by.

Another interesting development may be to consider using Support Vector Machines (SVMs) (Cherkassky and Mulier 1998, Vapnik 2000) to replace models $l_i(\mathbf{x})$. SVMs are able to turn nonlinear problems into optimal linear ones by projection of the problem into a very high dimensional space. This could mean that each RBF can more adequately cover a greater space since the filtering process need not be approximately linear in the vicinity of each RBF to produce an accurate enough overall model. Using the SVM approach may then obviate the need for more complex nonlinear optimisation solutions for nonlinear sections.

References

Applebaum SP, Chapman DJ (1976) Adaptive arrays with main beam constraints. IEEE Transactions on Antennas and Propagation. Vol. AP-24, 650-662.

Åström KJ, Wittenmark B (1995) Adaptive control. Addison-Wesley.

Bendat JS, Piersol AG (1971) Random data: Analysis and measurement procedures. Wiley-Interscience.

Benesty J, Gansler T, Morgan DR, Sondhi MM, Gay SL (2001) Advances in network and acoustic echo cancellation. Springer.

Benesty J, Huang Y, Editors (2003) Adaptive signal processing: Applications to real-world problems. Springer.

Bishop CS (1995) Neural networks for pattern recognition. Clarendon Press.

Box GEP, Jenkins GM (1970) Time series analysis forecasting and control. Holden-Day, San Francisco, 423-428.

Bozic SM (1994) Digital and Kalman Filtering. E. Arnold.

Brandstein M, Ward D, Editors (2001) Microphone arrays. Springer.

Bräunl T (2003) Embedded Robotics: Mobile robotic design and applications with embedded systems. Springer.

Bremaud P (2002), Mathematical principles of signal processing. Springer.

Broomhead DS, Lowe D (1988) Radial basis-functions, multi-variable functional interpolation and adaptive networks, Royal Signals and Radar Establishment Memorandum 4148, 28th March.

Burns B (1996) Training artificial neural networks with genetic algorithms to play games. Honour Thesis, The Department of Electrical and Electronic Engineering, The University of Western Australia.

Capon J (1969) High-resolution frequency-wavenumber spectrum analysis. Proceedings of the IEEE, 57:1408-1418.

Chen CH (1988) Signal processing handbook. Marcel Dekker Inc.

Cherkassky V, Mulier F. (1998) Learning from data. John Wiley and Sons, Inc.

Coley DA (1999) An introduction to genetic algorithms for scientists and engineers. World Scientific Publishing.

Cooley JW, Tukey J W (1965) An algorithm for the machine computation of complex Fourier series. Math. Comput., 19th April, 297-301.

Cross PA (1981) The computation of position at sea. The Hydrographic Journal, 20:7-16.

Cross PA (1982) Prediction, filtering and smoothing of offshore navigation data. The Hydrographic Journal, 25:5-16.

Cross PA (1983) Advanced least squares applied to position fixing. Department of Land Surveying, North East London Polytechnic, Working paper No. 6, 205pp.

Cross PA (1987) Kalman filtering and its application to offshore position-fixing. The Hydrographic Journal, 44:19-25.

Darwin C (1859) On the origin of species by means of natural selection, or preservation of favoured races in the struggle for life. John Murray, London

Diniz PSR (1997) Adaptive filtering: Algorithms and practical implementation. Kluwer Academic Publishers.

Durbin J (1960) The fitting of time-series models. Rev. Inst. Statist., 28:233-243.

Eberhart R, Dobbins B (1990) Neural network PC tools: A practical guide. Academic Press.

Falconer DD, Ljung L (1978) Application of fast Kalman estimation to adaptive equalization. IEEE Transactions on Communications, COM-26:1439-1446.

Fogel LR, Owens AJ, Walsh MJ (1966) Artificial intelligence through simulated evolution. John Wiley & Sons, New York.

Frisch R (1934) Statistical confluence analysis by means of complete regression systems. University of Oslo, Economic Institute, Publication n. 5 ed.

Gabel R, Roberts R (1987) Signals and linear systems. Third Edition, John Wiley and Sons.

Garas J (2000) Adaptive 3D sound systems. Kluwer Academic.

Gelfand SB, Krogmeier JV, Balasubramanian R (1995) A tree-structured piecewise linear filter with recursive least-squares adaptation. Proceedings of the 29th Asilomar Conference on Signals Systems and Computers, 1:673-675.

Gelfand SB, Ravishankar CS (1993) A tree-structured piecewise linear adaptive filter. IEEE Transactions on Information Theory, 39:6:1907-1922.

Gershenfeld N (1999) The Nature of mathematically modelling. Cambridge University Press.

Godard DN (1974) Channel equalization using a Kalman filter for fast data transmission. IBM J. Res. Dev. 18:267-273.

Gold B, Morgan N (2000) Speech and audio signal processing: Processing and perception of speech and music. John Wiley & Sons Inc..

Golub GH, Van Loan CF (1983) Matrix computations. John Hopkins University Press.

Gupta MM, Rao DH, Editors (1994) Neuro-control systems. IEEE Press.

Hassibi B, Sayed AH, Kailath T (1996) H^∞ optimality of the LMS algorithm. IEEE Transactions on Signal Processing, 44:267-280.

Hassoun MH (1995) Fundamentals of artificial neural networks. MIT Press.

Hayes MH (1996) Statistical digital signal processing and modelling. John Wiley & Sons, Inc.

Haykin S (1994) Neural networks, a comprehensive foundation. Macmillan College Publishing Co. Inc.

Haykin S (1996) Adaptive filter theory. Prentice-Hall.

Haykin S (1999) Neural networks, a comprehensive foundation. Second Edition, Upper Saddle River, NJ, Prentice-Hall Inc.

Haykin S, Kosko B, Editors (2001), Intelligent Signal Processing. Wiley-IEEE Press, First Edition.

Haykin S, Van Veen B (1999) Signals and systems. John Wiley and Sons, Inc..

Hebb DO (1949) The organization of behaviour. Wiley.

Hecht-Nielson R (1990) Neurocomputing. Addison-Wesley Pub. Co.

Heredia A, Arce GR (2000) Nonlinear filters based on combinations of piecewise polynomials with compact support. IEEE Transactions on Signal Processing, 48:10:2850-2863.

Hinton G, Sejnowski TJ, Editors (1999) Unsupervised learning: Foundations of neural computation. The MIT Press.

Hlawatsch F (1998) Time-frequency analysis and synthesis of linear signal spaces. Kluwer Academic Publishers.

Holland JH (1962) Outline for a logical theory of adaptive systems. Journal of the Association for Commuting Machinery.

Holland JH (1995) Adaptation in natural and artificial systems: An introductory analysis with applications to biology, control, and artificial intelligence. The MIT Press (First Edition 1975).

Hopfield JJ (1982) Neural networks and physical systems with emergent collective computational abilities. Proceedings of the National Academy of Sciences, 79:2554-2558.

Hopfield JJ (1984) Neurons with graded response have collective computational properties like those of two state neurons. Proceedings of the National Academy of Sciences, 81:3088-3092.

James W (1890) Psychology (Briefer Course). Holt.

Kailath T, Editor, (1977) Linear least-squares estimation. Benchmark Papers in Electrical Engineering and Computer Science, Prentice-Hall, Englewood Cliffs, N.J.

Kalman RE (1960) A new approach to linear filtering and prediction problems. Transactions of the ASME, Journal of Basic Engineering, 82: 35-45.

Kalman RE, Bucy RS (1961) New results in linear filtering and prediction theory. Transactions of ASME, J. Basic Eng., 83:95-108.

Kammler DW (2000) A first course in Fourier analysis. Prentice Hall.

Kartalpoulos SV (1996) Understanding neural Networks and Fuzzy Logic. IEEE Press.

Kohonen T (1990) Self-organising map. Proceedings of the IEEE, 78:1464-1480.

Kolmogorov AN (1939) Sur l'interpolation et extrapolation des suites stationaries. C. R. Acad. Sci. Paris 208:2043-2045. (English translation in (Kailath 1977))

Koza JR (1992) Genetic programming: On the programming of computers by means of natural selection. MIT Press, Cambridge, MA.

Koza JR, Bennett III FH, Andre D, M A (1996) Four problems for which a computer program evolved by genetic programming is competitive with human performance. Proceedings of the 1996 IEEE International Conference on Evolutionary Computing, 20-22 May:1-10.

Koza JR, Keane MA, Streeter MJ, Mydlowec W, Yu J, Lanza G (2003) Genetic programming: Routine human-competitive machine intelligence. Kluwer Academic Publishers.

Krein MG (1945) On a problem of extrapolation of A. N. Kolmogorov. C. R. (Dokl.) Akad. Nauk SSSR, 46:306-309. (Reproduced in (Kailath 1977))

Kurzweil R (1999) The age of spiritual machines: When computers exceed human intelligence. Allen and Unwin.

Lau SM, Leung SH, Chan BL (1992) A reduced rank second-order adaptive Volterra filter. ISSPA 92, Signal Processing and its Applications, Gold Coast, Australia, 16-21st August:561-563.

Lawson CL, Hanson RJ (1974) Solving least squares problems. Prentice-Hall.

Leung SH, Chan B L, Lau SM (1992) A second-order adaptive Volterra filter using three level sign algorithm. ISSPA 92, Signal Processing and its Applications, Gold Coast, Australia, 16-21st August:569-572.

Levine WS, Editor (1996) The control handbook. CRC Press and IEEE Press, 847-857, Petros I, Model reference adaptive control.

Levinson N (1947) The Wiener RMS (root mean square) error criterion in filter design and prediction. Journal of Math. Phys, 25:261-278.

Lim JS, Oppenheim AV (1988) Advanced topics in signal processing. Prentice Hall Signal Processing Series.

Loy NJ (1988), An engineer's guide to FIR digital filters. Prentice-Hall.

Lucky RW (1965) Automatic equalization for digital communications. Bell Syst. Tech. J., 44:547-588.

Makhoul J (1977) Stable and efficient methods for linear prediction. IEEE Transactions on Acoustics, Speech, and Signal Processing, ASSP-25:423-428.

Mamdani (1974) Applications of fuzzy logic algorithms for control of a simple dynamic plant. Proceedings of the IEE, 121:12:1585-1588.

Marks II RJ, Editor (1994) Fuzzy logic technology and applications. IEEE TAB.

Mathews JJ (1991) Adaptive polynomial filters. IEEE Signal Processing Magazine, 8:3:10-26.

Mattavelli M, Vesin JM, Amaldi E, Gruter R, (1996) A new approach to piecewise linear modelling of time series. IEEE Digital Signal Processing Workshop Proceedings, 502-505.

McCulloch JL, Pitts WA (1943) A logical calculus of the ideas immanent in nervous activity. Bulletin of Mathematics and Biophysics, 5:115-133.

McWhirter JG (1983) Recursive least-squares minimization using a systolic array. Proceedings of SPIE, Real-Time Signal Processing VI, Vol. 1152, San Diego, California.

Miller AR (1981) Pascal programs, for scientists and engineers. Sybex.

Mitchell TM (1997) Machine learning. McGraw-Hill.

Moody J, Darken C (1989) Fast learning in networks of locally-tuned processing units. Neural Computation, 1:2:281-294.

Morgan DR, Thi J (1995) A delayless subband filter. IEEE Transactions on Signal Processing, 43:1819-1830.

Mukaidono M (2001) Fuzzy logic for beginners. World Scientific Publishing.

Nadaraya EA (1964) On estimating regression. Theory Probability Applications, 9: 141-142.

Oppenheim AV, Schafer RW (1975) Digital signal processing. Prentice-Hall.

Parker DB (1985) Learning-logic. M.I.T. Cen. Computational Res. Economics Management Sci., Cambridge, MA, TR-47.

Plackett RL (1950) Some theorems in least squares. Biometrika, 37:149.

Poularikas A, Editor in Chief (1996) The transforms and applications handbook. CRC and IEEE Press.

Powell MJD (1985) Radial basis functions for multivariate interpolation: A review. Technical Report DAMPT 1985/NA12, Department of Applied Mathematics and Theoretical Physics, Cambridge University, England.

Press WH, Flannery BP, Teukolsky SA, Vetterling WT (1986) Numerical recipes: The art of scientific computing. Cambridge University Press.

Principe JC, Euliano NR, Lefebvre WC (1999) Neural and adaptive systems: Fundamentals through simulations. John Wiley and Sons, Inc.

Proakis JG, Manolakis DG (1996) Digital signal processing. Prentice-Hall.

Rabiner LR, Gold B (1975) Theory and application of digital signal processing. Prentice-Hall.

Robbins H, Munro S (1951) A stochastic approximation method. Ann. Math. Stat., 22:400-407.

Robinson EA. (1964) Wavelet composition of time series, in econometric model building. Edited by H. O. Wold. North-Holland, Amsterdam, 37-106.

Rosenblatt F (1958) The perceptron: A probabilistic model for information storage and organization in the brain. Psychoanalytic Review, 65:386-408.

Rumelhart DE, Hinton DE, Williams RJ (1986) Learning representation by backpropagating errors. Nature 323(9), 533-536.

Sayed AH, Kailath T (1994) A state-space approach to adaptive digital filters. IEEE Signal Processing Magazine 11:18-60.

Schalkoff RJ (1997) Artificial neural networks. McGraw-Hill.

Shynk JJ (1992) Frequency-domain and multirate adaptive filtering. IEEE Signal Processing Magazine, 9:1:14-37.

Simani S, Fantuzzi C, Rovatti R, Beghelli S (2000) Non-linear dynamic modelling in noisy environment using multiple model approach. Proceedings of the American Control Conference, 4:2332-2336.

Special issue on fuzzy and neural networks. IEEE Communications Magazine, September, 1992.

Specht DF (1988) Probabilistic neural networks for classification, mapping, or associative memory, IEEE Conference on Neural Networks, San Diego, July, 1:525-532.

Specht DF (1990) Probabilistic neural networks. International Neural Network Society, Neural Networks, 3:109-118.

Specht DF (1991) A general regression neural network. IEEE Transactions on Neural Networks, 2:6:568-576.

Stanley WD, Dougherty GR, Dougherty R (1984) Digital signal processing. Reston Publishing Company.

Stergiopoulos S (2001) Advanced signal processing handbook – Theory and implementation for radar, sonar, and medical imaging real-time systems. CRC Press.

Sutton RS, Barto AG (1999) Reinforcement learning: An introduction. The MIT Press.

Tanaka K (1997) An introduction to fuzzy logic for practical applications. Springer.

Turing AM (1950) Computing machinery and intelligence. Mind, LIX:433-460.

Urick RJ (1983) Principles of underwater sound. Third Edition, McGraw-Hill Book Company.

Van Den Boss A (1971) Alternative interpretation of maximum entropy spectral analysis. IEEE Transactions on Information Theory, IT-17:493-494.

Vapnik VN (1998) Statistical learning theory. Wiley.

Vapnik VN (2000) The nature of statistical learning theory. Springer-Verlag, Second Edition.

Vaseghi SG (1996) Advanced signal processing and digital noise reduction. Wiley and Teubner.

Verselinovic P, Leenaerts D (1996) A method for automatic generation of piecewise linear models. Proceedings of the 1996 IEEE International Symposium on Circuits and Systems "Connecting the World," 3:24-27.

Watson GS (1964) Smooth regression analysis. Sankhya Series A, 26:359-372.

Werbos PJ (1974) Beyond regression: new tools for prediction and analysis in the behavioural sciences. PhD dissertation, Committee on Appl. Math., Harvard Univ., Cambridge, M. A., November.

Werbos PJ (1990) Backpropagation through time: What it does and how it to do it. Proceedings of the IEEE, Vol. 78, No. 10.

Whitaker HP, Yamron J, Kezer A (1958) A design of model reference adaptive control systems for aircraft. Report R-164, Instrument Laboratory, Massachusetts Institute of Technology, Cambridge.

Widrow B, Hoff ME (1960) Adaptive switching circuits. WESTCON Convention, Record Part IV, 96-104.

Widrow B, et al. (1967) Adaptive antenna systems. Proceedings of the IEEE, 55:2143-2159.

Widrow B, Glover, McCook, Hauritz, Williams, Hearn, et al. (1975) Adaptive noise cancelling: Principles and applications. Proceedings of the IEEE, 63:12:1692-1716.

Widrow B, Stearns S (1985) Adaptive signal processing. Englewood Cliffs, NJ: Prentice Hall.

Widrow B, Winter R (1988) Neural networks for adaptive filtering and adaptive pattern recognition. IEEE Computer, March:25-39.

Wiener N (1949) Extrapolation, interpolation, and smoothing of stationary time series, with engineering applications. MIT Press, Cambridge, Mass.

Wiener N (1958) Nonlinear problems in random theory. Wiley, New York.

Yen J, Langari R, Zadeh LA (1995) Industrial applications of fuzzy logic and intelligent systems. IEEE Press.

Young J, Zaknich A, Attikiouzel Y (2001) Center reduction algorithm for the Modified Probabilistic Neural Network equalizer. IEEE International Joint Conference on Neural Networks (IJCNN), Washington, DC, USA, 1966-1970.

Yule GU (1927) On a method of investigating periodicities in disturbed series, with special reference to Wolfer's sunspot numbers. Philos. Trans. Royal Soc. London, A226:267-298.

Zadeh L (1965) Fuzzy sets. Inform. Control, 8:338-353.

Zaknich, A, Attikiouzel, Y., (1993) Automatic optimisation of the modified probabilistic neural network for pattern recognition and time series analysis. Proceedings of the First Australian and New Zealand Conference on Intelligent Information Systems, Perth, Western Australia, 1-3rd December, 152-156.

Zaknich A, Attikiouzel Y (1995a) Detection of oxalate needles in optical images using neural network classifiers. Proceedings of the First Australian and New Zealand Conference on Intelligent Information Systems, Perth, Western Australia, 1-3rd December, 1699-1702.

Zaknich A, Attikiouzel Y (1995b) Application of the modified probabilistic neural network to the enhancement of noisy short wave radio time and Morse code signals. Australian Journal of Intelligent Information Processing Systems (AJIIPS), 2:3:9-14.

Zaknich A (1997) Characterisation of aluminium hydroxide particles from the Bayer process using neural network and Bayesian classifiers. IEEE Transactions on Neural Networks, 8:4:919-931.

Zaknich A, deSilva CJ (1997) Adaptive learning schemes for the modified probabilistic neural network. Proceedings of the IEEE Third International Conference on Algorithms and Architectures for Parallel Processing, Melbourne, Australia, 597-610.

Zaknich A, Baker SK (1998) A real-time system for the characterisation of sheep feeding phases from acoustic signals of jaw sounds. Australian Journal of Intelligent Information Processing Systems (AJIIPS), 5:2:103-110.

Zaknich, A (1998) Introduction to the modified probabilistic neural network for general signal processing applications. IEEE Transactions on Signal Processing, 46:7:1980-1990.

Zaknich A, Attikiouzel Y (2000) A tuneable approximate piecewise linear model derived from the modified probabilistic neural network. IEEE Signal Processing Workshop on Neural Networks for Signal Processing (NNSP), Sydney, Australia, 1:45-53.

Zaknich A (2003a) Neural networks for intelligent signal processing. World Scientific Publishing, Series on Advanced Biology and Logic-Based Intelligence, Vol. 4.

Zaknich A (2003b) A practical sub-space adaptive filter. Neural Networks, 16:5/6:833-839.

Zaknich A (2003c) An integrated sensory-intelligent system for underwater acoustic signal-processing applications. IEEE Journal of Oceanic Engineering, 28:4:750-759.

Zaknich A (2004) A loudspeaker response model using tuneable approximate piecewise linear regression. Proceedings of the International Joint Conference on Neural Networks (IJCNN), Budapest, Hungary, 4:2711-2716.

Zames G (1981) Feedback and optimal sensitivity: model reference transformations, multiplicative seminorms, and approximate inverses. IEEE Trans. Autom. Control, AC-26:301-320.

Zeidler JR (1990) Performance analysis of LMS adaption filters. Proceedings of the IEEE, 78:12:1781-1806.

Zelniker G, Taylor F (1994) Advanced digital signal processing: Theory and applications. Marcel Dekker.

Index

Kalman and Wiener Filters, Neural Networks, Genetic Algorithms and Fuzzy Logic Systems Together in One Text Book

How can a signal be processed for which there are few or no *a priori* data?

Professor Zaknich provides an ideal textbook for one-semester introductory graduate or senior undergraduate courses in adaptive and self-learning systems for signal processing applications. Important topics are introduced and discussed sufficiently to give the reader adequate background for confident further investigation. The material is presented in a progression from a short introduction to adaptive systems through modelling, classical filters and spectral analysis to adaptive control theory, nonclassical adaptive systems and applications.

Features:

- Comprehensive review of linear and stochastic theory.

- Design guide for prac tical application of the least squares estimation method and Kalman filters.

- Study of classical adaptive systems together with neural networks, genetic algorithms and fuzzy logic systems and their combination to deal with such complex problems as underwater acoustic signal processing.

- Tutorial problems and exercises which identify the significant points and demonstrate the practical relevance of the theory.

- PDF Solutions Manual, available to tutors from springeronline.com, containing not just answers to the tutorial problems but also course outlines, sample examination material and project assignments to help in developing a teaching programme and to give ideas for practical investigations.

ISBN 1-85233-984-5

ISBN: 1-85233-984-5

9 781852 339845

springeronline.com

DATE DUE

GAYLORD No. 2333   PRINTED IN U.S.A.